Global warming and climate change are growing environmental concerns which are much in the scientific, governmental and public eye at present. The potential impact on freshwater and marine fish is immense, because most fish have no physiological ability to regulate their body temperature. This volume focuses on the effects of temperature at all levels of organization in fish, with particular emphasis on physiological function: cells, epithelia, organ systems, the whole organism, growth and development, reproduction, behaviour, pollutant inter-actions, ecology and population dynamics, with each chapter written by experts in the field. Many chapters also speculate on the long-term physiological and ecological implications to fish of a 2–4 °C global warming scenario over the next half century.

Researchers and graduate students in the areas of animal physiology and behaviour, environmental toxicology, population ecology, and fisheries biology and management will find this volume of particular interest.

T0275748

SOCIETY FOR EXPERIMENTAL BIOLOGY
SEMINAR SERIES: 61

GLOBAL WARMING: IMPLICATIONS FOR
FRESHWATER AND MARINE FISH

SOCIETY FOR EXPERIMENTAL BIOLOGY SEMINAR SERIES

A series of multi-author volumes developed from seminars held by the Society for Experimental Biology. Each volume serves not only as an introductory review of a specific topic, but also introduces the reader to experimental evidence to support the theories and principles discussed, and points the way to new research.

GLOBAL WARMING
Implications for freshwater and marine fish

Edited by

C.M. Wood and **D.G. McDonald**

Department of Biology,
McMaster University, Ontario, Canada

CAMBRIDGE
UNIVERSITY PRESS

PUBLISHED BY THE PRESS SYNDICATE OF THE UNIVERSITY OF CAMBRIDGE
The Pitt Building, Trumpington Street, Cambridge CB2 1RP, United Kingdom

CAMBRIDGE UNIVERSITY PRESS
The Edinburgh Building, Cambridge CB2 2RU, United Kingdom
40 West 20th Street, New York, NY 10011-4211, USA
10 Stamford Road, Oakleigh, Melbourne 3166, Australia

First published 1997

Typeset in Linotronic 300 Times 10/12 pt

A catalogue record for this book is available from the British Library

Library of Congress Cataloguing in Publication data

Global warming: implications for freshwater and marine fish / edited
by C.M. Wood and D.G. McDonald.
    p. cm. – (Society for Experimental Biology seminar series; 61)
    Includes index.
    ISBN 0 521 49532 6
    1. Fishes – Effect of temperature on. 2. Global warming –
Environmental aspects. I. Wood, Chris M. II. McDonald, D.G. (D.
Gordon) III. Series: Seminar series (Society for Experimental
Biology (Great Britain)); 61.
    QL639.1.G58   1997
    597.176–dc21   96-39001   CIP

ISBN 0 521 49532 6 hardback

Transferred to digital printing 2003

wv

Contents

Contributors

BALL, D.
Gatty Marine Laboratory, School of Biological and Medical
Sciences, University of St Andrews, St Andrews, Fife KY16 8LB,
UK

BRANDER, K.
Ministry of Agriculture, Fisheries and Food, Directorate of Fisheries
Research, Fisheries Laboratory, Pakefield Road, Lowestoft, Suffolk
NR33 0HT, UK

BUTLER, P.J.
School of Biological Sciences, University of Birmingham, Edgbaston,
Birmingham B15 2TT, UK

CRAWSHAW, L.I.
Department of Biology and Environmental Sciences and Resources
Program, Portland State University, Portland, Oregon 97207, USA

DeSTASIO, B.T.
Biology Department, Lawrence University, Appleton, Wisconsin
54911, USA

EGGINTON, S.
Department of Physiology, University of Birmingham, Edgbaston,
Birmingham B15 2TT, UK

FARRELL, A.P.
Biological Sciences, Simon Fraser University, Burnaby, British
Columbia V5A 1S6, Canada

HAZEL, J.R.
Department of Zoology, Arizona State University, Tempe, Arizona
85287-1501, USA

HOFMANN, G.E.
Hopkins Marine Station, Stanford University, Pacific Grove,
California 93950-3094, USA

HOULIHAN, D.F.
Department of Zoology, University of Aberdeen, Tillydrone
Avenue, Aberdeen AB9 2TN, UK

JOBLING, M.
NFH, University of Tromsø, 9037 Tromsø, Norway
JOHNSTON, I.A.
Gatty Marine Laboratory, School of Biological and Medical
Sciences, University of St Andrews, St Andrews, Fife KY16 8LB,
UK
KENNEDY, C.J.
Department of Biological Sciences, Simon Fraser University,
Burnaby, British Columbia V5A 1S6, Canada
MAGNUSON, J.J.
Center for Limnology, University of Wisconsin-Madison, Madison,
Wisconsin 53709, USA
McCARTHY, I.D.
Division of Environmental and Evolutionary Biology, Graham Kerr
Building, IBLS, University of Glasgow, Glasgow G12 8QQ, UK
McCORMICK, S.D.
Anadromous Fish Research Center, National Biological Service, PO
Box 796, Turners Falls, Massachusetts 01376, USA
McDONALD, D.G.
Department of Biology, McMaster University, 1280 Main Street
West, Hamilton, Ontario L8S 4K1, Canada
O'CONNOR, C.S.
Department of Biology and Environmental Sciences and Resources
Program, Portland State University, Portland, Oregon 97207, USA
PANKHURST, N.W.
Department of Aquaculture, University of Tasmania at Launceston,
PO Box 1214, Launceston, Tasmania 7250, Australia
REID, S.D.
Biology Department, Okanagan University College, 333 College
Way, Kelowna, British Columbia V1V 1V7, Canada
ROBERTSON, J.C.
Department of Zoology, Arizona State University, Tempe, Arizona
85287–1501, USA
ROMBOUGH, P.J.
Department of Zoology, Brandon University, Brandon, Manitoba
R7A 6A9, Canada
SHRIMPTON, J.M.
Anadromous Fish Research Center, National Biological Service, PO
Box 796, Turners Falls, Massachusetts 01376, USA
SOMERO, G.N.
Hopkins Marine Station, Stanford University, Pacific Grove,
California 93950–3094, USA

TAYLOR, E.W.
School of Biological Sciences, University of Birmingham, Edgbaston, Birmingham B15 2TT, UK
TAYLOR, S.E.
School of Biological Sciences, University of Birmingham, Edgbaston, Birmingham, B15 2TT, UK
VAN DER KRAAK, G.
Department of Zoology, University of Guelph, Guelph, Ontario N1G 2W1, Canada
WALSH, P.J.
Division of Marine Biology and Fisheries, Rosenstiel School of Marine and Atmospheric Science, University of Miami, Miami, Florida 33149, USA
WOOD, C.M.
Department of Biology, McMaster University, 1280 Main Street West, Hamilton, Ontario L8S 4K1, Canada
ZYDLEWSKI, J.D.
Anadromous Fish Research Center, National Biological Service, PO Box 796, Turners Falls, Massachusetts 01376, USA

Preface

While the details of climate change and global warming remain subjects of intense debate, there is a growing awareness that the earth's temperature regime is changing, as a result of either natural variation or anthropogenic emissions of 'greenhouse gases'. The effects of warming scenarios on agriculture, human health and vegetation have been extensively studied over the past decade, but there has been remarkably little research on the anticipated effects on fish and fisheries. This is somewhat surprising, given the fact that temperature is undoubtedly the most thoroughly studied environmental variable in fish biology! Fish are poikilotherms living in a medium of high heat capacity and conductance; their body temperature is normally within a few fractions of a degree of the water, and therefore the rates of all of their biological functions are critically dependent on environmental temperature.

A few years ago, as we started our own research programme in this area, we came to realize that there existed only a modest amount of information on the long-term effects of small temperature increments on fish. Nonetheless, there was also clearly available a wealth of classical data on 'temperature effects' from which extrapolation and speculation could be made. We thought it would be useful to bring together experts from all areas of fish biology, and ask them to summarize the existing information, to identify significant gaps in current understanding, and to speculate, based on their knowledge of the field, on the responses of fish to chronic small increments in temperature superimposed on natural regimes. We also felt that it would be useful for others active in global warming issues to have access to this collective wisdom. The result was a two-day symposium entitled 'Global Warming – Implications for freshwater and marine fish' and the present book of the same title with chapters contributed by each invited speaker or group.

We were intrigued to learn how very complex the task of predicting the effects of global warming on fish proteins, membranes, cells, tissues, organisms and communities might be. Firstly, thermal responses have to be considered in the context of thermal optima. Most, if not all, fish species have an optimum temperature, usually at the mid-point of their zone of tolerance, a temperature they prefer if given a choice, and at which such physiological processes as aerobic scope, cardiovascular function and growth reach maxima. Thus, cellular, physiological and behavioural responses will differ depending on whether the temperature increase is towards or away from the optimum. Moreover, not all processes within a species have the same optimum temperature. For example, the optimum temperature for embryonic growth in salmonids is lower than for juvenile growth which, in turn, is lower than the optimum temperature for smolting.

A second complicating factor is that the zone of thermal tolerance varies with species (i.e. stenothermal vs eurythermal) and, within species, with life stage and developmental process. For example, the lethality range is greater than the range permitting juvenile to adult growth, which in turn is larger than the range permitting successful gametogenesis and embryo/larval growth and development.

A third complicating factor is that, in fact, optimum thermal habitat is in short supply for most temperate zone freshwater species. Lake-dwelling species for much of the year can only find habitat that is below their optimum temperature while stream-dwelling fishes are forced to frequent habitats that are above their temperature optimum in summer and autumn, the approaching spawning season for most. There are also key information gaps:

- Research to date has focused much more on cold acclimation than warm acclimation. Processes of acclimation to temperatures that are colder than the optimum temperature undoubtedly involve a different suite of adjustments than those involved with acclimation to warmer temperatures.
- While the effects of temperature have been better studied at the biochemical, cellular and membrane level than at other levels; there is relatively little known of the impact of such changes on the physiology and behaviour of fish species. There is a real need, as well, to look at cellular processes in field populations.
- Eurythermal species have been studied much more often than stenothermal species, and the capacity of the former for thermal adaptation is likely to be greater. Moreover, the total number of species studied is quite small. For example, reproductive studies have typically involved cyprinids and salmonids only, groups which are relatively eurythermal and stenothermal, respectively.
- Studies have rarely used a 'global warming' temperature regime. Temperature acclimation studies usually maintain fish at two or more constant temperatures, with at least a 10 °C gap between, but fish often experience large daily fluctuations in temperature.

What are the effects of a small and gradual increase in temperature
against this large background fluctuation?
- There are often inadequate data on the actual thermal experience
  of fish, especially marine species, which complicates the interpretation of field studies.
- There is very little known of the indirect effects of global warming,
  let alone the examination of responses to such effects. Indirect
  effects include changes in water levels, salinity, amplitude of temperature fluctuations, and temperature–contaminant interactions.

What better venue could we ask for this symposium than the Annual
Meeting of The Society for Experimental Biology (SEB), and what
better site than St Andrews University, Scotland during a week of
glorious weather, 3–7 April 1995? We thank the Society of Experimental
Biology (SEB) for hosting and financially supporting the symposium,
and both the Osmoregulation and the Respiration Groups of the SEB
for additional financial support. We thank Drs Chris Bridges, Mike
Thorndyke, Ted Taylor, Neil Hazon, Ian Johnston and Ms Vicky
Wragg for the various important contributions they made in helping
us put together this successful symposium. We thank the Local Secretary
Dr Neil Hazon in particular for organizing such an enjoyable Annual
Meeting. We thank Maria Murphy, Alan Crowden, Barnaby Willitts,
Jane Bulleid and Miranda Fyfe at Cambridge University Press for
helping us see this book through to completion. Finally, we thank our
contributors for their enthusiasm and dedication to this project and
for their courage in speculating about issues we do not yet fully
understand, and to the many reviewers for their constructive guidance.

Chris M. Wood
D. Gordon McDonald

GEORGE N. SOMERO and
GRETCHEN E. HOFMANN

Temperature thresholds for protein adaptation: when does temperature change start to 'hurt'?

Introduction

Changes in species composition are commonly associated with disconti-
nuities or gradients in habitat temperature. It is common for one
species of a genus to replace its congener as a function of change in
latitude, altitude, depth or some other environmental variable which
is associated with a shift in temperature (Graves & Somero, 1982;
Dahlhoff & Somero, 1993a; Fields et al., 1993; Barry et al., 1995).
Data accumulated by biogeographers studying marine, freshwater and
terrestrial habitats suggest that even relatively minor differences in
habitat temperature may be adequate to drive species replacements,
especially in the case of ectotherms. With few exceptions, however,
the physiological and biochemical adaptations that may be instrumental
in establishing and maintaining these species replacement patterns are
unknown (Graves & Somero, 1982; Dahlhoff & Somero, 1993a, b).
Indeed, the complex interplay between physical and biological factors
in establishing distribution patterns has led to uncertainty about the
importance of fine-scale adaptation to temperature in setting biogeographical patterning. A shift in habitat temperature may lead to
pervasive effects on community structure because some critical link in
the food web, for example a plant which serves as the nutritional
source for one or more dominant animal species, may be severely
impacted. Perturbation of this key plant species could lead to major
shifts in occurrence or abundance of animals which depend on the
plant, even if the animals themselves are not directly impacted by the
temperature change (cf. Lubchenco et al., 1993). Understanding the
importance of temperature change in biogeography thus requires a
broad experimental approach, one which may begin with a detailed
examination of temperature effects on individual physiological and

Society for Experimental Biology Seminar Series 61: *Global Warming: Implications for freshwater and marine fish*, ed. C. M. Wood & D. G. McDonald. © Cambridge University Press 1996, pp. 1–24.

biochemical processes, yet which ultimately must progress to a broadly integrative analysis at the community level.

Concern about the effects of temperature changes of only a few degrees Celsius on organismal function, species distribution limits, and community structure has become more than an abstract issue of only academic concern. The possibility of global warming mandates that a clearer understanding be achieved of how temperature changes in the range of 1.5–4.5 °C (Intergovernmental Panel on Climate Change, 1992) affect the basic physiological and biochemical functions of organisms, especially when these changes in temperature occur at or near the upper limits of the thermal optima of the species. This review examines one aspect of this issue by focusing on acute and evolutionary effects of small changes in temperature on protein structure and function. The central questions addressed are as follows. Does thermal stress under natural environmental conditions, notably stress due to small increases in temperature at the upper limit of thermal tolerance ranges, lead to damage to protein systems? If so, is this damage readily reversible, or does irreversible protein denaturation also occur? What traits of proteins are most perturbed by changes in temperature and, therefore, most demanding of adaptive modification? What types of changes in protein sequence are required to modify a protein for activity under a new thermal regime?

As background for each of these questions is the central issue of the minimal amount of change in temperature, the 'threshold' for temperature effects, that is sufficient to perturb protein systems enough to favour an adaptive response. Issues of 'threshold' effects will be discussed in two temporal contexts: (i) the magnitudes of acute changes in temperature that are sufficient to cause protein unfolding and, as a result, trigger the heat shock response and/or proteolytic removal of damaged proteins, and (ii) the amount of temperature change that is sufficient over many generations to favour selection for evolutionary changes in protein thermal sensitivity.

Analysis will demonstrate that at the upper limits of thermal tolerance ranges, protein function and structure may be impaired significantly. Minor and readily reversible changes in protein structure, for example minor shifts in conformation, may substantially perturb critical functional traits like substrate binding ability. Larger, yet still reversible changes in conformation may require the activities of heat shock proteins (hsp's) to prevent inappropriate aggregation of partially unfolded proteins and to restore native conformation. Irreversible damage to proteins is also demonstrated through analysis of temperature-dependent changes in concentrations of ubiquitinated proteins. Irreversibly damaged proteins to which ubiquitin is covalently bound are tagged for proteolysis by non-lysosomal pathways. Over evolutionary time-spans, i.e. times allowing modification in protein amino acid sequences, temperature differences of only a few degrees Celsius are sufficient to select for adaptive changes in structure and function. Taken together, these data show that the levels of temperature change predicted by some models of global warming could have direct and significant effects on ectothermic organisms, as well as complex higher order effects due to influences on currents, upwelling, predator–prey interactions and other broader scale environmental processes (IPCC, 1992; Fields *et al.*, 1993; Kareiva, Kingsolver & Huey, 1993; Lubchenco *et al.*, 1993; Barry *et al.*, 1995; Roemmich & McGowan, 1995).

Thermal perturbation of protein structure: damage, rescue and degradation

From the *in vitro* to the *in situ*: do laboratory denaturation studies reflect 'real world' phenomena?

The literature of biochemistry contains hundreds of examples of thermal perturbation of protein structure and function. As discussed later, proteins are only marginally stable at physiological temperatures, and this means that their structures can be either reversibly or irreversibly damaged by relatively small increases in temperature. Although protein denaturation has been studied in exquisite detail, the great majority of studies of temperature–protein interactions tell us little about the consequences of temperature change for proteins under *in situ* conditions, where the variables of protein concentration, solute composition, and substrate concentrations can be expected to lead to substantially different sensitivities to temperature from those seen under more simplified, and often physiologically unrealistic, *in vitro* conditions. It is important, then, to begin this analysis by asking, 'Is there evidence that proteins *in situ* are significantly perturbed, in function and structure, by physiologically realistic conditions of heat stress?'

The heat shock response: roles of molecular chaperones

The most direct evidence that proteins are damaged by heat stress is the appearance in thermally-stressed organisms of elevated concentrations of heat shock (stress-induced) proteins (Lindquist & Craig, 1988; Parsell & Lindquist, 1993). Since the discovery made by R.M. Ritossa (1962) more than three decades ago that heat stress induces unique chromosomal puffing patterns in *Drosophila*, large numbers of investigators have been examining the nature of the proteins induced by heat stress and the genetic regulatory mechanisms that govern their expression. Even though the heat shock genes quickly became an important system for studying control of gene expression, the function of the heat shock proteins was unclear for many years after their initial discovery. Significant evidence has now accumulated to show that stress-induced proteins are members of a broad family of proteins, the molecular chaperones, that play a vital role in the normal processes of protein synthesis and maturation. Molecular chaperones are defined as proteins that assist in the proper folding, compartmentation and assembly of other proteins, but do not remain as part of the final, mature complex (Ellis & van der Vies, 1991; Craig, 1993). As their name implies, molecular chaperones play a key role in preventing improper complexing of proteins which, because they are not properly folded, may have a strong tendency to adhere to other proteins in inappropriate ways. For example, members of the hsp70 family bind to and stabilize nascent polypeptides being translated on ribosomes and prevent formation of complexes between the partially synthesized, not yet folded polypeptide and other proteins. Molecular chaperones, when functioning as stress proteins, perform a similar role (Becker & Craig, 1994). They bind to exposed sites on partially denatured proteins, which most likely are hydrophobic patches that are buried in the native conformation. Stress protein binding prevents aggregation of the denatured proteins, and affords an opportunity for refolding to the native conformation of the protein.

The expression of stress proteins is induced when cells are exposed to any conditions that can lead to denaturation of proteins (Hightower, 1980). Thus the generic term 'stress protein' is often used in place of the more specific term, 'heat shock protein'. Denaturing conditions include not only exposure to high temperatures, but also exposure to certain chemicals, e.g. ethanol, arsenate, and heavy metals, osmotic stress, and oxygen deprivation (Parsell & Lindquist, 1993). The common signal for stress protein induction is the presence of denatured, mis-folded or misaggregated proteins in the cell (Goff & Goldberg, 1985; Ananthan, Golberg & Voellmy, 1986; Parsell & Lindquist, 1993). Different stress proteins play different roles in the 'rescue' of unfolded proteins or misaggregated proteins. For example, hsp70 is thought to prevent the formation of aggregations of denatured proteins, whereas hsp104 is capable of dissolving aggregates and rescuing protein conformation (Parsell *et al.*, 1994).

The diverse functions of hsp's, and the variation in patterns of hsp expression found in different organisms and tissues (Parsell & Lindquist, 1993) suggest that the stress protein response and the strategy and pattern of stress protein expression may have a large impact on organismal function. However, as in the case of the literature on *in vitro* thermal denaturation of isolated proteins, the large literature on the heat shock response is focused more strongly on mechanistic questions related to understanding gene expression than on issues of environmental effects. Despite the appearance in the current literature of several hundred publications each year on the heat shock response, very few studies have examined hsp synthesis in organisms exposed to the normal heat stresses encountered in their natural habitats. For very few species can we state with any assurance that the heat shock response occurs under conditions of typical heat stress, e.g. during peak summer temperatures. Field studies are critical for another reason: stresses in addition to those due to temperature *per se* may activate the heat shock response. For example, desiccation stress due to prolonged exposure to high temperatures and/or low humidities might also affect protein stability and elicit the heat shock response. For this reason, field studies are absolutely essential if meaningful predictions are to be possible about thermal stress due to global warming. Laboratory studies of the heat shock response in which only temperature is varied can be but a first step in unravelling the complex effects of temperature stress under natural conditions.

The heat shock response in field populations

To illustrate adaptive patterns in the expression of hsp's under natural environmental conditions, two cases will be reviewed in which hsp concentrations and threshold induction temperature, defined as the temperature at which elevated synthesis of hsp's is first observed, have been measured in field populations. In studies of two species of goby fish (genus *Gillichthys*), Dietz and Somero (1992) showed that summer-acclimatized fish had higher levels of a 90 kDa class of hsp (hsp90) in brain tissue than did winter-acclimatized fish. For both winter and summer fishes, threshold induction temperatures were within the upper range of physiological temperatures. However, the threshold induction temperature for hsp90 synthesis was significantly higher in summer. Dietz (1994) demonstrated that in the eurythermal goby *Gillichthys mirabilis*, the threshold induction temperature could be shifted by as much as 8 °C as a result of laboratory temperature acclimation. These data reveal a high degree of plasticity in the heat shock response: both induction temperatures and concentrations of hsp's differ between conspecifics acclimatized to different temperatures. Most importantly, however, these data show that the threshold temperatures for induction of increased hsp synthesis lie well within the normal temperature ranges of the organisms. Thus, heat stress under natural habitat conditions can lead to significant perturbation of protein structure.

The conclusions from work with eurythermal goby fish were supported and extended by studies of sessile intertidal invertebrates that experience extreme thermal stress and temperature- and wind-derived stresses arising from desiccation. Hofmann and Somero (1995) examined the effects of acute temperature changes, seasonal acclimatization, and microhabitat location on the mussel *Mytilus trossulus*. Mussels from a rocky intertidal site (British Camp) on San Juan Island, Washington, USA, that encountered emersion periods of several hours duration exhibited large and rapid changes in body temperature during midday low tides in summer. Body temperatures climbed from ambient sea water temperatures (10 °C in August) to over 30 °C in approximately 3 hours during low tides on clear summer days (Fig. 1). With *in vitro* labelling protocols that exposed gill fragments to ^{35}S-labelled amino acids (methionine and cysteine), induction of hsp70 was observed by 25 °C (Fig. 2), a temperature well below the peak tissue temperatures encountered by the British Camp specimens in summer. By 30 °C, a sharp attenuation of protein synthesis was apparent, and by 35 °C, protein synthesis was essentially fully blocked. Mussels from a second site, Argyle Creek, which were not exposed to air during low tides, also exhibited hsp70 induction at 25–28 °C (Hofmann & Somero, 1995). Induction temperatures for hsp70 synthesis were also subject to seasonal acclimatization, as in the case of goby fish (D. Roberts, G.E. Hofmann & G.N. Somero, unpublished observations).

Using quantitative antibody analysis, it was demonstrated that levels of hsp's of the 70 kDa class, e.g. hsp68, were significantly higher in gill tissue in summer than in winter. Mussels exposed to prolonged periods of emersion (British Camp site) contained higher hsp68 concentrations than mussels that were continuously immersed (Argyle Creek site; Fig. 3). For mussels at the British Camp site in August, the levels of hsp68 reflected the duration of emersion; by 5 and 8 hours of emersion, hsp68 levels were significantly higher than at the beginning of emersion. In February, hsp68 levels did not change during an 8 hour emersion period.

Comparisons of laboratory-acclimated mussels and field-collected mussels that had experienced similar thermal histories showed that the

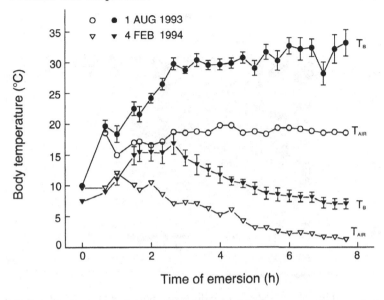

Fig. 1. Body temperatures (T_B) (T_B) of *Mytilus trossulus* recorded during a tidal cycle at the British Camp site on San Juan Island, Washington, USA. Body temperatures were measured, using an implanted thermistor, every 20 min on five mussels during emersion on summer (August) and winter (February) daytime low tides. (Data from Hoffmann & Somero, 1995.)

heat shock response was significantly reduced in the laboratory-held group (D. Roberts, G. E. Hofmann & G. N. Somero, unpublished observations). This finding suggests that under laboratory acclimation conditions in which the mussels were continually immersed, the absence of stress due to desiccation, or some other factor related to aerial exposure, reduced perturbation of proteins and, therefore, reduced the magnitude of the stress protein response. The differences observed between field-acclimatized and laboratory-acclimated individuals provide a *caveat* about the dangers of relying on laboratory-based studies to make predictions about effects occurring in natural habitats. (See the chapter by McCarthy and Houlihan for a further consideration of the difference between acclimatization and acclimation.)

These and other studies of heat stress on natural populations (Bosch *et al.*, 1988; Sanders *et al.*, 1991; Dietz & Somero, 1992; Sharp *et al.*, 1994; White, Hightower & Schultz, 1994) show unequivocally that body temperatures normally encountered under seasonal or diurnal extremes of temperature can sufficiently perturb protein structure to elicit a

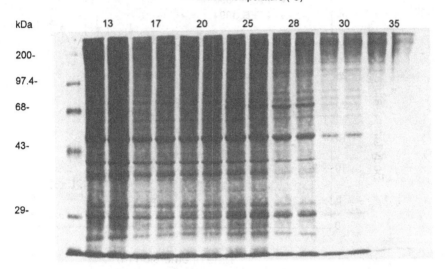

Fig. 2. Heat shock protein expression profile in gill of the mussel *M. trossulus*. The fluorograph illustrates patterns of protein synthesis in isolated gill tissue exposed to different incubation temperatures. Duplicate samples were run from a single individual at each of seven incubation temperatures. Protein molecular weight standards are indicated on the left of the figure. The dark band first appearing at 25 °C with an approximate molecular weight of 70 kDa is hsp68 (see text). (Data from Hofmann & Somero, 1995.)

strong heat shock response. However, studies of hsp synthesis by themselves do not indicate whether irreversible damage is done to proteins under conditions of natural heat stress. It is conceivable that the heat shock response is sufficient to rescue the majority, and perhaps all, of the proteins whose structures are perturbed by heat stress.

The ubiquitination response: evidence for irreversible protein denaturation under natural habitat conditions

To determine whether heat stress under natural habitat conditions is sufficient to lead to irreversible denaturation of proteins another experimental approach is required, one that quantifies the amounts of irreversibly denatured proteins that accumulate under thermal stress. To this end, Hofmann and Somero (1995) measured levels of ubiquitinated proteins in tissues of *M. trossulus* from the same populations used in

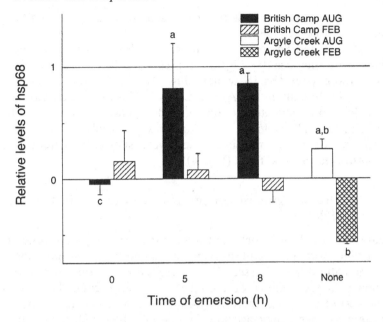

Fig. 3. Western analysis of relative levels of an hsp70 isoform, hsp68, in gill tissue of *M. trossulus* individuals from two sites on San Juan Island, Washington. British Camp site: animals emersed for a long period during low tide; Argyle Creek: animals remained immersed during full tidal cycle ('None'). Levels of hsp68 were measured using laser densitometry and standardized using the intensity of a common weight marker run with each gel. The zero point line represents equal intensity of the signals of the molecular weight marker band and hsp68 band, and the bars show increases or decreases relative to the intensity of the marker band. The horizontal line at 1 represents 2-fold higher intensity of hsp68 relative to the molecular weight marker. Each band represents the mean ± SEM for 5 mussels, except for British Camp in August, for which $n = 4$ for 0 hours and $n = 3$ for the 5 h time point. Letters next to error bars indicate significant differences: a, significantly different from corresponding winter collection (ANOVA; $p < 0.05$); b, significantly different from samples collected at British Camp (ANOVA; $p < 0.05$); c, significantly different from 5 h and 8 h time points. (Data from Hofmann & Somero, 1995.)

the hsp experiments discussed above. Ubiquitin is a small protein (approximately 8 kDa) which, when covalently bound to a denatured protein, 'tags' the protein for proteolysis by non-lysosomal proteases (Rechsteiner, 1987). Therefore, the accumulation of ubiquitinated proteins can serve as an index of the amount of irreversible denaturation

caused by exposure of cells to a particular stress, such as high temperature (Bond *et al.*, 1988).

Hofmann and Somero (1995) found that levels of ubiquitinated proteins were significantly higher in mussels collected in summer than
from those collected in winter (Fig. 4). Furthermore, mussels that
occurred subtidally in the Argyle Creek habitat, i.e. in an environment
with low and stable temperatures, had significantly less ubiquitinated
protein, as well as lower standing-stocks of hsp's, than conspecifics
from the British Camp site, which were subject to extended aerial
exposure during low tides (Fig. 4).

The energy costs of protein damage *in situ* may be high

Data on the heat shock and ubiquitination responses suggest that
reversible unfolding and irreversible denaturation of proteins occur
during natural heat stresses. The consequences of both forms of protein
damage in the context of ecological energetics could be substantial.
The cost of replacing irreversibly damaged proteins is likely to be high.
Protein synthesis may account for 20–30% of cellular ATP turnover
under basal conditions (Hawkins, 1985, 1991; Houlihan, 1991). The
increased cost of replacing damaged proteins could add significantly to
this already large share of the cellular energy budget.

Even in situations where molecular chaperones can rescue heat-
damaged proteins, major energy costs may be added to the cells' energy
budgets. The synthesis and function of hsp's requires substantial energy.
In some organisms heat stress is followed immediately by reduction or
elimination of synthesis of proteins other than hsp's (Parsell & Lindquist, 1993). Refolding of an unfolded protein by chaperones may
require hundreds of ATP equivalents (Parsell & Lindquist, 1993). Thus,
the preservation of the cellular protein pool in the face of temperature
change may represent a significant energy cost to the cell. Elevated
temperatures may increase these costs sufficiently to threaten survival
or reproduction of organisms that are confronted with elevated cell
temperatures that cause large increases in protein unfolding. Increasing
environmental (= body) temperatures for ectotherms therefore may
require the evolutionary development of proteins with higher thermal
stabilities. The occurrence of homologues of proteins with different
thermal stabilities among species adapted evolutionarily to different
temperatures indicates the importance of this type of molecular
evolution.

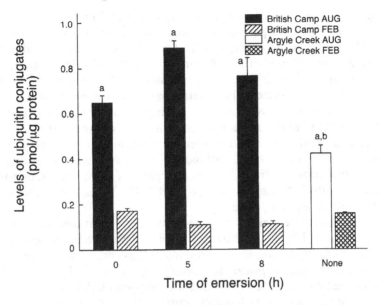

Fig. 4. Comparative levels of ubiquitin conjugates in gill tissue of summer- and winter-acclimatized *M. trossulus*. Ubiquitin conjugates in gill tissue extracts were quantified using a solid-phase immunoassay as described by Haas and Bright (1985), with a polyclonal rabbit anti-ubiquitin conjugate antibody provided by Dr A. L. Haas. Bars represent the mean ± SEM for gill samples from 5 mussels, except for the following: British Camp (August) 0 h, $n = 4$; 5 h, $n = 3$. Letters next to error bars indicate significant differences. a, significantly different from corresponding winter collection (ANOVA, $p < 0.05$); b, significantly different from specimens collected at British Camp (ANOVA, $p < 0.05$). (Data from Hofmann & Somero, 1995.)

Protein thermal stability and evolutionary adaptation to temperature

Heat stability is strongly correlated with adaptation temperature

In view of the evidence that thermal denaturation of proteins can occur under normal environmental conditions, it is not surprising to find that orthologous homologues of proteins (variations on a protein theme that are encoded by a common gene locus in different species and populations) from species adapted to different temperatures exhibit highly regular variation in thermal stability. This trend is shown in

Fig. 5 for eye lens proteins (crystallins) isolated from vertebrates whose maximal body temperatures span a range of approximately 50 °C (McFall-Ngai & Horwitz, 1990). Species studied include the highly cold-adapted stenothermal Antarctic fish *Pagothenia borchgrevinski*, whose upper lethal temperature is near 4 °C (Somero & DeVries, 1967), and the thermophilic desert iguana, *Dipsosaurus dorsalis*, whose core temperature may reach 47 °C. There is a regular increase in heat denaturation temperature with maximal body temperature across all taxa examined. The trend shown for crystallins reflects patterns seen for several other sets of orthologous homologues of enzymatic and structural proteins from a wide diversity of organisms (Johnston & Walesby, 1977; Swezey & Somero, 1982; Hochachka & Somero, 1984; Dahlhoff & Somero, 1993a).

Selection for enhanced thermal stability in heat-tolerant organisms can be viewed as an adaptive response for minimizing thermal damage to proteins under *in situ* conditions. However, it might seem even more reasonable, *a priori*, to maximize protein stability in all species and thereby reduce, or even eliminate, the potential for thermal damage to proteins and its potentially important consequences for cellular function and energy budgets. In view of the comparative ease with which molecular biologists are able to alter protein thermal stability using methods of site-directed mutagenesis (Matthews, 1987; Powers

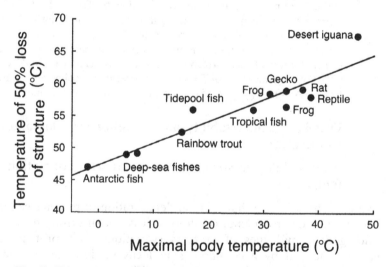

Fig. 5. Thermal denaturation (temperature of 50% loss of structure) of lens proteins (crystallins) of vertebrates with different maximal body temperatures. (Data from McFall-Ngai & Horwitz, 1990.)

et al., 1993), one can ask why natural selection has not led to stronger proteins, which are more able to withstand thermal stress.

The importance of maintaining marginally stable protein structures

The answer to this question about this putative 'shortfall' of selection involves one of the most important quantities in protein biochemistry, the net stabilization free energies of protein structure. Globular proteins such as enzymes have net free energies of stabilization of only approximately 30–65 kJ mol^{-1} (8–17 kcal mol^{-1}) (Jaenicke, 1991). Despite the involvement of hundreds of non-covalent ('weak') bonds in the formation of the native, folded protein molecule, the net free energy of stabilization is equivalent to only a few 'weak' bonds (Jaenicke, 1991). The reason why proteins are less stable than they could be, at least in theory, is that the maintenance of a capacity for undergoing rapid conformational changes is critical for most aspects of protein function. For example, during the binding of ligands (substrates and cofactors), a protein typically undergoes a change in conformation. The requirement for flexibility of structure has led to selection for proteins that are marginally stable at physiological temperature, i.e. selection favours a 'trade-off' between structural stability and functional capacity (Jaenicke, 1991; Somero, 1995).

The strong correlation that exists between adaptation temperature and protein thermal stability suggests that marginal stability is strongly conserved among homologous forms of proteins from organisms adapted to different temperatures. Because the net stabilization free energy of a protein is only of the order of a few 'weak' bonds, interspecific, temperature-adaptive differences in net stabilization free energy are extremely small indeed. Although in most cases it is not clear how naturally occurring homologues with different thermal stabilities differ in sequence, studies of genetically engineered proteins have demonstrated that amino acid substitutions as simple as single glycine to alanine replacements can effect a significant change in structural stability (Matthews, 1987). Thus, adaptation of proteins to temperature is a subtle process, one in which even seemingly minor changes in sequence can lead to adaptively important changes in structural stability and, as we now consider, in functional (kinetic) properties.

Protein function and adaptation to temperature

The kinetic properties of enzymes, notably ligand binding events, are strongly affected by temperature. The sources of these temperature effects include the non-covalent interactions that occur between enzyme and ligand at the binding site, and the conformational changes that occur during binding and catalysis. Thus, adaptive differences in temperature effects on binding could be due to (i) alterations in the residues that interact with ligands, or (ii) modifications in the conformational changes that accompany binding. Based on available data on sequences of homologous proteins, it appears that active sites are generally fully conserved among orthologous homologues (Powers *et al.*, 1993; Tsuji *et al.*, 1994; Fields, 1995). Therefore, adaptation does not appear to involve substitutions in the active sites of proteins. This conclusion leads to the following conjecture: interspecific, temperature-adaptive differences in the kinetic properties of orthologous homologues are likely to reside in differences in the conformational flexibility of the protein. This conjecture links the observed differences in thermal stability with correlated differences in kinetic properties.

K_m values are highly conserved among species

Enzyme–ligand interactions are at once highly sensitive to temperature, as well as other physical and chemical variables, and highly conserved among species (Yancey & Somero, 1978; Yancey & Siebenaller, 1987; Dahlhoff & Somero, 1993a; Somero, 1995). Conservation of ligand binding is illustrated by the extent to which Michaelis–Menten constants (K_m values) are conserved among differently temperature-adapted species at their normal body temperatures (Fig. 6). For the A-type ('muscle isozyme') of lactate dehydrogenase, LDH-A, the K_m of pyruvate is conserved within the range of *c.* 0.15–0.35 mM pyruvate across species, including Antarctic fishes adapted to -1.9 °C and desert reptiles that encounter temperatures near 47 °C. Differences in the effects of temperature on K_m values are noted in comparisons of stenothermal and eurythermal species. Enzymes of the latter species generally exhibit smaller changes with measurement temperature than do the homologous enzymes of stenotherms (Coppes & Somero, 1990). Perturbations of enzymatic function by increases in habitat temperature may have a greater impact on stenothermal species than on eurytherms. These differential effects may be one way in which global warming will have differential impacts on different organisms.

'Threshold' temperature increases favouring selection for new protein variants are only a few degrees Celsius

The broad interspecific patterns noted for LDH-As are reflected on a fine temperature scale by orthologous homologues of LDH-A in con-

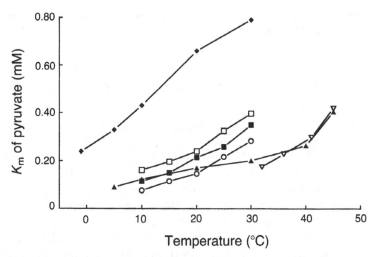

Fig. 6. Effect of temperature on apparent Michaelis–Menten constants (K_m) of pyruvate for A-type lactate dehydrogenases (LDH-A) purified from skeletal muscle of vertebrates with different body temperatures (physiological temperatures ranges are given in parentheses). ◆, Stenothermal Antarctic fish, *Trematomus pennelli* (−1.9 °C). Barracuda congeners: □, *Sphyraena idiastes* (14–22 °C); ■, *S. lucasana* (16–28 °C); ○, *S. ensis* (22–30 °C). ▲, Eurythermal goby fish, *Gillichthys seta* (9–38 °C); ▽, Desert iguana (30–47 °C). (Data from Graves & Somero (1982), Donahue (1982), and J. Podrabsky, W. Winter and G.N. Somero, unpublished.)

generic fishes adapted to slightly different thermal minima and maxima. For example, LDH-As of barracuda congeners (genus *Sphyraena*) that differ in maximal body temperature by only 3–8 °C reflect the differences noted in the broad comparisons among vertebrates adapted to temperatures between −1.9 °C and 47 °C (Fig. 6). The K_m conservation patterns noted in these studies of vertebrate LDH-As are seen as well in malate dehydrogenases of invertebrates from several taxa (Dahlhoff & Somero, 1991, 1993a). The patterns noted in comparisons of congeneric species from habitats that differ only slightly in temperature provide a basis for the conjecture that differences in maximal habitat temperature of only a few degrees Celsius can, given sufficient evolutionary time, favour changes in amino acid sequence that lead to restoration of kinetic properties and the appropriate balance between flexibility and stability of protein structure. Thus, even though it is impossible to determine the absolute minimal amount of change in body temperature that is adequate to favour selection for adaptive changes in proteins, it seems increasingly clear that changes in maximal habitat temperature of only a few degrees Celsius are above the 'threshold' at which perturbation is sufficient to favour adaptation.

Minor changes in sequence can effect adaptation to temperature

Studies of sequences of orthologous homologues of proteins from species (or populations of conspecifics) adapted to different temperatures have revealed that only very minor changes in sequence are sufficient to cause shifts in thermal stability and function. Here, 'minor' refers to two different aspects of sequence: the number of substitutions *per se*, and the chemical nature of these substitutions. Powers and colleagues (1993) showed that differences in thermal stability between allozymes of LDH-B in populations of the killifish *Fundulus heteroclitus* from different latitudes can result from a single amino acid substitution: a shift between alanine and serine. This is an extremely minor change, chemically, in that it represents only the removal or addition of a single hydroxyl group from a protein with *c*. 330 residues. Nonetheless, this hydroxyl group is involved in a hydrogen bond that stabilizes the enzyme's structure. Recalling the low net stabilization free energies of globular proteins, it should not be surprising that a single 'weak' chemical bond can have such a marked influence on protein structural stability.

Studies of LDH-A homologues from barracudas have likewise shown that small differences in sequence may account for differences in stability and kinetic properties (L. Z. Holland, M. McFall-Ngai & G. N. Somero, unpublished data). As in the case of the LDH-B homologues, a single amino acid substitution may be sufficient to explain the differences in thermal responses between LDH-A homologues of congeneric barracudas.

The minor amounts of change in sequence that appear adequate to alter protein stability and function have important implications in the context of rates of evolutionary adaptation to temperature. If only minor changes in sequence are required to adapt a protein to a new temperature range, then rates of protein evolution in the face of thermal stress could be rapid. Because the impact of global warming is a function of both the total change in temperature and the rate at which warming occurs (IPCC, 1992), the rate at which organisms can adapt to rising temperatures is a critical issue in predicting the effects of warming.

Currently, it is difficult to predict quantitatively how rapidly proteins could evolve in the face of the temperature changes predicted by global warming models. It is pertinent to note, however, that laboratory thermal selection studies indicate that rapid changes in thermal optima can occur in bacteria (Bennett, Dao & Lenski, 1990) and animals (Huey, Partridge & Fowler, 1991). These studies may provide crucially important insights into potential effects of global climate change.

Conclusions and suggestions for future research

Because of proteins' low net free energies of stabilization, which are necessitated by the need for proteins to undergo reversible changes in conformation during function, proteins are only marginally stable at physiological temperatures. This fundamental characteristic of proteins has several critical implications for protein evolution and for the potential effects of global warming on ectothermic organisms.

Perhaps of greatest significance is the fact that low net stabilization free energies render proteins susceptible to structural perturbation by temperatures within the upper reaches of the physiological temperature range. The induction of hsp's at temperatures well within the normal body temperature range of a species is a clear indication that the marginal stabilities of proteins render them susceptible to heat-induced structural damage under natural conditions. Moreover, even though some amount of perturbation of protein structure may be reversed by the activities of hsp's, the accumulation of ubiquitinated proteins under heat stress indicates that irreversible, as well as reversible, damage can occur under natural conditions. Future studies must characterize more fully the nature of the heat shock and ubiquitination responses in field populations. Characterizing the temperatures at which induction of hsp synthesis and buildups of ubiquitinated proteins occur, and quantifying the standing-stock concentrations of hsp's and ubiquitinated proteins in different species at different seasons, may provide a quantitative measure of thermal stress. The findings that changes in hsp threshold induction temperatures and in cellular concentrations of hsp's and ubiquitinated proteins occur seasonally (Dietz & Somero, 1992; Hofmann & Somero, 1995) necessitate that careful field studies of differently acclimatized natural populations be carried out. Laboratory studies alone are not likely to provide an adequate basis for evaluating *in situ* thermal perturbation, as shown by the attenuated heat shock response of laboratory-acclimated mussels (D. Roberts, G.E. Hofmann & G.N. Somero, unpublished observations). In the field, thermal stress is likely to be experienced in conjunction with other stresses, for example those due to desiccation and anaerobiosis. Future studies must be strongly field-orientated if the true impact of environmental change is to be evaluated quantitatively.

A second important implication of marginal protein stabilities that merits close study in natural populations relates to energy budgets. The costs of protein synthesis, including production of hsp's, chaperone-mediated protein renaturation, and protein degradation, comprise a large share of the cellular energy budget. In light of the demonstration that natural populations experience reversible and irreversible denaturation of cellular proteins, one potential consequence of global warming is an increase in the amount of energy that will be required to maintain the pool of cellular proteins (see McCarthy & Houlihan, this volume). Additional energy directed towards maintaining the cellular protein pool may be lost to key processes like reproduction. Future studies must quantify as precisely as possible the impacts of global warming on energy budgets under natural environmental conditions to determine whether thermal stress exerts a negative impact on reproductive capacities and rates of growth.

A third implication of the marginal stabilities of proteins that is relevant for studies of global warming is that even subtle temperature-induced shifts in protein conformation are likely to be manifested in physiologically significant changes in function. The finding that enzyme–ligand interactions are so highly temperature-sensitive, especially in cold-adapted stenothermal species like Antarctic fishes, is an indication that perturbations in structure that are far less than those required to fully denature a protein can have large effects on the protein's function. In effect, 'physiological denaturation' of a protein, defined as thermal disruption of protein function at extremes of the physiological temperature range, may take place at temperatures much lower than those at which full structural unfolding occurs. *In vitro* studies of thermal perturbation of proteins therefore must examine physiologically relevant effects such as disruption of ligand binding to allow predictions to be made about the effects of global warming.

A fourth implication of the inherent structural flexibility of proteins has implications for studies of protein evolution, specifically of the types of adaptive differences in sequence that can effect adaptation to temperature. For example, because the energy costs of reversible changes in conformation contribute strongly to binding energies for enzymes like LDH, sequence changes that influence conformational rigidity could alter ligand binding properties. It follows that temperature adaptation could involve residues distant from the active site, as long as these residues play a role in establishing conformational flexibility. The finding that active site sequences of orthologous homologues of proteins tend to be fully conserved among species suggests that differences in the responses of kinetic properties to change in temperature are not due to different types of non-covalent bonds between ligand and enzyme. Rather, these differences must be due to different strengths of the same types of 'weak' bonds. Minor changes in conformation that lead to shifts in the position of amino acid residues important in binding could lead to major changes in kinetic properties like K_m of substrate and cofactor. Studies of the structures of orthologous homologues of enzymes from differently thermally-adapted species are likely to provide important new insights into structure–function relationships of proteins. Comparisons of congeneric species whose proteins are extremely similar in sequence and may differ only by adaptively important amino acid substitutions may be especially useful in this regard.

Congeneric species hold promise in another important arena of study: elucidation of 'threshold' temperature changes that are sufficient to favour perturbation of protein function and structure and, hence, to influence biogeographical patterning. Studies of congeners that live in environments that differ in maximal temperature by only a few degrees Celsius support the conjecture that temperature differences as small as 3–8 °C are sufficient to favour selection for new orthologous forms of proteins with different thermal sensitivities. Comparative studies of this nature cannot, of course, show what the minimal amount of temperature change is that could favour selection. However, these studies can provide evidence about the amount of change in temperature that is sufficient over evolutionary periods to favour adaptive change in proteins. The prediction from studies of congeners is that temperature increases of the magnitude predicted by some models of global climate change (IPCC, 1992) could be enough to exert significant perturbation of protein structure and function, especially in the case of species that already are encountering temperatures near the upper limits of their optimal temperature ranges. Future comparisons of temperature effects on proteins of congeneric ectotherms may further refine our understanding of the linkages between biogeographical patterning and protein adaptation to temperature.

A fifth implication of marginal protein stability concerns rates of protein evolution. The demonstration in evolutionary studies of sequences of orthologous homologues and in experimental manipulation of sequences using site-directed mutagenesis, that only very minor changes in sequence are sufficient to alter protein thermal stability and kinetic properties, suggests that adaptation of proteins to a new thermal regime could be rapid. Studies in which temperature is being used in laboratory evolution studies (e.g. Bennett *et al.*, 1990; Huey *et al.*, 1991) could reveal a great deal about the rates of protein evolution in response to thermal stress and, therefore, of the rates at which ectothermic species may be able to respond adaptively to anthropogenic changes in habitat temperatures. It will be critical in these studies to find the molecular 'needles in the haystack' that are responsible for the induced changes in organismal thermal preferences and tolerances.

A final implication from this discussion of protein–temperature interactions is that effects of global warming may be different on stenothermal and eurythermal species. The latter species may have substantially more ability for acclimatization because they possess (i) additional isoforms of enzymatic proteins with different thermal optima (Johnston, 1983; Lin & Somero, 1995a, b), and (ii) greater abilities to restructure their membrane systems adaptively when confronted with changes in temperature (Dahlhoff & Somero, 1993b). These and other differences between stenotherms and eurytherms suggest that the effects of global warming will not be uniform among ectothermic species (cf. Somero, Dahlhoff & Lin, 1996). Future studies must include species with widely different thermal tolerance ranges in order to provide a comprehensive picture of how global warming might affect diverse taxa.

In conclusion, to predict the effects of global warming, biologists will need to maintain a broad experimental focus, one which demands collaborative efforts among biochemists, physiologists, biogeographers and community ecologists. Studies of protein–temperature interactions, while only a small segment of this broad experimental approach, are starting to reveal some of the critical effects on metabolic function, cellular energy budgets, and biogeographical patterning that could befall ectothermic organisms if even the relatively modest predicted increases in temperature due to global warming were to occur. Future studies must build on this initial experimental base and also deal with the suite of complex, indirect, secondary effects of temperature that may result from community-level perturbations: for example, from strong effects of temperature on particular sites in food webs. If mechanistic laboratory studies are paired with appropriate field studies, then biologists may succeed in developing models that not only can predict, but which can also suggest mechanisms for ameliorating, the effects of global warming.

Acknowledgements

Portions of this work were supported by National Science Foundation grant IBN-9206660 to GNS and a National Science Foundation Marine Biotechnology and Oceans Sciences Postdoctoral Fellowship to GEH.

References

Ananthan, J., Goldberg, A. L. & Voellmy, R. (1986). Abnormal proteins serve as eukaryotic stress signals and trigger the activation of heat shock genes. *Science*, **232**, 522–4.

Barry, J. P., Baxter, C. H., Sagarin, R. D. & Gilman, S. E. (1995). Climate-related, long-term faunal changes in a California rocky intertidal community. *Science*, **267**, 672–5.

Becker, J. & Craig, E. A. (1994). Heat shock proteins as molecular chaperones. *European Journal of Biochemistry*, **219**, 11–23.

Bennett, A. F., Dao, K. M. & Lenski, R. E. (1990). Rapid evolution in response to high temperature selection. *Nature*, **346**, 79–81.

Bond, U., Agell, N., Haas, A. L., Redman, K. & Schlesinger, M. J. (1988). Ubiquitin in stressed chicken embryo fibroblasts. *Journal of Biological Chemistry*, **263**, 2384–8.

Bosch, T. C. G., Krylow, S. M., Bode, H. R. & Steele, R. E. (1988). Thermotolerance and synthesis of heat shock proteins: these responses are present in *Hydra attenuata* but absent in *Hydra oligactis*. *Proceedings of the National Academy of Sciences USA*, **85**, 7927–31.

Coppes, Z. L. & Somero, G. N. (1990). Temperature-adaptive differences between the M_4-lactate dehydrogenases of stenothermal and eurythermal Sciaenid fishes. *Journal of Experimental Zoology*, **254**, 127–31.

Craig, E. A. (1993). Chaperones: helpers along the pathways to protein folding. *Science*, **260**, 1902–3.

Dahlhoff, E. & Somero, G. N. (1991). Pressure and temperature adaptation of cytosolic malate dehydrogenases of shallow- and deep-living marine invertebrates: Evidence for high body temperatures in hydrothermal vent animals. *Journal of Experimental Biology*, **159**, 473–87.

Dahlhoff, E. & Somero, G. N. (1993a). Kinetic and structural adaptations of cytoplasmic malate dehydrogenases of eastern Pacific abalones (genus *Haliotis*) from different thermal habitats: biochemical correlates of biogeographical patterning. *Journal of Experimental Biology*, **185**, 137–50.

Dahlhoff, E. & Somero, G. N. (1993b). Effects of temperature on mitochondria from abalone (genus *Haliotis*): adaptive plasticity and its limits. *Journal of Experimental Biology*, **185**, 151–68.

Dietz, T. J. (1994). Acclimation of the threshold induction temperatures for 70-kDa and 90-kDa heat shock proteins in the fish *Gillichthys mirabilis*. *Journal of Experimental Biology*, **188**, 333–8.

Dietz, T. J. & Somero, G. N. (1992). The threshold induction temperature of the 90-kDa heat shock protein is subject to acclimatization in eurythermal goby fishes (genus *Gillichthys*). *Proceedings of the National Academy of Sciences USA*, **89**, 3389–93.

Donahue, V. E. (1982). Lactate dehydrogenase: structural aspects of environmental adaptation. PhD dissertation, University of California, San Diego.

Ellis, R. J. & van der Vies, S. M. (1991). Molecular chaperones. *Annual Review of Biochemistry*, **60**, 321–47.

Fields, P. (1995). Adaptation to temperature in two genera of coastal fishes, *Paralabrax* and *Gillichthys*. PhD dissertation, University of California, San Diego.

Fields, P., Graham, J. B., Rosenblatt, R. H. & Somero, G. N. (1993). Effects of expected global change on marine faunas. *Trends in Ecology and Evolution*, **8**, 30–7.

Goff, S. A. & Goldberg, A. L. (1985). Production of abnormal proteins in *E. coli* stimulates transcription of *lon* and other heat shock genes. *Cell*, **41**, 587–95.

Graves, J. E. & Somero, G. N. (1982). Electrophoretic and functional enzymic evolution in four species of eastern Pacific barracudas from different thermal environments. *Evolution*, **36**, 97–106.

Haas, A. L. & Bright, P. M. (1985). The immunochemical detection and quantification of intracellular ubiquitin-protein conjugates. *Journal of Biological Chemistry*, **260**, 12464–73.

Hawkins, A. J. S. (1985). Relationships between the synthesis and breakdown of protein, dietary absorption and turnover of nitrogen and carbon in the blue mussel, *Mytilus edulis* L. *Oecologia*, **66**, 42–9.

Hawkins, A. J. S. (1991). Protein turnover: a functional appraisal. *Functional Ecology*, **5**, 222–33.

Hightower, L. E. (1980). Cultured animal cells exposed to amino acid analogues or puromycin rapidly synthesize several polypeptides. *Journal of Cell Physiology*, **102**, 407–24.

Hochachka, P. W. & Somero, G. N. (1984). *Biochemical Adaptation*. Princeton, NJ: Princeton University Press.

Hofmann, G. E. & Somero, G. N. (1995). Evidence for protein damage at environmental temperatures: seasonal changes in levels of ubiquitin conjugates and hsp70 in the intertidal mussel *Mytilus trossulus*. *Journal of Experimental Biology*, **198**, 1509–18.

Houlihan, D. F. (1991). Protein turnover in ectotherms and its relationship to energetics. In *Advances in Comparative and Environmental Physiology*, Vol. 7, ed. R. Gilles, pp. 1–43. Berlin: Springer-Verlag.

Huey, R. B., Partridge, L. & Fowler, K. (1991). Thermal sensitivity of *Drosophila melanogaster* responds rapidly to laboratory natural selection. *Evolution*, **45**, 751–6.

Intergovernmental Panel on Climate Change (IPCC) (1992). *Climate Change 1992: The Supplementary Report to the IPCC Assessment*, ed. J. T. Houghton, B. A. Callander & S. K. Varney. Cambridge: Cambridge University Press.

Jaenicke, R. (1991). Protein stability and molecular adaptation to extreme conditions. *European Journal of Biochemistry*, **202**, 715–28.

Johnston, I. A. (1983). Cellular responses to an altered body temperature: the role of alterations in the expression of protein isoforms. In: *Cellular Acclimatisation to Environmental Change*, ed. A. R. Cossins & P. Sheterline. *Society for Experimental Biology Seminar Series* 17, pp. 121–43. Cambridge: Cambridge University Press.

Johnston, I. A. & Walesby, N. J. (1977). Molecular mechanisms of temperature adaptation in fish myofibrillar adenosine triphosphatases. *Journal of Comparative Physiology B*, **119**, 195–206.

Kareiva, P. M., Kingsolver, J. G. & Huey, R. B. (1993). *Biotic Interactions and Global Change*. Sunderland, Mass.: Sinauer Associates.

Lin, J. J. & Somero, G. N. (1995a). Temperature-dependent changes in expression of thermostable and thermolabile isozymes of cytosolic malate dehydrogenase in the eurythermal goby fish *Gillichthys mirabilis*. *Physiological Zoology*, **68**, 114–28.

Lin, J. J. & Somero, G. N. (1995b). Thermal adaptation of cytoplasmic malate dehydrogenases of eastern Pacific barracuda (*Sphyraena* spp.): the role of differential isoenzyme expression. *Journal of Experimental Biology*, **198**, 551–60.

Lindquist, S. & Craig, E. A. (1988). The heat-shock proteins. *Annual Review of Genetics*, **22**, 631–77.

Lubchenco, J., Navarette, S. A., Tissot, B. N. & Castilla, J. C. (1993). Possible ecological responses to global climate change: nearshore benthic biota of northeastern Pacific ecosystems. In *Earth Systems Responses to Global Change*, ed. H. Mooney, E. R. Fuentes & B. I. Kronberg, pp. 147–66, London: Academic Press.

McFall-Ngai, M. J. & Horwitz, J. (1990). A comparative study of the thermal stability of the vertebrate eye lens: Antarctic ice fish to the desert iguana. *Experimental Eye Research*, **50**, 703–9.

Matthews, B. W. (1987). Genetic and structural analysis of the protein stability problem. *Biochemistry*, **26**, 6885–8.

Parsell, D. A., Kowal, A. S., Singer, M. A. & Lindquist, S. (1994). Protein disaggregation mediated by heat-shock protein hsp104. *Nature*, **372**, 475–8.

Parsell, D. A. & Lindquist, S. (1993). The function of heat-shock proteins in stress tolerance: degradation and reactivation of damaged proteins. *Annual Review of Genetics*, **27**, 437–96.

Powers, D. A., Smith, M., Gonzalez-Villasenor, I., DiMichele, L., Crawford, D., Bernardi, G. & Lauerman, T. (1993). A multidisciplinary approach to the selectionist/neutralist controversy using the model teleost, *Fundulus heteroclitus*. In *Oxford Surveys in Evolutionary Biology*, Vol. 9, ed. D. Futuyma & J. Antonovics, pp. 43–107. Oxford: Oxford University Press.

Rechsteiner, M. (1987). Ubiquitin-mediated pathways for intracellular proteolysis. *Annual Review of Cell Biology*, **3**, 1–30.

Ritossa, R. M. (1962). A new puffing pattern induced by heat shock and DNP in *Drosophila*. *Experientia*, **18**, 571–3.

Roemmich, D. & McGowan, J. (1995). Climatic warming and the decline of zooplankton in the California current. *Science*, **267**, 1324–6.

Sanders, B. M., Hope, C., Pascoe, V. M. & Martin, L. S. (1991). Characterization of the stress protein response in two species of *Collisella* limpets with different temperature tolerances. *Physiological Zoology*, **64**, 1471–89.

Sharp, V. A., Miller, D., Bythell, J. C. & Brown, B. E. (1994). Expression of low molecular weight HSP 70 related polypeptides from the symbiotic sea anemone *Anemonia viridis* Forskall in response to heat shock. *Journal of Experimental Marine Biology and Ecology*, **179**, 179–93.

Somero, G. N. (1995). Proteins and temperature. *Annual Review of Physiology*, **57**, 43–68.

Somero, G. N., Dahlhoff, E. & Lin J. J. (1996). Stenotherms and eurytherms: mechanisms establishing thermal optima and tolerance ranges. In *Phenotypic and Evolutionary Adaptation to Temperature*, ed. I. A. Johnston & A. F. Bennett. *Society for Experimental Biology Seminar Series*. Cambridge: Cambridge University Press (in press).

Somero, G. N. & DeVries, A. L. (1967). Temperature tolerance of some Antarctic fishes. *Science*, **156**, 257–8.

Swezey, R. R. & Somero, G. N. (1982). Polymerization thermodynamics and structural stabilities of skeletal muscle actins from vertebrates adapted to different temperatures and pressures. *Biochemistry*, **21**, 4496–503.

Tsuji, S., Qureshi, M. A., Hou, E. W., Fitch, W. M. & Li, S. S.-L. (1994). Evolutionary relationships of lactate dehydrogenases (LDHs) from mammals, birds, an amphibian, fish, barley, and bacteria: LDH cDNA sequences from *Xenopus*, pig, and rat. *Proceedings of the National Academy of Sciences USA*, **91**, 9392–6.

White, C. N., Hightower, L. E. & Schultz, R. J. (1994). Variation in heat-shock proteins among species of desert fishes (Poeciliidae, *Poeciliopsis*). *Molecular Biology and Evolution*, **11**, 106–19.

Yancey, P. H. & Siebenaller, J. F. (1987). Coenzyme binding ability of homologues of M_4-lactate dehydrogenase in temperature adaptation. *Biochimica et Biophysica Acta*, **924**, 483–91.

Yancey, P. H. & Somero, G. N. (1978). Temperature dependence of intracellular pH: its role in the conservation of pyruvate apparent K_m values of vertebrate lactate dehydrogenase. *Journal of Comparative Physiology B*, **125**, 129–34.

JOHN C. ROBERTSON
and JEFFREY R. HAZEL

Membrane constraints to physiological function at different temperatures: does cholesterol stabilize membranes at elevated temperatures?

Introduction – impact of global warming on membrane physiology

Consequences of poikilothermy: perturbation of membrane structure

Because temperature determines the rates of biological processes as well as the distribution of conformations assumed by both molecules and molecular aggregates (e.g. membranes) in cells (Hochachka & Somero, 1984), physiological function is markedly perturbed by fluctuations in body temperature. Thermal perturbation of function is particularly acute for aquatic poikilotherms such as fish, for the high heat capacity and relatively low oxygen content of aquatic environments combine to ensure that, as a consequence of respiration, the body tissues remain within 1 °C of ambient water temperature (Carey et al., 1971). One of the primary consequences of poikilothermy in fishes is the perturbation of membrane organization when body temperature changes, for the structural lipids of biological membranes display complex phase behaviours and physical properties that are exquisitely sensitive to temperature (Hazel, 1995). Temperature effects are most evident as altered physical properties of the phospholipid acyl domain, which in turn determine the order (time-averaged position in space) and rates of motion of membrane constituents, the molecular geometries of membrane phospholipids, and the dynamic phase behaviour of the membrane. At normal physiological temperatures (i.e. the temperature to which an animal is either acclimated or adapted), the fatty acyl chains of membrane phospholipids are moderately disordered and

Society for Experimental Biology Seminar Series 61: *Global Warming: Implications for freshwater and marine fish*, ed. C. M. Wood & D. G. McDonald. © Cambridge University Press 1996, pp. 25–49.

relatively fluid due to the presence of 3–7 gauche rotamers (rotations about carbon–carbon single bonds) per acyl chain (Casal & Mantsch, 1984); however, in spite of this disorder, the average molecular geometry of most phospholipid molecules is essentially cylindrical. Since cylindrically-shaped phospholipids (such as phosphatidylcholine (PC), which possesses a relatively large headgroup compared to its hydrophobic volume) pack regularly to form stable bilayers, they are referred to as bilayer-stabilizing lipids and the state of the membrane at normal physiological temperatures is referred to as a **lamellar** (i.e. bilayer) **liquid-crystalline** or **fluid phase** (Fig. 1).

In the context of global warming, effects of temperatures higher than the normal physiological temperature on membrane structure and function are most relevant. At temperatures just above the physiological range, membranes become increasingly disordered as the hydrophobic

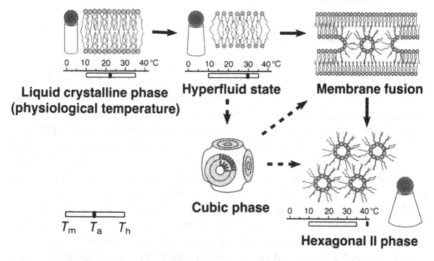

Liquid crystalline phase (physiological temperature)

Hyperfluid state

Membrane fusion

T_m T_a T_h

Cubic phase

Hexagonal II phase

Fig. 1. Effects of elevated temperatures on membrane physical properties and phase state. With warming from the physiological temperature, phospholipid geometry begins to deviate from the cylindrical shape characteristic of the liquid crystalline phase and the membrane becomes less ordered (hyperfluid). With further warming, phospholipid geometry becomes conical and there is a transition (occurring at temperature T_h) to the hexagonal II (H_{II}) phase, possibly proceeding through a cubic phase intermediate. As a consequence of warming, the operational temperature (T_a) is shifted away from its normal position between the gel/fluid (T_m) and fluid/hexagonal phase transitions (T_h) to a point in closer proximity to or above T_h, thus altering the dynamic phase behaviour (fusogenicity) of the membrane.

domain of phospholipids expands and the average molecular geometry begins to deviate from cylindrical, resulting in a 'hyperfluid' state of the membrane. With additional increases in temperature, disorder in the acyl domain of membrane phospholipids is further increased to a point where some phospholipids (most notably phosphatidylethanol-amine (PE), which possesses a relatively small headgroup relative to its hydrophobic volume) assume a conical rather than a cylindrical geometry. Although conical lipids can accommodate increased disorder in the acyl domain, they cannot, in the absence of cylindrically-shaped lipids, pack to form a stable lamellar phase; consequently, at some critical temperature (T_h), conical or bilayer-destabilizing lipids assume an **inverted hexagonal (H_{II}) phase**, which destroys the integrity (i.e. barrier properties) of the membrane bilayer (Seddon, 1990). Thus, in addition to increasing the rates and extent of acyl chain motions, elevated temperatures also reduce the temperature interval between the temperature at which the membrane is functioning and the hexagonal phase transition. Since the proximity of the membrane to the H_{II} phase transition is an important determinant of the dynamic phase behavior of a membrane (i.e. the ability to undergo the regulated fusion required for endo- and exocytosis and intracellular membrane trafficking), elevated temperatures will destabilize the lamellar phase (Hazel, 1995).

Consequences of poikilothermy: perturbation of membrane function

The membrane properties reviewed above are sufficiently sensitive to temperature that even the moderate warming (2–4 °C in average temperatures within the next century) predicted by climate change models (Intergovernmental Panel on Climate Change, 1992), and certainly the excursions in temperature from average values, are anticipated to have a significant impact on membrane structure and function. For example, membrane activities as diverse as Na^+/K^+-ATPase (Harris, 1985), active chloride transport (Gasser, Goldsmith & Hopfer, 1990), receptor-mediated uptake of low density lipoprotein (Kuo, Weinfeld & Loscalzo, 1990), the rotational mobility of integral membrane proteins (Squier, Bigelow & Thomas, 1988), the passive permeability of fluid phase membranes, and the collision coupling between membranous components of signal transduction pathways (Rimon, Hanski & Levitzki, 1980) are all positively correlated with membrane fluidity, which is increased at elevated temperatures. The phase state and/or physical properties of membrane lipids also appear to define the thermal limits of physiological function. For example, the leakage of potassium ion from muscle fibres at elevated temperatures has been implicated as a cause of heat death in crayfish (Gladwell, Bowler & Duncan, 1975), and in microorganisms the maximum growth temperature can be decreased by growth on low-melting-point fatty acids (Kruuv *et al.*, 1983). In addition, recent experiments indicate that *E. coli* regulate the phospholipid composition of their membranes so that the transition to the H_{II} phase occurs approximately 10 °C above the growth temperature (Rietveld *et al.*, 1993). Collectively, these data suggest that increases in membrane permeability or the formation of non-lamellar phases (with the consequent loss of bilayer integrity), may constrain physiological function at high temperatures, just as formation of the gel phase does at low temperatures.

The problem of global warming: stabilizing the fluid phase bilayer

It is clear from the foregoing discussion that global warming challenges the ability of poikilothermic organisms to maintain critical physical properties and phase behaviours of cellular membranes. In order to deal successfully with this challenge, organisms must be able to offset the reduction in membrane order and increased propensity to form non-lamellar phases induced by elevated temperatures. Thus, from a membrane perspective, the problem of global warming can be reduced to finding suitable mechanisms for stabilizing the fluid phase bilayer at elevated temperatures.

Does cholesterol stabilize biological membranes at elevated temperatures?

The inherent thermal sensitivity of the phase behaviour and physical properties of membrane lipids limits the temperature range over which a given set of membrane constituents can function effectively. It is therefore not surprising that eurythermal poikilotherms, of which freshwater and marine fish are prime examples, restructure their membranes so that lipids of appropriate physical and regulatory properties are matched to the prevailing thermal conditions. Based on a diverse literature describing the effects of growth temperature on membrane lipid composition (Hazel & Williams, 1990), any one or a combination of the following adjustments to membrane lipid composition could contribute to the stabilization of membrane structure and function at temperatures above the normal physiological range: (1) increased proportions of saturated fatty acids (Hazel, 1988); (2) elevated proportions of bilayer-stabilizing (e.g. PC) to bilayer-destabilizing (e.g. PE) phospholipids (Pruitt, 1988; Thompson, 1989); (3) elevated proportions of plasmalogen (i.e. alk-1-enyl ether phospholipids) compared to diacyl phospholipids (Matheson, Oei & Roots, 1980); (4) an increased content of neutral lipids such as cholesterol. Although there are considerable data to support all of these possible mechanisms of membrane stabilization at warm temperatures, and all are likely to contribute to some degree to the adaptation of membrane structure and function to high temperatures, the remainder of this chapter will focus solely on the role of neutral lipids in this process.

Effects of cholesterol on biological membranes

Because of their well-established biophysical effects on membranes, sterols (such as cholesterol) or sterol analogues appear to be ideally suited to stabilize membrane structure at elevated temperatures. Furthermore, modulation of membrane cholesterol content may occur more quickly and at less metabolic cost than adjustments to other aspects of membrane lipid composition. Cholesterol is an amphipathic steroid possessing a 3-β-hydroxyl group at one end of the molecule and a flexible alkyl tail at the other (Yeagle, 1985). The steroid nucleus lies parallel to and buried within the acyl domain of the membrane with the 3-β-hydroxyl group exposed near the bilayer surface (in the vicinity of the ester carbonyl linkages of the phospholipids) and the alkyl tail projecting toward the bilayer interior. Cholesterol can be readily incorporated into most membranes up to molar ratios of 1:1 and possibly higher relative to phospholipids, and its actions on membrane physical and functional properties have been extensively studied. Cholesterol exerts the following effects on fluid phase lipids: (i) increases the orientational order of membrane constituents, particularly in that portion of the acyl chains closest to the headgroup region and in direct contact with the steroid nucleus, by inhibiting the formation of gauche rotamers in the phospholipid acyl chains (Davies *et al.*, 1990). This effect is reflected in an increased membrane thickness and the 'condensation' of membrane lipids (i.e. the actual surface area occupied at an air–water interface in binary mixtures of phospholipid and cholesterol is less than the summed areas of the individual components); (ii) reduces the area compressibility of a membrane, leading to greater cohesion and reduced permeability (Bloch, 1991); (iii) broadens, and in some cases, eliminates the gel/

fluid transition, due, in part, to its ability to disrupt the packing of gel phase lipids while ordering fluid phase lipids (Yeagle, 1985); (iv) has minimal impact on the rotational diffusion coefficients of membrane constituents, thus reflecting the maintenance of a fluid-like microviscosity (Bloch, 1991); (v) increases the space between lipid molecules in the headgroup region, permitting water to penetrate further into the lipid–water interface (Subczynski et al., 1994); (vi) reduces the free volume in the region of the bilayer adjacent to the aqueous interface, but creates free volume in the centre of the bilayer (Subczynski et al., 1994); and (vii) may either increase or decrease the transition temperature to the H_{II} phase, depending on the chemical composition of the phospholipids in the bilayer (Seddon, 1990). Thus, cholesterol exerts a predominantly stabilizing effect on fluid phase membranes, primarily by promoting the trans configuration about single bonds in the acyl chains of phospholipids and by filling free volumes (hydrophobic pockets) in the bilayer matrix. Based on the majority of these actions, cholesterol appears ideally suited to the stabilization of fluid phase membranes at elevated temperatures. However, the vast majority of the cholesterol paradigm has been developed from studies of the interactions between cholesterol and predominantly saturated phospholipids, whereas weaker interactions occur between cholesterol and diunsaturated phospholipids. In fact, interactions between cholesterol and unsaturated phospholipids are in some cases markedly diminished. This is evidenced by the failure of 30–50 mol% cholesterol to reduce membrane permeability, to be freely miscible with unsaturated phospholipids (resulting in fluid phase immiscibility of cholesterol-rich and poor membrane domains), and to reduce the enthalpies of the gel/fluid transition (Keough, 1992).

In summary, results derived from the interactions of cholesterol with phospholipids in model lipid membranes lead to the prediction that, in poikilotherms, the elevation of membrane cholesterol content may be a particularly effective mechanism for stabilizing plasma membranes (where the vast majority of cellular cholesterol is located; Lange et al., 1989) against the perturbing effects of warm temperatures. Nevertheless, until recently, the potential involvement of sterols in the thermal adaptation of membrane function in poikilotherms has not been extensively evaluated. Furthermore, the physical effects which cholesterol has in model membrane systems of relatively simple composition may not apply to natural mixtures of phospholipids. This is especially true in regard to membranes rich in unsaturated phospholipids, as is typical of the membranes of many freshwater and marine fish.

Stabilization of plasma membranes by sterol analogues in prokaryotes

Because most prokaryotes lack the capacity to produce sterols (see below), cholesterol is not typically a component of cell membranes in bacteria and related organisms. Absence of cholesterol from prokaryotic membranes, however, does not reflect a phylogenetic limit to the presence of stabilizing compounds in biological membranes. In fact, such components are widely distributed, critical elements of prokaryotic cellular membranes. Thus, while prokaryotes do not contain cholesterol, they do contain compounds which (1) play a structural role in membranes analogous to that of cholesterol, and (2) may themselves represent phylogenetic precursors of cholesterol in terms of their molecular evolution and role in membrane stabilization.

As proposed by Rohmer, Bouvier and Ourisson (1979), these prokaryotic sterol analogues are polyterpenes derived from the isoprenoid biosynthetic pathway which yields the tetracyclic triterpenoid cholesterol in eukaryotes. Hopanoids are amphipathic pentacyclic triterpenoids which, like cholesterol, possess a planar, rigid polycyclic ring structure and the molecular dimensions and polarity to intercalate into one hemilayer of a membrane. Interestingly, hopanoids like bacteriohopanetetrol (Fig. 2) have their polar groups on the flexible hydrocarbon tail, and thus have an opposite ring-tail membrane orientation to that of cholesterol. On the other hand, tetraterpene carotenoids are typically long, rigid hydrocarbons which could traverse the bilayer and maintain a stabilizing orientation via hydrophobicity or terminal hydrophilic groups (Fig. 2).

Hopanoids are widely distributed among prokaryotes, often in quantities similar to the sterols of eukaryotes (Rohmer, Bouvier-Nave & Ourisson, 1984; Sahm *et al.*, 1993), and membrane association of hopanoids has been demonstrated in several bacterial species (Hermans, Neuss & Sahm, 1991; Jurgens, Simonin & Rohmer, 1992). In model systems, hopanoids condense membranes (Poralla, Kannenberg & Blume, 1980; Kannenberg *et al.*, 1983, 1985), decrease liposome permeability (Kannenberg *et al.*, 1985) and broaden the gel to fluid phase transition (Kannenberg *et al.*, 1983), effects all similar to those described for cholesterol. The suppression of growth accompanying inhibition of hopanoid synthesis in several species of bacteria (Flesch & Rohmer, 1987) illustrates the physiological importance of these compounds. The high hopanoid content of *Zymomonas moblis*, an ethanol-producing bacterium, has been suggested to reflect adaptation to a membrane-destabilizing alcohol environment (Hermans *et al.*, 1991).

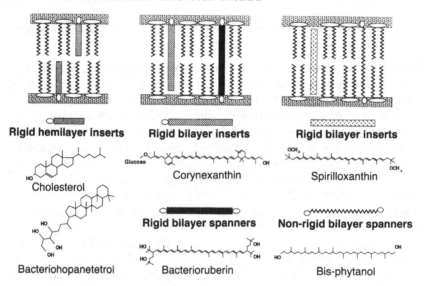

Fig. 2. Interaction of various stabilizing elements with membrane bilayers is dictated by molecular dimensions and structural characteristics. Amphipathic rigid inserts could order membranes by intercalating into one hemilayer of the membrane (e.g. cholesterol and bacteriohopanoids, top left) and interacting with other lipid components. Alternatively, longer inserts (e.g. carotenoids, top centre and right) could have a stabilizing effect by extending across the bilayer; presence of terminal polar groups or hydrophobicity would maintain the reinforcing orientation of these compounds in the membrane. Structures of compounds exemplifying each type of proposed interaction are given.

Moreover, the content of hopanoids in the thermoacidophilic bacterium *Bacillus acidocaldarius* was found to increase with lowered pH and especially with higher temperatures (Poralla, Hartner & Kannenberg, 1984).

Carotenoids are present in membranes of both prokaryotic and eukaryotic organisms. While the membranes of sterol-requiring mycoplasmas contain sterol but no carotenoids, membranes of *Acholeoplasma laidlawii* (which does not require sterol) contain carotenoids instead of sterols (Huang & Haug, 1974). Biophysical studies have demonstrated that polar carotenoids intercalate into and stabilize (i.e. order) both model phospholipid bilayers (Milon *et al.*, 1986; Subczynski *et al.*, 1992) and membranes of extracted *Halobacterium* and *A. laidlawii* lipids (Rottem & Markowitz, 1979; Lazrak *et al.*, 1988).

Thus, there is substantial evidence in support of polyterpenes acting as reinforcing components in prokaryotic membranes, a role analogous to that of cholesterol in eukaryotic plasma membranes. Furthermore, the membrane content of these stabilizing elements appears to be regulated in an adaptive fashion in response to changes in environmental conditions.

From an evolutionary perspective, the analogous membrane function of cholesterol and other polyterpenoids in phylogenetically diverse organisms is particularly intriguing in view of the biosynthetic relationship of these compounds. It has been suggested (Rohmer *et al.*, 1979; Ourisson, Rohmer & Poralla, 1987) that the isoprenoid biosynthetic pathway represents a phylogeny which reflects the molecular evolution of membrane-reinforcing elements (Fig. 3). In this scheme, tetra-

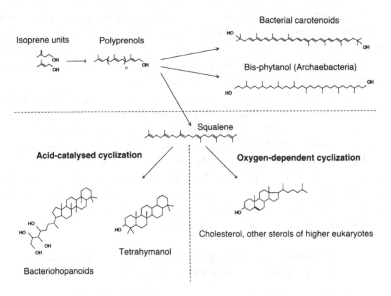

Fig. 3. A proposed phylogeny of membrane-stabilizing components is reflected in the isoprenoid biosynthetic pathway. The earliest membrane-ordering compounds may be long rigid polyterpenes, such as carotenoids, which could arise via polymerization of polyprenols. The more advanced polycyclic triterpenoids of some prokaryotes or lower eukaryotes (e.g. hopanoids, tetrahymanol) are produced through proton-catalysed cyclization of squalene. Alternatively, synthesis of cholesterol and other sterols of eukaryotes proceeds via an oxygen-dependent squalene cyclase mechanism, which presumably evolved after the photosynthesis-dependent development of an oxygenated atmosphere on Earth.

terpenoids like carotenoids represent the most 'primitive' membrane stabilizers; these compounds could arise via polymerization of polyprenols. The presence of carotenoids and derivatives of bis-phytanol in archaebacteria has been suggested to reflect the ancestral nature of these membrane elements (Rohmer et al., 1979). Production of terpenes containing ring structures suggests greater complexity, requiring evolution of a squalene cyclase system and additional steps to ensure a proper molecular fit of membrane components. The proton-catalysed cyclization of squalene in prokaryotes gives rise to a family of polycyclic triterpenoids which include hopanoids. Sterol biosynthesis, however, proceeds via a different pathway, dependent on molecular oxygen for generation of the cyclization substrate (squalene epoxide) and for subsequent sterol-modifying reactions. Lack of cholesterol in prokaryotes has thus been attributed to absence of the oxygen-based cyclase system, reflecting evolution in the oxygen-free atmospheric conditions predating photosynthesis (Bloch, 1991). Progression along this phylogeny has further been interpreted as indicating increasing utility as membrane components, with cholesterol thus representing the most effective membrane stabilizer (Bloch, 1991).

According to this view, terpenoids represent an ancient and conserved mechanism for regulating the physical properties of biological membranes. Involvement of cholesterol in adaptive plasma membrane restructuring may then have evolutionary roots in ancestors which lacked a complex organelle network and utilized carotenoids or hopanoids to adjust the physical properties of their cellular membranes. Studies of the metabolism and membrane interactions of terpenoids in prokaryotes thus might contribute to understanding of the cellular physiology of cholesterol in eukaryotes.

Evidence supporting a role for cholesterol in stabilizing membranes at elevated temperatures

Cholesterol in eukaryotic cell membranes

Although the possible involvement of cholesterol in adaptive membrane restructuring in lower eukaryotes has not been directly investigated, the ciliate *Tetrahymena pyriformis* has been the subject of a number of relevant studies. By way of proton-catalysed cyclization of squalene, this eukaryote produces tetrahymanol, a pentacyclic triterpenoid related to hopanoids; surface membranes of *Tetrahymena*, particularly the ciliary membranes, are enriched in tetrahymanol (Ramesha & Thompson, 1982). However, if provided with exogenous cholesterol, synthesis of tetrahymanol is suppressed and cholesterol is incorporated into membranes. Evidence also indicates that sterols other than cholesterol, as well as hopanoids and carotenoids, can all support growth when tetrahymanol production is chemically inhibited (Sahm *et al.*, 1993). These observations thus support the proposed functional analogy of different terpenoids as structural elements of membranes. Increased proportions of tetrahymanol (in relation to phospholipids) were reported in ciliary membranes of 39 °C-acclimated as compared to 13 °C-acclimated cells (Ramesha & Thompson, 1982), suggesting this cholesterol-like compound may be important in the adaptation of some *Tetrahymena* membranes to elevated temperatures.

The sterol:phospholipid ratios of two other lower eukaryotes, *Neurospora crassa* (Aaronson, Johnston & Martin, 1982) and *Saccharomyces cerevisiae* (Hunter & Rose, 1972), have also been reported to increase at higher growth temperatures. However, these studies analysed total cell lipids, not isolated membranes, and include sterols which may have characteristics unlike those of cholesterol.

Somewhat surprisingly, work with mammalian cells in culture provides some compelling experimental support for involvement of cholesterol in adaptive regulation of membrane order. For example, in Chinese hamster ovary (CHO) fibroblasts, total membrane cholesterol:phospholipid molar ratios were positively correlated with both culture temperature (ranging from 32 to 41 °C) and membrane order (Anderson *et al.*, 1981). When CHO cell plasma membranes were chemically fluidized (thereby mimicking the disordering effects of temperature increase) up-regulation of cholesterol synthesis was suggested by the resulting 2–3-fold increases in activity of 3-hydroxy-3-methylglutaryl coenzyme A (HMGCoA) reductase and microsomal membrane cholesterol content (Sinensky & Kleiner, 1981). And in a mutant CHO cell line unable to regulate sterol metabolism, elevating plasma membrane cholesterol content by nutritional manipulation led to a parallel decrease in plasma membrane fluidity (Sinensky, 1978).

Data in the literature relating membrane cholesterol content to growth temperature are conflicting, in part, because many studies were performed with intracellular membranes. These investigations predate the seminal work of Lange and colleagues (Lange *et al.*, 1989), demonstrating that the vast majority of cellular cholesterol is present in the plasma membrane. Thus, while direct positive correlations between growth temperature and cholesterol content of organelle membranes from metazoan poikilotherms have been reported (e.g. Chang & Roots, 1989), these data may reflect some degree of impurity of membrane fractions. It now appears clear that the plasma membrane is the most appropriate location for examining the possibility that cholesterol modulation is involved in an adaptive membrane response.

Cholesterol levels in the plasma membranes of various tissues of thermally-acclimated trout

To ascertain if cholesterol might play a role in temperature-evoked membrane restructuring in a vertebrate poikilotherm, cholesterol levels were quantified in plasma membranes isolated from several different tissues of 20 °C-acclimated and 5 °C-acclimated rainbow trout (Robertson & Hazel 1995). Cholesterol content, expressed as the cholesterol:phospholipid (C:P) molar ratios, exhibited a consistent pattern of change in plasma membranes from trout in the two temperature acclimation groups (Fig. 4). In liver, kidney and gill, 20 °C-acclimated fish had significantly higher ($p < 0.05$) plasma membrane C:P ratios than fish maintained at 5 °C. In erythrocyte plasma membranes the mean C:P ratio was again higher in warm- versus cold-acclimated fish,

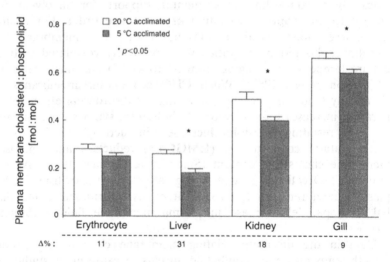

Fig. 4. Cholesterol:phospholipid molar ratios of plasma membranes isolated from tissues of rainbow trout acclimated to either 5 °C or 20 °C. Data are means of 7–10 preparations per tissue at each temperature; bars indicate SEM. Asterisks denote significant (p < 0.05) difference between warm- and cold-acclimation groups as judged by t-test of arcsin transformed ratio data. Magnitude of temperature-associated change in relative cholesterol content ($\triangle$%) of plasma membranes for each tissue is calculated as:

$$\triangle\% = [(C:P \text{ at } 20° - C:P \text{ at } 5°) \div (C:P \text{ at } 20°)] \times 100.$$

but this difference was not significant ($p = 0.25$). These results thus (1) constitute correlative support for cholesterol modulation as a component of temperature-induced plasma membrane reorganization in a number of trout tissues, and (2) are consistent with the prediction that cholesterol may stabilize plasma membranes in the face of the disordering effects of increased temperature. The physiological significance of changes seen in trout membranes is further supported by studies of the effects of cholesterol and temperature on the physical properties of lipids extracted from trout liver plasma membranes (see below). Thus, thermal modulation of cholesterol levels in plasma membranes of a number of trout tissues appears to occur in a manner consistent with homeoviscous adaptation.

Analysis of a number of tissues also allows evaluation of any tissue differences in membrane cholesterol composition or acclimatory response. Gross differences were found in the relative plasma membrane cholesterol content of the trout tissues examined (Fig. 4). Gill had the highest mean C:P ratio, followed in descending order by kidney, erythrocyte and liver; gill and kidney C:P ratios were 2 to 2.5 times higher than those of liver and red blood cells. Similar differences were seen in cholesterol:protein ratios, with values for both gill and kidney plasma membranes again being 2-fold greater than those of either liver or erythrocytes. Heterogeneity in plasma membrane cholesterol level has been noted among tissues of other species and presumably reflects functional or structural differences in the membranes of various cell types.

In relation to relative cholesterol content, then, the degree of the temperature-associated shift in plasma membrane cholesterol content follows an inverse pattern in the different trout tissues. The greatest proportional rise in plasma membrane C:P ratio with warm-acclimation (indicated as $\Delta\%$ in Fig. 4) was seen in liver, followed by kidney, and was least in gill and erythrocytes. These data thus suggest that plasma membrane acclimatory responses involving cholesterol may not be of equal physiological significance in cells of different tissues. For example, while liver and erythrocyte plasma membranes had similar gross C:P ratios, warm-acclimation resulted in a 3-fold greater increase in the C:P ratio in liver as compared to red blood cells. Again, the tissue-specific functional requirements of cells could account for variation in membrane adaptive responses involving cholesterol. This relationship may extend to the level of discrete plasma membrane domains within a polarized cell: for example, the basolateral plasma membrane domain of hepatocytes contains less cholesterol and a different phospholipid composition compared with the apical region (Meier *et al.*, 1984). These structural differences are likely to support domain-specific membrane activities and suggest that adaptive changes in plasma membrane composition, including those involving cholesterol, may occur independently in each domain. Studies with trout intestinal epithelia, described below, illustrate the domain-specific nature of thermally-evoked changes in plasma membrane composition.

Interspecific comparisons of plasma membrane cholesterol indicate a positive correlation between body temperature and cholesterol content

Based on comparison with data from other species, trout plasma membrane C:P ratios further suggest that a general relationship may exist between the thermal physiology of vertebrates and plasma membrane cholesterol concentrations. The C:P molar ratios for plasma membranes from liver, kidney, erythrocytes and enterocytes of trout are lower than published values for corresponding tissues from homeotherms (Fig. 5). This interspecific difference is greatest in erythrocyte and liver membranes, where trout C:P ratios are 2–4-fold less than those of organisms maintaining higher body temperatures. The relatively low cholesterol contents reported for plasma membranes from intestinal epithelial cells of another poikilotherm, the tortoise (Chappelle & Gilles-Baillien, 1989; Fig. 5), are also consistent with a phylogenetic pattern. Thus, in addition to the acclimatory changes seen in some trout plasma membranes, comparative data suggest a possible adaptive, evolutionary relationship between plasma membrane cholesterol concentrations and thermal status.

Any broad correlations drawn from interspecific comparisons of membrane cholesterol concentrations must be tempered in view of the present lack of data for non-mammalian species. Another tissue for which acclimatory and interspecific data are available is sperm. No difference was found in sperm plasma membrane cholesterol concentrations in trout acclimated to 8 or 18 °C (mean C:P molar ratios of 0.33 and 0.35, respectively; Labbe et al., 1995); furthermore, the reported trout sperm C:P ratios fall within the range of values reported for mammalian and avian species (Parks & Lynch, 1992). However, differences in phospholipid class, fatty acid composition and membrane fluidity, all responses typical of temperature acclimation in membranes of other trout tissues, were also not seen in sperm plasma membranes from warm- versus cold-acclimated trout (Labbe et al., 1995). Thus, lack of a thermally-adaptive plasma membrane response in sperm might

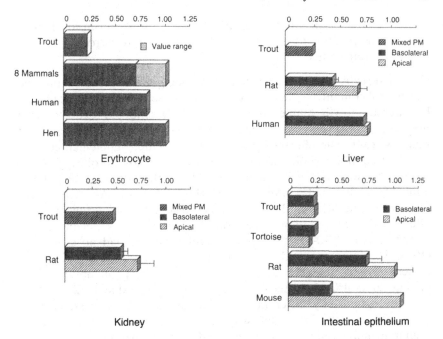

Fig. 5. Comparison of cholesterol:phospholipid (C:P) molar ratios of plasma membranes isolated from different tissues of trout with corresponding published values for homeotherms. Consistently lower values for the trout suggest a correlation may exist between membrane cholesterol concentrations and the thermal physiology of vertebrates. Data for trout are from 20 °C-acclimated animals. Bars indicate SEM where values from independent studies were combined to generate a mean C:P ratio. Trout intestinal epithelia data from Crockett & Hazel, 1995. Homeotherm data from: erythrocytes: Nelson, 1967; Kleinig *et al.*, 1971; Flamm & Schachter, 1982; liver: Meier *et al.*, 1984; Wolters *et al.*, 1991; intestinal epithelia: Kawai, Fujita & Nakao, 1974; Brasitus & Schachter, 1980; Chappelle & Gilles-Baillien, 1983, 1989; kidney: Perez-Albarsanz *et al.*, 1991; Imai *et al.*, 1992; Ramsammy *et al.*, 1993.

reflect the highly specialized nature of this membrane and cell type. It is also notable that C:P ratios from trout kidney and, particularly, gill plasma membranes approach magnitudes similar to those encountered in mammalian tissues. This may indicate an overriding functional role, perhaps in permeability regulation, for cholesterol in plasma membranes of certain tissues.

Cholesterol supplementation orders phospholipids from the plasma membranes of cold-acclimated trout hepatocytes

To test directly the potential impact of changes in plasma membrane cholesterol content similar in magnitude to those induced in hepatocytes by altered growth temperature (Fig. 4), phospholipid extracts were prepared from liver plasma membranes isolated from 5 °C-acclimated rainbow trout. The phospholipid extracts were then supplemented with 10, 15 or 20 mol% cholesterol and the effects of cholesterol supplementation on membrane order determined from the fluorescence polarization of 1,6-diphenyl-1,3,5-hexatriene (DPH; Fig. 6). At all temperatures tested, addition of 20 mol% cholesterol had a significant ordering effect (higher polarization values) on the membranes; supplementation to the extent of 10–15 mol% cholesterol generally had no significant effect on membrane order, although some degree of ordering was evident at most temperatures. Of particular interest, membranes containing 20 mol% cholesterol displayed an order value at 20 °C that was not significantly different from that of membranes possessing no cholesterol at 5 °C. This result suggests that alterations in the C:P ratio similar in magnitude to those observed with temperature acclimation in plasma membranes of trout liver are sufficient to largely offset the disordering effects of elevated temperatures on membrane physical properties. Cholesterol thus orders the relatively unsaturated phospholipids of hepatic plasma membranes, supporting the hypothesis that modulation of membrane cholesterol content is a significant feature of the temperature-induced compositional adjustments that account for the thermal adaptation of membrane structure and function in multicellular poikilotherms such as fish.

Exceptions to the cholesterol and thermal adaptation hypothesis

The data from intraspecific and interspecific comparisons of trout plasma membrane cholesterol content, as well as the biophysical evidence, are generally consistent with a role for cholesterol in the stabilization of membrane structure at elevated temperatures. However, as seen in the cases of trout erythrocytes and sperm, it cannot be concluded that the cholesterol content of all plasma membranes varies directly with growth temperature. Similarly, variation in membrane cholesterol content to achieve membrane stability is not necessarily always driven by changes in temperature. A case in point is illustrated by the opposing

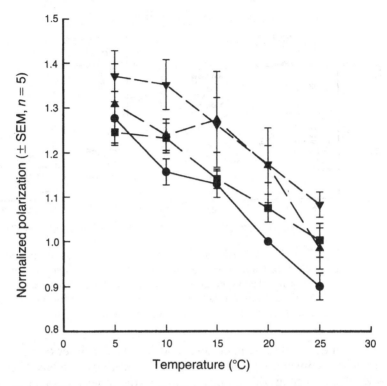

Fig. 6. Effect of cholesterol addition to plasma membrane phospholipid extracts of 5 °C-acclimated rainbow trout. Fluorescence polarization values for DPH (1,6-diphenyl-1,3,5-hexatriene) were normalized to the value for no cholesterol at 20 °C. Data are presented as mean ± SEM for 5 experiments. ●, 0 mol% cholesterol; ■, 10 mol% cholesterol; ▲, 15 mol% cholesterol; ▼, 20 mol% cholesterol.

responses of the apical (brush border) and basolateral membranes of trout enterocytes to temperature acclimation (Crockett & Hazel, 1995; Fig. 7). The basolateral domains (Fig. 7a) display nearly perfect thermal compensation of membrane order: although an excursion by 20 °C-acclimated trout into water at 5 °C orders the membrane to the extent indicated by a rise in polarization from point A to B, subsequent acclimation to 5 °C disorders the membrane by an amount equivalent to the drop in polarization between points B and C. Because lipid order is similar in 20 °C- and 5 °C-acclimated trout when compared at the respective acclimation temperatures, membrane order is conserved. Nevertheless, cholesterol contents in the basolateral domain do not

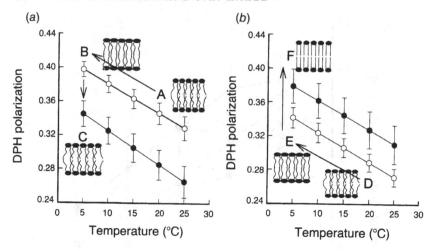

Fig. 7. Effect of temperature on the order (as reflected in the fluorescence polarization of DPH) of (a) basolateral and (b) apical or brush border plasma membrane domains prepared from (●) 5 °C- and (○) 20 °C-acclimated fish. Presented as mean ± SD for 5 experiments. Refer to text for an explanation of the figure.

vary with acclimation temperature, and the greater order of membranes in warm-acclimated fish can be attributed entirely to reduced levels of acyl chain unsaturation. Conversely, in the apical domain of trout enterocytes (Fig. 7b), lipid order displays an inverse compensation to temperature acclimation. Not only does an acute drop in temperature from 20 to 5 °C order the membrane by an amount equivalent to the interval between D and E, but acclimation to 5 °C further orders the membrane to an extent indicated by the interval from E to F – consequently, the apical domain is significantly more ordered in 5 °C- than 20 °C-acclimated trout. In this case, however, the increased order characteristic of cold-acclimated fish can be attributed entirely to elevated concentrations of membrane cholesterol (C:P ratios are 0.31 vs 0.25 in cold- and warm-acclimated trout, respectively), since acyl chain composition is not significantly influenced by temperature acclimation. Thus, in the apical plasma membrane domains of trout enterocytes, elevated cholesterol concentrations do result in increased membrane order, consistent with the paradigm of cholesterol action on membrane lipids. However, this response cannot be attributed to thermal adaptation since the membranes become more ordered at lower temperatures. Presumably, in this example, the lipid composition of the apical membrane domain is responding to some environmental factor other than temperature (possibly the concentration of intestinal bile salts).

Conclusions and suggestions for future research

Sterols or structurally similar molecules are required by all eukaryotic and many prokaryotic organisms as stabilizing components of the plasma membrane. In some cases (e.g. liver, kidney and gill plasma membranes of temperature-acclimated rainbow trout) increased proportions of cholesterol relative to phospholipid are sufficient in magnitude to offset the effects of elevated temperature on membrane order. In addition, interspecific comparisons of cholesterol:phospholipid molar ratios in the plasma membranes of various tissues of animals characterized by different body temperatures are also consistent with a role for cholesterol in stabilizing membranes at elevated temperatures. However, in some membrane systems (e.g. plasma membranes of fish sperm and erythrocytes, and the basolateral domains of trout enterocytes), cholesterol content is not significantly influenced by temperature acclimation, even though thermal compensation of membrane order may be observed. Collectively, these results indicate that although modulation of membrane cholesterol content may be a common mechanism for adjusting the physical properties of a membrane to prevailing thermal conditions, it is not the sole mechanism by which such adaptations can occur. Furthermore, although modulation of membrane order is a consistently observed outcome of variations in cholesterol content, in some cases (e.g. the brush border domain of trout enterocytes) cholesterol content apparently responds to stimuli other than temperature.

Several directions for future research into the possible role of cholesterol in thermal adaptation of biological membranes are suggested by the data presented in this chapter. (i) It would be useful to extend the scope of current interspecific comparisons of plasma membrane cholesterol contents to include a wider range of species from a broader range of body temperatures. Such studies should employ common protocols of membrane isolation and analysis so that data are more readily comparable than comparisons inferred from literature data using differing methods of isolation and analysis. (ii) Most studies, to date, have focused on the effects of cholesterol on membrane physical properties and little attention has been paid to possible effects on the dynamic phase behaviour of real membrane systems. It may be that an equally important aspect of cholesterol action is to stabilize the bilayer phase of membranes and reduce the proximity to the H_{II} phase transition

(Hazel, 1995). This possibility needs to be explored. (iii) A better understanding of the regulatory mechanisms that result in altered concentrations of membrane cholesterol might provide insights into why cholesterol content appears to be a primary mechanism of thermal adaptation in some membranes, whereas in other membranes alternate adaptive mechanisms are more predominant. This question applies not only to differences between tissues but also to variability in acclimatory membrane responses at the cellular level (i.e. between discrete plasma membrane domains or between the plasma membrane and cholesterol-poor intracellular membranes). (iv) The significance of cholesterol as a component of thermally-adaptive membrane restructuring may be temporally dynamic. Studies relating cholesterol and phospholipid-based plasma membrane compositional changes over the course of thermal acclimation could address this possibility. Manipulating membrane cholesterol concentrations using chemicals which suppress cholesterol synthesis might also be useful in investigating the role of cholesterol in adaptive plasma membrane responses.

It is clear that a variety of lipid compositional changes can contribute to the preservation of membrane physical properties in response to temperature change. Modulation of cholesterol content is one of the thermally-adaptive processes which operate in the plasma membrane. However, this mechanism is not active in the plasma membranes of all cells. The specific functional requirements and local environments of each cell define exactly how plasma membrane physical properties are regulated. Thus the availability of cholesterol, or other membrane components, to participate in plasma membrane adaptive regulation may reflect accommodation of functional demands. This may, in turn, limit the adaptive capacity of the membrane by restricting the restructuring response. Global warming may then challenge the ability of cells, acquired over an evolutionary time-scale, to maintain this adaptive balance.

Acknowledgement

This work was supported by National Science Foundation Grant IBN 9205234.

References

Aaronson, L. R., Johnston, A. M. & Martin, C. E. (1982). The effects of temperature acclimation on membrane sterols and phospholipids of *Neurospora crassa*. *Biochimica et Biophysica Acta*, **713**, 456–62.

Anderson, R. L., Minton, K. W., Li, G. C. & Hahn, G. M. (1981). Temperature-induced homeoviscous adaptation of Chinese hamster ovary cells. *Biochimica et Biophysica Acta*, **641**, 334–348.

Bloch, K. (1991). Cholesterol: evolution of structure and function. In: *Biochemistry of Lipids, Lipoproteins and Membranes*, ed. D. E. Vance & J. Vance, pp. 363–81. New York: Elsevier.

Brasitus, T. A. & Schachter, D. (1980). Lipid dynamics and lipid–protein interactions in rat enterocyte basolateral and microvillus membranes. *Biochemistry*, **19**, 2763–9.

Carey, F. G., Teal, J. M., Kanwisher, J. W., Lawson, K. D. &. Beckett, J. S. (1971). Warm-bodied fish. *American Zoologist*, **11**, 137–45.

Casal, H. L. & Mantsch, H. H. (1984). Polymorphic phase behavior of phospholipid membranes studied by infrared spectroscopy. *Biochimica et Biophysica Acta*, **779**, 381–401.

Chang, M. C. J. & Roots, B. I. (1989). The lipid composition of mitochondrial outer and inner membranes from the brains of goldfish acclimated at 5 and 30 °C. *Journal of Thermal Biology*, **14**, 191–4.

Chappelle, S. & Gilles-Baillien, M. (1983). Phospholipids and cholesterol in brush border and basolateral membranes from rat intestinal mucosa. *Biochimica et Biophysica Acta*, **753**, 269–71.

Chappelle, S. & Gilles-Baillien, M. (1989). Variation in the lipids in the intestinal membranes of active and hibernating tortoises. *Biochemical Systematics and Ecology*, **9**, 233–40.

Crockett, E. L. & Hazel, J. R. (1995). Cholesterol levels explain inverse compensation of membrane order in brush border but not homeoviscous adaptation in basolateral membranes from the intestinal epithelia of rainbow trout. *Journal of Experimental Biology*, **198**, 1105–13.

Davies, M. A., Schuster, H. F., Brauner, J. W. & Mendelsohn, R. (1990). Effect of cholesterol on conformational disorder in dipalmitoylphosphatidylcholine bilayers. A quantitative IR study of the depth dependence. *Biochemistry*, **29**, 4368–73.

Flamm, M. & Schachter, D. (1982). Acanthocytosis and cholesterol enrichment decrease lipid fluidity of only the outer human erythrocyte membrane leaflet. *Nature*, **298**, 290–2.

Flesch, G. & Rohmer, M. (1987). Growth inhibition of hopanoid synthesizing bacteria by squalene cyclase inhibitors. *Archives of Microbiology*, **147**, 100–4.

Gasser, K. W., Goldsmith, A. & Hopfer, U. (1990). Regulation of chloride transport in parotid secretory granules by membrane fluidity. *Biochemistry*, **29**, 7282–8.

Gladwell, R. T., Bowler, K. & Duncan, C. J. (1975). Heat death in the crayfish *Austropotamobius pallipes* – ion movements and their effects on excitable tissues during heat death. *Journal of Thermal Biology*, **1**, 79–94.

Harris, W. E. (1985). Modulation of (Na⁺, K⁺)-ATPase activity by the lipid bilayer examined with dansylated phosphatidylserine. *Biochemistry*, **24**, 2873–83.

Hazel, J. R. (1988). Homeoviscous adaptation in animal cell membranes. In *Advances in Membrane Fluidity – Physiological Regulation of Membrane Fluidity*, ed. R. C. Aloia, C. C. Curtain & L. M. Gordon, **6**, 149–88. New York: Liss.

Hazel, J. R. (1995). Thermal adaptation in biological membranes: Is homeoviscous adaptation the explanation? *Annual Review of Physiology*, **57**, 19–42.

Hazel, J. R. & Williams, E. E. (1990). The role of alterations in membrane lipid composition in enabling physiological adaptation of organisms to their physical environment. *Progress in Lipid Research*, **29**, 167–227.

Hermans, M. A. F., Neuss, B. & Sahm, H. (1991). Content and composition of hopanoids in *Zymomonas moblis* under various growth conditions. *Journal of Bacteriology*, **173**, 5592–5.

Hochachka, P. W. & Somero, G. N. (1984). *Biochemical Adaptation*. Princeton, NJ: Princeton University Press.

Huang, L. & Haug, A. (1974). Regulation of membrane lipid fluidity in *Acholeplasma laidlawii*: effect of carotenoid pigment content. *Biochimica et Biophysica Acta*, **352**, 361–70.

Hunter, K. & Rose, A. H. (1972). Lipid composition of *Saccharomyces cerevisiae* as influenced by growth temperature. *Biochimica et Biophysica Acta*, **260**, 639–53.

Imai, Y., Scoble, J. E., McIntyre, N. & Owen, J. S. (1992). Increased Na⁺-dependent D-glucose transport and altered lipid composition in renal cortical brush-border membrane vesicles from bile duct-ligated rats. *Journal of Lipid Research*, **33**, 473–83.

Intergovernmental Panel on Climate Change (IPCC) (1992). *Climate Change 1992: The Supplementary Report to the IPCC Assessment*, ed. J. T. Houghton, B. A. Callander & S. K. Varney. Cambridge: Cambridge University Press.

Jurgens, U. J., Simonin, P. & Rohmer, M. (1992). Localization and distribution of hopanoids in membrane systems of the cyanobacterium *Synechocystis* PCC 6714. *FEMS Microbiology Letters*, **92**, 285–8.

Kannenberg, E., Blume, A., Geckeler, K. & Poralla, K (1985). Properties of hopanoids and phosphatidylcholines containing ω-cyclohexane fatty acid in monolayer and liposome experiments. *Biochimica et Biophysica Acta*, **814**, 179–85.

Kannenberg, E., Blume, A., McElhaney, R. N. & Poralla, K. (1983). Monolayer and calorimetric studies of phosphatidylcholines containing branched chain fatty acids and their interactions with cholesterol and with a bacterial hopanoid in model membranes. *Biochimica et Biophysica Acta* **733**, 111–16.

Kawai, K., Fujita, M. & Nakao, M. (1974). Lipid components of two different regions of an intestinal epithelial cell membrane of mouse. *Biochimical et Biophysica Acta*, **369**, 222–33.

Keough, K. M. W. (1992). Unsaturation and the interactions of phospholipids with cholesterol and proteins. In *Structural and Dynamic Properties of Lipids and Membranes*, ed. P. J. Quinn & R. J. Cherry, pp. 19–28. London: Portland Press.

Kleinig, H., Zentgraf, H., Comes, P. & Stadler, J. (1971). Nuclear membranes and plasma membranes from hen erythrocytes. II. Lipid composition. *Journal of Biological Chemistry*, **246**, 2996–3000.

Kruuv, J, Glofcheski, D., Cheng, K.-H., Campbell, S. D., Al-Qysi, H. M. A., Nolan, W. T. & Lepock, J. R. (1983). Factors influencing survival and growth of mammalian cells exposed to hypothermia. I. Effects of temperature and membrane lipid perturbers. *Journal of Cellular Physiology*, **115**, 179–85.

Kuo, P., Weinfeld, M. & Loscalzo, J. (1990). Effect of membrane fatty acyl composition on LDL metabolism in Hep G2 hepatocytes. *Biochemistry*, **29**, 6626–32.

Labbe, C., Maisse, G., Muller, K., Zachowski, A., Kaushik, S. & Loir, M. (1995). Thermal acclimation and dietary lipids alter the composition, but not the fluidity, of trout sperm plasma membrane. *Lipids*, **30**, 23–33.

Lange, Y., Swaisgood, M. H., Ramos, B. V. & Steck, T. L. (1989). Plasma membranes contain half the phospholipid and 90% of the cholesterol and sphingomyelin in cultured human fibroblasts. *Journal of Biological Chemistry*, **264**, 3786–93.

Lazrak, T., Wolff, G., Albrecht, A., Nakatani, Y., Ourisson, G. & Kates, M. (1988). Bacterioruberins reinforce reconstituted *Halobacterium* lipid membranes. *Biochimica et Biophysica Acta*, **939**, 160–2.

Matheson, D. F., Oei, R. & Roots, B. I. (1980). Changes in the fatty acyl composition of phospholipids in the optic nerve of temperature-acclimated goldfish. *Physiological Zoology*, **53**, 57–69.

Meier, P. J., Sztul, E. S., Reuben, A. & Boyer, J. L. (1984). Structural and functional polarity of canalicular and basolateral plasma membrane vesicles isolated in high yield from rat liver. *Journal of Cell Biology*, **98**, 991–1000.

Milon, A., Wolff, G., Ourisson, G. & Nakatani, Y. (1986). Organization of carotenoid–phospholipid bilayer systems. *Helvetica Chimica Acta*, **69**, 12–24.

Nelson, G. J. (1967). Lipid composition of erythrocytes in various mammalian species. *Biochimica et Biophysica Acta*, **144**, 221–32.

Ourisson, G., Rohmer, M. & Poralla, K. (1987). Prokaryotic hopanoids and other polyterpenoid sterol surrogates. *Annual Review of Microbiology*, **41**, 301–33.

Parks, J. E. & Lynch, D. V. (1992). Lipid composition and thermotropic phase behavior of boar, bull, stallion, and rooster sperm membranes. *Cryobiology*, **29**, 255–66.

Perez-Albarsanz, M. A., Lopez-Aparicio, P., Senar, S. & Recio, M. N. (1991). Effects of lindane on fluidity and lipid composition in rat renal cortex membranes. *Biochimica et Biophysica Acta*, **1066**, 124–30.

Poralla, K., Hartner, T. & Kannenberg, E. (1984). Effect of temperature and pH on the hopanoid content of *Bacillus acidocaldarius*. *FEMS Microbiology Letters*, **23**, 253–6.

Poralla, K., Kannenberg, E. & Blume, A. (1980). A glycolipid containing hopane isolated from the acidophilic, thermophilic *Bacillus acidocaldarius*, has a cholesterol-like function in membranes. *FEBS Letters*, **113**, 107–10.

Pruitt, N. L. (1988). Membrane lipid composition and overwintering strategy in thermally-acclimated crayfish. *American Journal of Physiology*, **254**, R870–6.

Ramesha, C. S. & Thompson, G. A. (1982). Changes in the lipid composition and physical properties of *Tetrahymena* ciliary membranes following low temperature acclimation. *Biochemistry*, **21**, 3612–17.

Ramsammy, L. S., Boos, C., Josepovitz, C. & Kaloyanides, G. J. (1993). Biophysical and biochemical alterations of renal cortical membranes in diabetic rat. *Biochimica et Biophysica Acta*, **1146**, 1–8.

Rietveld, A. G., Killian, J. A., Dowhan, W. & de Kruijff, B. (1993). Polymorphic regulation of membrane phospholipid composition in *Escherichia coli*. *Journal of Biological Chemistry*, **268**, 12427–33.

Rimon, G., Hanski, E. & Levitzki, A. (1980). Temperature dependence of β receptor, adenosine receptor, and sodium fluoride stimulated adenylate cyclase from turkey erythrocytes. *Biochemistry*, **19**, 4451–60.

Robertson, J. C. & Hazel, J. R. (1995). Cholesterol content of trout plasma membranes varies with acclimation temperature. *American Journal of Physiology*, **269**, R1113–17.

Rohmer, M., Bouvier, P. & Ourisson, G. (1979). Molecular evolution of biomembranes: Structural equivalents and phylogenetic precursors of sterols. *Proceedings of the National Academy of Sciences USA*, **76**, 847–51.

Rohmer, M., Bouvier-Nave, P. & Ourisson, G. (1984). Distribution of hopanoid triterpenes in prokaryotes. *Journal of General Microbiology*, **130**, 1137–50.

Rottem, S. & Markowitz, O (1979). Carotenoids act as reinforcers of the *Acholeplasma laidlawii* lipid bilayer. *Journal of Bacteriology*, **140**, 944–8.

Sahm, H., Rohmer, M., Bringer-Meyer, S., Sprenger, G. A. & Welle, R. (1993). Biochemistry and physiology of hopanoids in bacteria.

In *Advances in Microbial Physiology*, vol.35, ed. A. H. Rose, pp.247–73. New York: Academic Press.

Seddon, J. M. (1990). Structure of the inverted hexagonal (H_{II}) phase, and non-lamellar phase transitions of lipids. *Biochimica et Biophysica Acta*, **1031**, 1–69.

Sinensky, M. (1978). Defective regulation of cholesterol biosynthesis and plasma membrane fluidity in a Chinese hamster ovary cell mutant. *Proceedings of the National Academy of Sciences USA*, **75**, 1247–9.

Sinensky, M. & Kleiner, J. (1981). The effects of reagents that increase membrane fluidity on the activity of 3-hydroxy-3-methylglutaryl coenzyme A reductase in the CHO-K1 cell. *Journal of Cell Physiology*, **108**, 309–16.

Squier, T. C., Bigelow, D. J. & Thomas, D. D. (1988). Lipid fluidity directly modulates the overall protein rotational mobility of the Ca-ATPase in sarcoplasmic reticulum. *Journal of Biological Chemistry*, **263**, 9178–86.

Subczynski, W. K., Markowska, E., Gruszecki, W. I. & Sielewiesiuk, J. (1992). Effects of polar carotenoids on dimyristolphosphatidylcholine membranes: a spin-label study. *Biochimica et Biophysica Acta*, **1105**, 97–108.

Subczynski, W. K., Wisniewska, A., Yin, J., Hyde, J. S. & Kusumi, A. (1994). Hydrophobic barriers of lipid bilayer membranes formed by reduction of water penetration by alkyl chain unsaturation and cholesterol. *Biochemistry*, **33**, 7670–81.

Thompson, G. A. (1989). Membrane acclimation by unicellular organisms in response to temperature change. *Journal of Bioenergetics and Biomembranes*, **21**, 43–59.

Wolters, H., Spiering, M., Gerding, A., Slooff, M. J. H., Kuipers, F., Hardonk, M. J. & Vonk, R. J. (1991). Isolation and characterization of canalicular and basolateral fractions from human liver. *Biochimica et Biophysica Acta*, **1069**, 61–9.

Yeagle, P. L. (1985). Cholesterol and cell membranes. *Biochimica et Biophysica Acta*, **822**, 267–87.

IAN D. McCARTHY
and DOMINIC F. HOULIHAN

The effect of temperature on protein metabolism in fish: the possible consequences for wild Atlantic salmon (*Salmo salar* L.) stocks in Europe as a result of global warming

Introduction

In fish, it is generally accepted that increased temperatures, within the thermal range for a particular species, will lead to increased growth rates providing that food is not limiting. Protein growth occurs when protein synthesis exceeds protein degradation. Therefore, under-standing the effect of temperature on protein synthesis and food/protein consumption is fundamental to predicting the potential effects of climatic change on the growth performance of fish. Within the preferred thermal range, long-term exposure of ectotherms to lower or higher environmental temperatures may lead to a degree of physio-logical and biochemical independence from the acute effects of the alteration in environmental temperature, a process generally termed temperature compensation. The limits within which such independence (or compensation) may fall are ultimately determined by the genetic potential (genotype) of an organism. Since gene expression is depen-dent upon the synthesis of proteins, it seems likely that metabolic adjustments may be accompanied by changes in protein synthesis. *In vivo* rates of protein synthesis have most commonly been studied by measuring radiolabelled amino acid incorporation rates using the 'flooding dose' technique (Garlick, McNurlan & Preedy, 1980). This technique, and its experimental application in fish biology, have been covered in depth in recent reviews (Houlihan, Carter & McCarthy, 1995a, b; Houlihan *et al.*, 1995c) and will not be considered here.

Society for Experimental Biology Seminar Series 61: *Global Warming: Implications for freshwater and marine fish*, ed. C. M. Wood & D. G. McDonald. © Cambridge University Press 1996, pp. 51–77.

The effect of various abiotic and biotic factors on protein synthesis in fish has been the subject of several reviews (Fauconneau, 1985; Houlihan, 1991; Houlihan, Mathers & Foster, 1993; Houlihan et al., 1995a, b, c) and earlier work on the adaptive response of protein metabolism in fish to reduced water temperature has also been covered (Haschemeyer, 1978).

The aim of this chapter is to review the recent data on the effects of temperature on protein synthesis in fish, particularly with the aim of predicting the possible effects of global warming on this process. This review will focus on temperate fish since most of the available data are for salmonid and cyprinid fish. The conclusions obtained will be placed in an ecological context by speculating about the possible effects of global warming on the freshwater stages of wild Atlantic salmon stocks in Europe. Before reviewing the effects of temperature on protein synthesis it is important to distinguish between temperature acclimatization and temperature acclimation. Under conditions where more than one environmental variable is changing simultaneously (commonly ration and temperature) the physiological change(s) are called 'acclimatization' (Hazel & Prosser, 1974). The term 'acclimation' has been defined as the physiological change(s) developed over a period of time in response to the alteration of a *single* environmental variable such as water temperature (Hazel & Prosser, 1974; Foster *et al.*, 1992). In this review we will distinguish between temperature effects and food effects and will compare the effects of acute and chronic exposure to elevated water temperatures on protein synthesis in fish.

The mechanism of protein synthesis

The mechanism of protein synthesis in eukaryotes is a complex process involving three types of ribonucleic acid, ribosomal (rRNA), messenger (mRNA) and transfer (tRNA) RNA, together with a number of macro-molecules (enzymes, initiation (eIF), elongation and release factors, ATP and GTP) (Haschemeyer, 1978; Alberts et al., 1983; Pain & Clemens, 1986; Reeds & Davies, 1992). The process by which proteins are synthesized, known as translation (or the ribosome cycle) can be divided up into three stages (Fig. 1). Initiation involves a ribosome and a molecule of a specific initiator, tmet-t-RNA (methionyl-tRNA$_f$), binding to a particular site on the mRNA at the beginning of the coding sequence. An amino-acyl tRNA is then able to bind and the synthesis of the first peptide bond takes place. During the elongation phase the ribosome moves relative to the mRNA and amino acids are

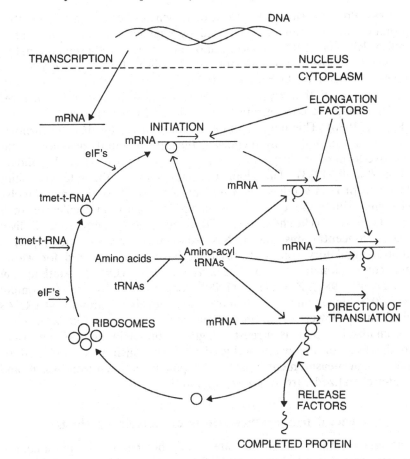

Fig. 1. Schematic outline of the ribosome cycle in eukaryotes (after Haschemeyer, 1978; Pain & Clemens, 1986). Abbreviations are defined in the text.

incorporated into the growing peptide chain. Once the ribosome reaches the end of the coding sequence on the mRNA, known as termination, it is released together with the completed protein chain. The ribosome is then available for reattachment on the same or another molecule of mRNA.

The overall rate of protein synthesis within eukaryotic tissues can be regulated by the number of ribosomes present in the cell/tissue/ organ (which will determine the maximum rate of protein synthesis possible), by the rate at which the ribosomes synthesize proteins (RNA activity) or by the supply of free amino acids as substrates for the process. Previous work has shown that each of these ways of regulating protein synthesis can be influenced by changes in water temperature in fish. Ribosomal RNA concentration increases as a thermoacclimatory response to low water temperature (Goolish, Barron & Adelmann, 1984; Ferguson & Danzmann, 1990; see next section). It has also been suggested that metabolic rate compensation might be achieved by increasing the concentrations of critical (if not all) cellular enzymes (Koban, 1986). This may be obtained through readjusting the turnover rate of an enzyme or by increasing the rate of interconversion of the inactive form of an enzyme to the active form by reversible phosphorylation (Sidell, 1983). The elongation step may be the rate-controlling factor in protein synthesis and temperature compensation may involve changing either the quantity of elongation factor molecules or their specific activity (Haschemeyer, 1978; Simon, 1987). Differences in liver tRNA concentrations and tRNA isoacceptor species, which play a central regulatory role in translation, have been reported for warm- and cold-acclimatized carp, *Cyprinus carpio* L. (Oñate, Amthauer & Krauskopf, 1987). Zuvic *et al.* (1980) report that the liver tRNA content was 70% higher and *in vivo* intracellular levels of amino-acyl tRNAs significantly lower in warm-acclimatized carp compared with cold-acclimatized carp. This suggested higher amino acid incorporation rates in the liver of the warm-acclimatized carp which was confirmed by subsequent measurement in isolated hepatocytes taken from warm- and cold-acclimatized carp (Saez *et al.*, 1982).

General temperature effects on protein synthesis

The relationship between the rate of any biochemical or physiological process and water temperature in fish is asymmetrical. Providing that resources are not limiting and the temperature increase occurs within the thermal limits for the species, then the rate of any biochemical or physiological process increases with increasing water temperature to a maximum, above which the rate declines near the upper limit for the species (Huey & Kingsolver, 1989; Atkinson, 1994). There is evidence to suggest that maximum rates of protein synthesis occur at the optimum temperature for any species. Measurements of oxygen consumption and *in vivo* rates of protein synthesis in isolated hepatocytes from rainbow trout, *Oncorhynchus mykiss* (Walbaum), measured at a range of incubation temperatures between 5 and 20 °C, have demonstrated an asymmetric relationship with temperature (Fig. 2; Pannevis & Houlihan, 1992), with the highest rates occurring around the optimum water temperature for growth in rainbow trout of 17 °C (Hokanson, Kleiner & Thorslund, 1977).

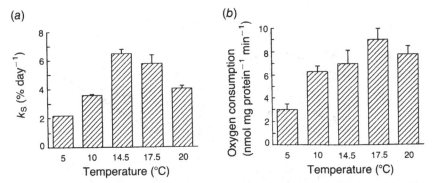

Fig. 2. The effect of acute temperature incubation at 5, 10, 14.5, 17.5 and 20 °C on (*a*) fractional rates of protein synthesis (k_s, % day^{-1}) and (*b*) oxygen consumption (nmol.mg^{-1} protein.min^{-1}) for hepatocytes isolated from rainbow trout (*Oncorhynchus mykiss*) acclimatized to 10 °C (data redrawn from Pannevis & Houlihan, 1992).

The effects of temperature on protein synthesis and RNA activity in the liver and muscle of ectotherms (fish, frog, lizard) and endotherms (birds and mammals) are shown in Fig. 3. The aim of this analysis is to examine the general effect of temperature on the process of protein synthesis. In compiling this summary figure the following criteria were used to select data: (i) all ectotherms were allowed long-term acclimatiz-ation to constant temperature prior to the measurement of protein synthesis; (ii) all ectotherms were reared at temperatures within the thermal preferenda for the species; (iii) all animals were allowed con-tinuous access to food or were fed high rations; (iv) the muscle data consists of epaxial white muscle in fish and predominately slow fibres in endotherms. The relationship between temperature and protein syn-thesis is exponential for both the liver and the muscle data (Fig. 3*a*), with significantly higher rates in the liver compared to muscle at any given temperature. The scatter in the data may be due to species differences or differences in nutritional status, rearing conditions or methodology. However, the slopes of the two regression lines are similar, suggesting a common temperature effect on protein synthesis in different tissues despite differences in relative rates. This exponential increase in tissue rates of protein synthesis with increasing temperature has also been found in single-species studies (see below).

Rates of protein synthesis in any particular tissue are largely pro-portional to the total RNA concentrations which are commonly expressed as the RNA:protein ratio or 'capacity for protein synthesis' (C_s, mg RNA.g^{-1} protein; Sugden & Fuller, 1991). Dividing the tissue fractional rate of protein synthesis by the tissue C_s value will give an

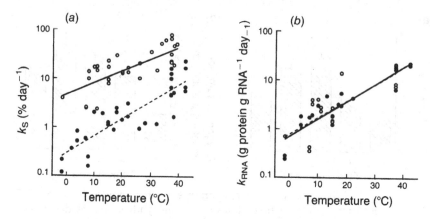

Fig. 3. (*a*) The effect of temperature acclimatization on the fractional rate of protein synthesis (k_s, % day^{-1}) in the liver (○, solid line) and muscle (●, broken line). The relationships are described by the following equations: liver, $\log_{10}k_s = 0.026T\,°C + 0.659$ ($R^2 = 0.581$, $n = 35$, $p < 0.001$); muscle, $\log_{10}k_s = 0.035T\,°C - 0.529$ ($R^2 = 0.738$, $n = 34$, $p < 0.001$). (*b*) The effect of temperature on RNA activity (k_{RNA}, g protein synthesized. g^{-1} RNA. day^{-1}) in the liver (○, solid line) and muscle (●, broken line). The relationships are described by the following equations liver, $\log_{10}k_{RNA} = 0.034T\,°C - 0.111$ ($R^2 = 0.617$, $n = 19$, $p < 0.001$); muscle, $\log_{10}k_{RNA} = 0.036T\,°C - 0.153$ ($R^2 = 0.858$, $n = 22$, $p < 0.001$). The data in (*a*) and (*b*) are from: Ashford & Pain (1985); Carter *et al.* (1993b); Fauconneau & Arnal (1985); Foster *et al.* (1992); Garlick, Burk & Swick (1976); Haschemeyer (1983); Haschemeyer, Persell & Smith (1979); Houlihan *et al.* (1994); Laurent *et al.* (1978); Lobley *et al.* (1980, 1994); McCarthy (1993); McCarthy *et al.* (unpublished); McMillan & Houlihan (1988); Millward *et al.* (1976); Peragón *et al.* (1992, 1993, 1994); Perera (1995); Pocrnjic *et al.* (1983); Preedy *et al.* (1988); Sayegh & Lajtha (1989); Smith & Haschemeyer (1980); Smith *et al.* (1996); Sugden & Fuller (1991); Tessauraud *et al.* (1992); Watt *et al.* (1988); Yacoe (1983).

estimation of RNA efficiency which has been termed 'RNA activity' (k_{RNA}, g protein synthesized.g^{-1} RNA.day^{-1}) (Sugden & Fuller, 1991). RNA activity will be dependent on the rates of peptide chain initiation, elongation and termination and will therefore exhibit temperature sensitivity. The number of studies in Fig. 3*a* where tissue C_s was measured is fewer than the synthesis measurements, but the liver and muscle data both indicate an exponential relationship between temperature and k_{RNA} (Fig. 3*b*). However, in contrast to the protein synthesis results, k_{RNA} values do not appear to be different between liver and muscle and there is a single temperature response. The concentration of RNA is higher in the liver compared to muscle (Smith & Haschemeyer, 1980; Preedy *et al.*, 1988; Peragón *et al.*, 1994) and hence it appears that this, and not the efficiency of RNA, may be responsible for differences in synthesis between tissues. Also a single line for ecto- and endotherms indicates a similarity in RNA performance controlled simply by temperature.

The influence of previous thermal history and period of acclimatization on protein synthesis

The effect of temperature on protein synthesis in fish has usually emphasized the compensatory adaptations exhibited by fish when they are exposed to reduced water temperature. Haschemeyer (1978) found that when assayed at 22 °C, liver cell-free preparations from toadfish acclimatized to 10 °C had 75% higher *in vitro* rates of incorporation of [³H]phenylalanine into liver protein compared with animals acclimatized to 22 °C. Similarly, Simon (1987) obtained cytosolic extracts from various tissues of rainbow trout acclimatized to 4 and 17 °C and measured *in vitro* elongation rates at a range of temperatures between 5 and 25 °C. Cold-acclimatized fish exhibited higher specific elongation rates for each of the tissues compared to warm-acclimatized fish at each experimental temperature (except red muscle). *In vivo* measurements of protein synthesis in the tissues of warm- and cold-acclimatized fish (Fig. 4*a*; Watt *et al.* 1988) show similar results to the *in vitro* data. Watt *et al.* (1988) transferred groups of common carp to a range of water temperatures (8–34 °C) after 5 weeks acclimatization to cold (8 °C) or warm water (18 °C) and measured protein synthesis in the red and white muscle after less than 7 days acclimatization to the new water temperature. The relationships between temperature and protein synthesis in the red and white muscle were exponential for both the warm- and cold-acclimatized groups (Fig. 4*a*). However, at each water temperature, cold-acclimatized fish exhibited significantly higher fractional rates of protein synthesis in both red and white muscle. This higher rate of synthesis was attributed to increased RNA activity as a result of increased elongation and/or initiation factor activity following cold acclimatization (Nielsen, Plant & Haschemeyer, 1977; Haschemeyer, 1978; Harmon, Proud & Pain, 1984).

The studies of Haschemeyer (1978), Simon (1987) and Watt *et al.* (1988) show that previous thermal history (cold or warm

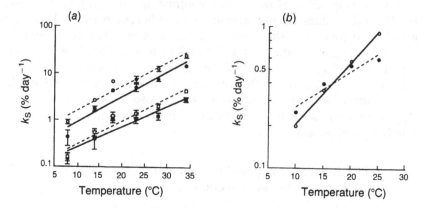

Fig. 4. (a) The effect of cold (8 °C) and warm (18 °C) acclimatization on fractional rates of protein synthesis (k_s, % day^{-1}) in the red and white muscle of carp (*Cyprinus carpio*), following acclimation to a range of water temperatures for 7 days (data redrawn from Watt *et al.*, 1988). The cold- and warm-acclimatized fish are represented by the open symbols/dotted lines and closed symbols/solid lines respectively. The data for the red and white muscle are indicated by the circles and squares respectively and are presented as mean ± SD ($n = 5$). Synthesis rates were measured using a single intraperitoneal flooding dose injection of [^{3}H]phenylalanine. (b) The effect of short- and long-term acclimatization to a range of water temperatures on fractional rates of protein synthesis (k_s, % day^{-1}) in the white muscle of carp (data redrawn from Loughna & Goldspink, 1985). The short-term and long-term acclimatized fish were transferred from 15 °C to the new water temperatures for 24 hours and 4 weeks respectively before synthesis rates were measured using a single intraperitoneal flooding dose injection of [^{3}H]phenylalanine. The data for the short-term and long-term acclimated fish are represented by the open symbols/solid line and closed symbols/dotted line respectively. Each data point represents the mean of 5 animals.

acclimatization) has a significant influence on the protein synthetic response when given a relatively short-term exposure (e.g. up to 7 days) to a range of water temperatures. However, the period of exposure to a new water temperature prior to measurement also has a significant influence on protein synthesis. Loughna & Goldspink (1985) acclimatized common carp to 15 °C, and then measured *in vivo* rates of protein synthesis in the white muscle 24 hours after transfer to a range of water temperatures between 10 and 25 °C and after 4 weeks acclimatization to the new water temperature (Fig. 4*b*). The results obtained following short-term (24 h) and long-term (4 weeks) acclimatization both gave an exponential relationship between temperature and protein synthesis. White muscle synthesis rates measured after 24 hours or 4 weeks were similar for fish transferred to both 15 and 20 °C. However, short-term exposure resulted in significantly higher white muscle synthesis rates at 25 °C and significantly lower synthesis rates at 10 °C compared to rates measured after 4 weeks acclimatization (Fig. 4*b*), indicating a degree of thermal compensation with long-term exposure.

Temperature and protein synthesis: influence of nutritional status

The studies of Loughna and Goldspink (1985) and Watt *et al.* (1988), reviewed in the previous section, were referred to by the authors as 'temperature acclimation'. This is misleading since the fish were fed to satiation twice a day and it is likely that temperature was not the only factor influencing protein synthesis. Food consumption rates vary with water temperature and therefore the term 'temperature acclimatization' (as defined earlier) is a more accurate description. Food consumption has a strong influence on rates of protein synthesis (Houlihan, Hall & Gray, 1989; Carter *et al.*, 1993a; McCarthy, Houlihan & Carter, 1994) and therefore it is important to quantify individual or group food consumption rates, for example using radiography (reviewed in McCarthy *et al.*, 1993) in studies of fish physiology. In this section the importance of separating temperature effects from feeding effects is highlighted.

The effect of temperature acclimation on protein synthesis has been studied by Foster *et al.* (1992), who reared juvenile cod (*Gadus morhua* L.) individually at 5 and 15 °C. Each fish was fed daily a single meal corresponding to 3% of its body weight and therefore consumption rates were similar at 5 and 15 °C (Fig. 5a). After 6–7 weeks, the cold (5 °C) and warm-acclimated (15 °C) groups showed similar specific growth rates (Fig. 5a) and tissue rates of protein synthesis (Fig. 5*b*). In each of the tissues examined, cold-acclimated fish had significantly higher total RNA concentrations (Fig. 5c) and the warm-acclimated fish maintained similar rates of protein synthesis by increasing RNA activity by 60% (Fig. 5*d*). The elevated total RNA concentration in the tissues of the cold-acclimated fish are further evidence of the thermal compensatory adaptation to low water temperatures discussed earlier. Translation rates were slower in the tissues of the cold-acclimated cod but under conditions of equal amino acid supply,

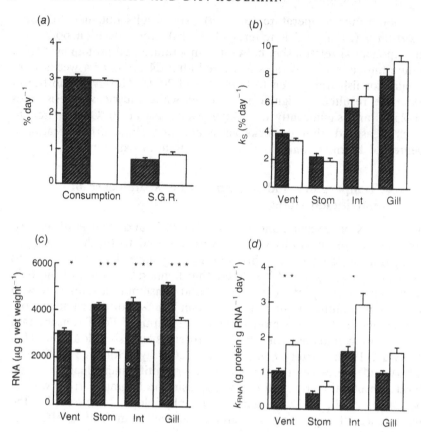

Fig. 5. The effect of temperature acclimation on the growth perform-ance and protein metabolism of Atlantic cod (*Gadus morhua*) fed similar rations and reared at 5 and 15 °C (data redrawn from Foster *et al.*, 1992). The data for the cold- (5 °C) and warm- (15 °C) acclimated fish are represented by hatched and white bars respectively (mean ± SEM, $n = 8$). (*a*) Consumption (% body weight day^{-1}) and whole body specific growth rate (S.G.R., % day^{-1}). (*b*) Fractional rates of protein synthesis (k_s, % day^{-1}), (*c*) tissue total RNA content (µg RNA.g^{-1} wet weight), (*d*) RNA activity (k_{RNA}, g protein synthe-sized. g^{-1} RNA.day^{-1}) for the ventricle (Vent), stomach (Stom), intestine (Int) and gill. Synthesis rates were measured using a single intravenous flooding dose injection of [^{3}H]phenylalanine via the caudal blood vessels. Mean values are compared using ANOVA (*$p <$ 0.05, **$p <$ 0.01, ***$p <$ 0.001).

increased RNA concentrations resulted in similar rates of protein synthesis and growth compared to fish acclimated to 15 °C.

Recently, we have reared juvenile rainbow trout under *ad libitum* feeding conditions at 7 and 17 °C for 4 weeks in order to examine the effect of temperature acclimatization (i.e. variable temperature and food supply) on protein synthesis (I. McCarthy *et al.*, unpublished data; Fig. 6). Food consumption (measured using radiography), growth, food conversion ratio, and white muscle and whole body fractional rates of protein synthesis were significantly higher in the warm-acclimatized fish (Fig. 6*a*, *b*). Thus elevated temperature together with increased food consumption resulted in an increase in protein synthesis (see also Fauconneau & Arnal, 1985). The total RNA concentration in the white muscle and the whole body were significantly higher in the cold-acclimatized fish as a result of thermal compensation (Fig. 6*c*), but RNA activities in the white muscle and whole body were significantly higher in the warm-acclimatized fish (Fig. 6*d*).

Most studies have focused on the effects of temperature on protein synthesis and there are very few data available for protein growth and degradation. Mathers *et al.* (1993) present data for juvenile rainbow trout (7 weeks post first feeding) reared at 5, 10 and 15 °C and either fed a high ration or starved for 4 weeks. Whole body rates of protein synthesis were significantly lower in the feeding fish at 5 °C but were similar at 10 and 15 °C (Fig. 7*a*). Protein growth increased significantly with increasing water temperature. The effects of temperature on protein degradation rates were variable with degradation rates increasing from 5 to 10 °C but decreasing again at 15 °C. For the starved fish, whole body fractional rates of protein synthesis, degradation and whole body protein loss all increased significantly with increasing water temperature (Fig. 7*b*). There were no differences in RNA concentration, expressed on a weight-specific basis (μg RNA.g^{-1} dry weight) or as the RNA:protein ratio between the fed animals at 5, 10 and 15 °C, and the increased synthesis rates at 10 and 15 °C were due to increased RNA activity (k_{RNA}). In larger fish, RNA concentrations decrease with increasing temperature (see above) but the available data suggest that RNA concentrations are much higher in larval and juvenile fish and at low water temperatures thermal compensation may not occur (Mathers *et al.*, 1993).

Possible effects of global warming on growth and physiology of wild Atlantic salmon stocks in Europe

Meteorological monitoring has indicated an increase in atmospheric CO_2 levels and a rise in global air temperatures. Global climatic change

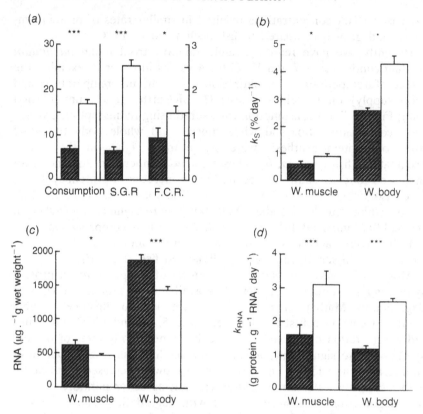

Fig. 6. The effect of temperature acclimatization on the growth performance and protein metabolism of rainbow trout, fed *ad libitum* rations and reared at 7 and 17 °C (I. McCarthy *et al.*, unpublished data). The data for the cold (7 °C) and warm (17 °C) acclimatized fish are represented by hatched and white bars respectively. The data are presented as mean ± SEM for the cold ($n = 13$) and warm ($n = 15$) acclimatized fish. (*a*) Consumption (mg dry food consumed. g^{-1} wet weight.day^{-1}), whole body specific growth rate (S.G.R., % day^{-1}) and food conversion ratio (F.C.R., g wet weight gain.g^{-1} dry food consumed). (*b*) Fractional rates of protein synthesis (k_s % day^{-1}), (*c*) Tissue total RNA content (μg RNA.g^{-1} wet weight), and (*d*) RNA activity (k_{RNA}, g protein synthesized.g^{-1} RNA.day^{-1}) for the white muscle (W.muscle) and whole body (W.body). Synthesis rates were measured using a single intraperitoneal flooding dose injection of [³H]phenylalanine. Mean values are compared using ANOVA (*$p < 0.05$, ***$p < 0.001$).

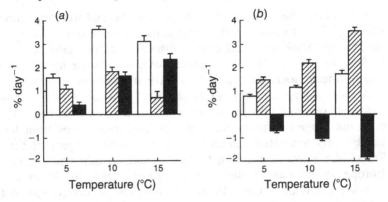

Fig. 7. Whole body fractional rates of protein synthesis (k_s, open bars), protein degradation (k_d, hatched bars) and protein growth (k_g, solid bars) for (*a*) fed and (*b*) starved juvenile rainbow trout reared at 5, 10 and 15 °C. The data are presented as means ± SEM. (Data redrawn from Mathers *et al.*, 1993.)

models have predicted that the average global air temperature may increase by up to 0.5 °C per decade (Haughton, Jenkins & Ephraums, 1990). In freshwater ecosystems, fish are already exposed to fluctuations in water temperature on an annual, seasonal and daily basis; these fluctuations can be quite large: however, the general trend is upward. For example, Schindler *et al.* (1990) report that the mean annual air and water temperature of the Experimental Lakes Area of northwestern Ontario has increased by 2 °C and the length of the ice-free season has increased by 3 weeks over the last 30 years. The effects of global warming on freshwater fish may vary according to latitude: it is possible that climatic change may result in a general year round shift in average water temperatures or it may accentuate the seasonal differences already present (see below). However, most of the models that seek to describe the possible effects of climatic change on freshwater fish have focused on lacustrine environments (e.g. Hill & Magnuson, 1990; Magnuson, Meisner & Hill, 1990; McLain, Magnuson & Hill, 1994; see also Magnuson & DeStasio, this volume) and there is a scarcity of data for riverine fish (Power, 1990; Jensen, 1991).

We now move on to speculate about the possible effects that climatic change may have on the growth and physiology of wild Atlantic salmon (*Salmo salar* L.) stocks in Europe. This species has a wide distribution in Europe (see below) and is dependent on the riverine environment for several key stages of its life cycle (spawning, embryo and juvenile development). Specifically, we will use the thermal tolerance model of Elliott (1991), together with conclusions on the effect of temperature on protein metabolism and growth drawn from the earlier part of this review, to predict the consequences of global warming for northern and southern stocks of anadromous Atlantic salmon.

Atlantic salmon are present as genetically distinct stocks in many of the rivers draining into the Atlantic Ocean and the Baltic Sea. Historically the range of the species has extended southwards from Iceland, the Barents Sea (River Pechora) and southwestern part of the Kara Sea (River Kara) in northern Russia through Scandinavia and northern Europe to the Bay of Biscay and northern Portugal (River Douros) (see MacCrimmon & Gott, 1979). The life cycle of anadromous Atlantic salmon has been the subject of many reviews (Metcalfe & Thorpe, 1990; Thorpe, Metcalfe & Huntingford, 1992; Gibson, 1993); here we concentrate on the effects of elevated water temperature on the freshwater (parr) stage. We will focus on two rivers, the Alta in northern Norway (70° N) and the Nivelle in southwest France (43° 30′ N) located near to the northern and southern limits respectively of the range for the species in Europe. The existing average monthly water temperatures in the Alta and Nivelle are shown in Fig. 8, together with the predicted temperature profile following global warming and the current timing of spawning, hatching and first feeding of the salmon in the rivers.

In the Alta River, the average water temperature is close to 0 °C for 6 months during the northern winter. In the spring the temperature rises rapidly, reaching around 12 °C in the summer, and declines rapidly with the onset of winter. The adult fish in the Alta spawn from mid-November to late December. Owing to the cold winter temperatures the developmental rate of the eggs is slow and the eggs hatch in early June and reach first feeding in mid-July. Salmon parr are visual feeders and reduce or cease feeding below 7 °C (Metcalfe & Thorpe, 1990; Elliott, 1991). The proportion of the year when conditions would be suitable for growth is indicated in Fig. 8 as the proportion of the year when the average monthly water temperature is above 7 °C. For the Alta River, the growing season is about 98 days (range 82–112) or 27% of the year. Growth in these conditions is slow; the average size of the parr by October is 43 mm and the mean smolt age is 3.85 years (range 3–5 years) (reviewed in Jensen, 1991). In contrast to the Alta, the average monthly water temperatures in the Nivelle are higher throughout the year with water temperatures of around 8 °C in the winter and 20 °C in the summer. The adult fish in the Nivelle spawn from mid-December to mid-January. Due to the warmer water temperature, development is faster, the eggs hatch in

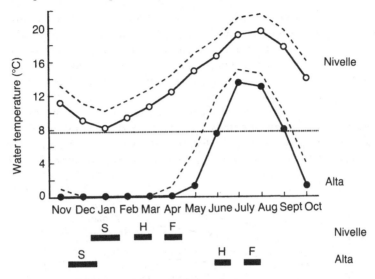

Fig. 8. Average monthly water temperatures for the River Nivelle in southwest France (○) and the River Alta in northern Norway (●). The solid line represents the present water temperatures and the broken lines represent the expected temperature regimes predicted from global warming models (data for the Alta are redrawn from Jensen, 1991). The present timing of spawning (S), hatching of the eggs (H) and first feeding of the alevins (F) are also indicated. Periods above the dotted line (7 °C) are suitable for growth. Data for Nivelle and Alta fish are taken from Beall *et al.* (1994) and Jensen (1991) respectively.

March and fry begin feeding in April; by October the average size is 99 mm (Beall *et al.*, 1994). Figure 8 would suggest that the opportunity for growth in the Nivelle river is year round. However, accounting for daily fluctuations in temperature, it is likely that the growing season for salmon in the Nivelle is similar to that of salmon in Asturias in northern Spain: about 330 days or 90% of the year (Garcia de Leániz & Martinez, 1988). As a result of this greater opportunity for feeding and growth, salmon in the Nivelle smolt after 1 or 2 years in freshwater (Beall *et al.*, 1994). The predicted effects of climatic change on the average monthly water temperatures in the Alta and Nivelle are also indicated in Fig. 8. Winter water temperatures in the Alta will still remain low. However in the spring, water temperatures will begin to rise earlier and faster and will peak around 14 °C in the summer, and the autumn decline to below 7 °C will be delayed by about 2–3 weeks (Jensen, 1991). In Fig. 8, we have suggested that the effect of climatic change will be a year round 2 °C rise in water temperature in the Nivelle. This would result in average water temperatures of around 10 °C in the winter and 22 °C in the summer. The opportunity for growth will continue to be at least 90% of the year in the Nivelle; however, the growing season in the Alta may increase to about 45% of the year.

Atlantic salmon from the rivers Lune and Leven in northwest England were used experimentally by Elliott (1991) to construct the thermal tolerance polygon in Fig. 9. Elliott reviewed the available laboratory and field data for thermal tolerance in Atlantic salmon from Europe

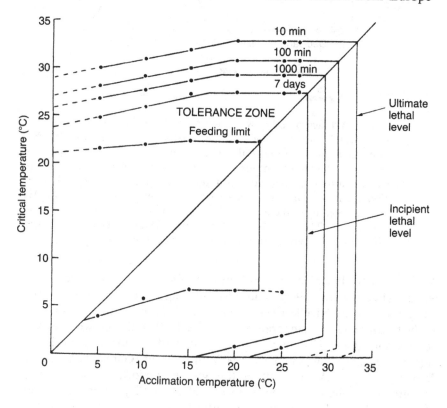

Fig. 9. Thermal tolerance polygon for juvenile Atlantic salmon (*Salmo salar*), showing the feeding zone, tolerance zone, incipient lethal level (survival over 7 days), ultimate lethal level (survival over 10 min) and intermediate levels (survival over 1000 min, 100 min): each point is the arithmetic mean of 16 values. (Figure taken from Elliott, 1991, used with permission, Blackwell Scientific Ltd.)

and Canada and found that these data were in general agreement with his results. It is not known whether salmon from the Alta and the Nivelle would show exactly the same thermal tolerance polygons as those defined by Elliott (1991). However, Jobling (this volume) reviews the available data for species with a broad latitudinal range and concludes that high latitude and low latitude fish show similar temperature preferenda and thermal tolerance. Therefore we will use Elliott's model to speculate on the possible effects of increased water temperature on the growth performance of Atlantic salmon from the Alta and Nivelle.

The predicted increase in the average peak summer water temperature in the River Alta from 12 to 14 °C is well within the feeding zone of Elliott (1991). Providing that food is available, this would result in an increase in the consumption rates of the salmon resulting in increases in protein synthesis and growth (Fig. 10). In order for the fish to stop feeding, the water temperature would have to rise 8 °C, to 22 °C (Fig. 9). Mortalities due to thermal stress would not occur unless the water temperature was maintained continuously at 26 °C for over 7 days or reached 28–30 °C in the short term. However, weekly and diurnal fluctuations about the summer mean of 14 °C are to be expected. Therefore, provided that food is not limiting, protein synthesis rates would be variable as a result of short-term fluctuations (hours to days) but overall climatic change would result in increased rates of protein synthesis and growth in the Alta salmon. As a result of increased growth rates during the summer the average age at smolting in the River Alta would decrease to 3.0 years (Jensen, 1991). It is possible that insufficient food would be consumed by the fish for maintenance on these warm days; therefore body reserves would need to be used to fuel maintenance metabolism (including increased protein degradation) and weight loss could occur. It is unlikely that the temperature would remain continuously elevated but might exceed 22 °C during the middle of the day. However, it is unlikely that these conditions would be maintained for a sufficient period of time for mortalities to occur.

In Fig. 8 we have predicted that the average peak summer water temperature in the River Nivelle in the summer would increase from 20 to 22 °C. This is the limit of the feeding zone in Elliott's model and any increase above this temperature in the short term would result in the cessation of feeding by the salmon (Fig. 9). Mortalities due to thermal stress would occur only if the water temperature was maintained continuously at 27 °C for over 7 days or reached 31–33 °C in the short term (minutes to hours). More probably the increase in summer water temperature would result in sustained periods of time when the consumption rates of the fish are sub-maintenance or the fish stop feeding altogether. The time-scale of this exposure might vary between years but overall we would predict an increase in the period of increased protein degradation and weight loss during the summer and an increase in summer mortality rates due to starvation (Fig. 10). The possibility of direct thermal mortality is unlikely but cannot be completely discounted. Increased summer mortality rates should result in a decrease in the number of fish smolting after 2 years (i.e. 2 summers) in fresh water and 1-year-old smolts will predominate. Data collected between 1942 and 1988 indicate that the average age at smolting has already decreased from 1.3 to 1.1 years in the Adour River in southwest France (Prouzet, 1990).

For the sake of simplicity, we have only considered the effect of a single parameter (temperature) on one stage of the life cycle (parr). We have not considered the movement of fish within the rivers to locate bodies of cooler water (see chapter by Crawshaw & O'Connor). The effects of climatic change on food supply, on egg survival and development, on the timing of first feeding, smolt migration and spawning, or on mortalities due to increased predation have not been considered. The effect of climatic change on wild salmonid stocks will be more complex. For example, any increase in growth opportunity can

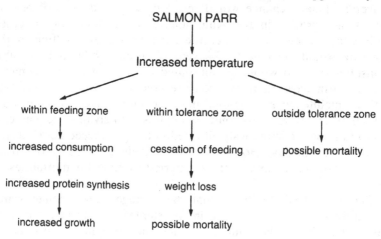

Fig. 10. A summary of the possible effects of increased water temperature on consumption, growth and protein metabolism of salmon parr based on the model of Elliott (1991) in Fig. 9.

only be exploited by the salmon if there is a corresponding increase in food supply. This may be a crucial factor for northern salmon stocks, since river productivity will be dependent on photoperiod as well as temperature and it is possible that survival rates at the first feeding stage may play a vital role in determining the effects of climatic change in these stocks. Increased summer mortalities in the southern rivers may result in the extinction of these stocks and a general shift in the range of the species. There is some evidence to suggest that this may occur since until recently the northeastern limit of the distribution was the Pechora River in western Russia. However, MacCrimmon & Gott (1979) report that as a result of climatic warming from 1919 to 1938, Atlantic salmon penetrated eastward into the Kara Sea and were in the Kara River as early as 1932. Climatic change may result in the colonization of new rivers in Siberia and the Arctic Islands by Atlantic salmon (MacCrimmon & Gott, 1979).

Summary and suggestions for future research

As a result of global warming we suggest that there will be a general northward shift in the geographic distribution of Atlantic salmon in Europe. It is likely that salmon stocks at the southern end of the range of the species in northern Spain and southwest France will become extinct. The opportunity for feeding and growth is likely to increase for salmon stocks at the existing northern end of the range of the species in northern Norway resulting in a decrease in the average age of smolting in these river stocks. It is also possible that new rivers in Siberia and the Arctic Islands may be colonized.

The mechanisms by which these changes in geographic distribution may occur can be summarized as follows.

1 Any increase in water temperature within the thermal preferenda for a species, provided that food supply is not limiting, will result in higher food consumption rates and increased rates of protein synthesis and growth. Therefore, providing food is available in the rivers, this will result in increased rates of protein synthesis and growth in the northern salmon stocks as the length of the annual growing season increases, resulting in a decrease in the average age at smolting.

2 The predicted increase in water temperatures during the summer may increase the period of time during the summer when the salmon parr in the southern salmon stocks stop feeding as a result of thermal stress. During these periods

of time, rates of protein degradation and weight loss will increase with increasing water temperature.

3   Both short-term (hours) and long-term (days to weeks) exposure to increased water temperature of fish acclimatized to a lower water temperature will result in increased rates of protein synthesis. The magnitude of increase in the rate of protein synthesis will be greater following short-term exposure to elevated water temperature.

It is important to point out that most of the data reviewed here have been obtained from short-term studies (maximum 7 weeks) under experimental conditions, such as controlled temperature and food availability. These studies have examined the temperature response of protein synthesis in fish reared under two constant water temperatures 10 °C apart, or a range of constant water temperatures spanning 20 to 30 °C. In the latter case, the temperature response of protein synthesis has been measured after 24 hours and 4 weeks acclimatization. The results obtained do allow some predictions to be made about the effects of elevated water temperature on protein synthesis in fish. However, this experimental design is not ideal when attempting to predict the effect of long-term climatic change on the physiology, growth and ecology of freshwater fish in the wild where changes in water temperature will be superimposed on existing fluctuations that already vary in magnitude and time-scale.

Specifically we would suggest further research in the following areas.

1   Further examination of the effects of temperature on translation processes in fish cells should be made. The effect of water temperature on protein synthesis appears to be tissue-specific (see Foster *et al.*, 1992) and the temperature response may vary between cell types (see Smith & Houlihan, 1995).

2   Tissue RNA concentrations are increased in fish exposed to reduced water temperatures; however, we still do not know the mechanism controlling this thermoacclimatory response. The effect of water temperature on the rates of synthesis and degradation of RNA should be investigated and the relationship between rates of RNA turnover and rates of protein synthesis should be examined.

3   The measurement of protein synthesis over a range of water temperatures throughout the thermal range for a particular species should be carried out to confirm whether maximum rates of protein synthesis occur at the optimum water tem-

perature for that species. The energetic cost per unit of protein synthesized should also be measured to confirm whether the lowest energetic costs correspond with the optimum temperature for growth. Rates of protein synthesis should be measured in fish reared at water temperatures above the incipient lethal level for the species (see Elliott, 1991) to examine how the protein metabolism of fish responds to thermal stress.

4   Studies should examine how short-term (hours to days) and seasonal (weeks to months) fluctuations in water temperature affect protein metabolism and growth of fish. The use of experimental designs that mimic natural thermal regime and the predicted temperature changes following global warming (for example, see Reid *et al.*, 1995) are to be encouraged as this will enable the effects of climate change on the bioenergetics of coldwater and temperate freshwater fish to be more accurately modelled and the effects of global warming to be more successfully predicted.

Acknowledgements

This research was conducted with a grant from the Biotechnology and Biological Sciences Research Council as part of their Global Environmental Change research programme. We thank Isadora Karamboula, Leyton Hackney, Stewart Owen, Richard Smith and Anne Farquhar for their assistance, Dr Edward Beall (INRA, St Pee-sur-Nivelle) for providing temperature data for the Nivelle River and Blackwell Scientific Ltd (Oxford) for copyright approval to reproduce Fig. 10.

References

Alberts, B., Bray, D., Lewis, J., Raff, M., Roberts, K. & Watson, J. D. (1983). *Molecular Biology of the Cell*, New York: Garland.

Ashford, A. J. & Pain, V. M. (1985). Effect of diabetes on the rates of synthesis and degradation of ribosomes in rat muscle and liver *in vivo*. *Journal of Biological Chemistry*, **261**, 4059–65.

Atkinson, D. (1994). Temperature and organism size – a biological law for ectotherms? *Advances in Ecological Research*, **25**, 1–58.

Beall, E., Dumas, J., Claireaux, D., Barrière, L. & Marty, C. (1994). Dispersal patterns and survival of Atlantic salmon (*Salmo salar* L.) juveniles in a nursery stream. *ICES Journal of Marine Science*, **51**, 1–9.

Carter, C. G., Houlihan, D. F., Brechin, J. & McCarthy, I. D. (1993a). The relationships between protein intake and protein accretion, synthesis and retention efficiency for individual grass carp, *Ctenopharyngodon idella* (Val.). *Canadian Journal of Zoology*, **71**, 392–400.

Carter, C. G., Houlihan, D. F., Brechin, J., McCarthy, I. D. & Davidson, I. (1993b). Protein synthesis in grass carp, *Ctenopharyngodon idella* (Val.) and its relation to diet quality. In *Fish Nutrition in Practice*, ed. S. J. Kaushik & P. Luquet, pp. 673–80. Paris: INRA.

Elliott, J. M. (1991). Tolerance and resistance to thermal stress in juvenile Atlantic salmon, *Salmo salar. Freshwater Biology*, **25**, 61–70.

Fauconneau, B. (1985). Protein synthesis and protein deposition in fish. In *Nutrition and Feeding in fish*, ed. C. B. Cowey, A. M. Mackie & J. G. Bell, pp. 17–45. London: Academic Press.

Fauconneau, B. & Arnal, M. (1985). In *vivo* protein synthesis in different tissues and the whole body of rainbow trout (*Salmo gairdneri* R.). Influence of environmental temperature. *Comparative Biochemistry and Physiology*, **82A**, 179–87.

Ferguson, M. M. & Danzmann, R. G. (1990). RNA/DNA ratios in white muscle as estimates of growth in rainbow trout held at different temperatures. *Canadian Journal of Zoology*, **68**, 1494–8.

Foster, A. R., Houlihan, D. F., Hall, S. J. & Burren, L. J. (1992). The effects of temperature acclimation on protein synthesis rates and nucleic acid content of juvenile cod (*Gadus morhua* L.). *Canadian Journal of Zoology*, **70**, 2095–102.

Garcia de Leániz, C. & Martinez, J. J. (1988). The Atlantic salmon in the rivers of Spain with particular reference to Cantabria. In *Atlantic Salmon: Planning for the Future*, ed. D. Mills & D. Piggins, pp. 179–209. London: Croom Helm.

Garlick, P. J., Burk, T. L. & Swick, R. W. (1976). Protein synthesis and RNA in tissues of the pig. *American Journal of Physiology*, **230**, 1108–12.

Garlick, P. J., McNurlan, M. A. & Preedy, V. R. (1980). A rapid and convenient technique for measuring the rate of protein synthesis in tissues by injection of [^{3}H] phenylalanine. *Biochemical Journal*, **192**, 719–23.

Gibson, R. J. (1993). The Atlantic salmon in freshwater: spawning, rearing and production. *Reviews in Fish Biology and Fisheries*, **3**, 39–73.

Goolish, E. M., Barron, M. G. & Adelmann, I. R. (1984). Thermoacclimatory response of nucleic acid and protein content of carp muscle tissue: influence of growth rate and relationship to glycine uptake by scales. *Canadian Journal of Zoology*, **62**, 2164–70.

Harmon, C. S., Proud, C. G. & Pain, V. M. (1984). Effects of starvation, diabetes and acute insulin treatment on the regulation

of polypeptide chain initiation in rat skeletal muscle. *Biochemical Journal*, **223**, 687–96.

Haschemeyer, A. E. V. (1978). Protein metabolism and its role in temperature acclimation. In *Biochemical and Biophysical Perspectives in Marine Biology*, Vol. 4, ed. D. C. Malins & J. R. Sargeant, pp. 329–84. London: Academic Press.

Haschemeyer, A. E. V. (1983). A comparative study of protein synthesis in nototheniids and icefish at Palmer Station, Antarctica. *Comparative Biochemistry and Physiology*, **76**B, 541–3.

Haschemeyer, A. E. V., Persell, R. & Smith, M. A. K. (1979). Effect of temperature on protein synthesis in fish of the Galapagos and Perlas Islands. *Comparative Biochemistry and Physiology*, **64**B, 91–5.

Haughton, J. T., Jenkins, G. J. & Ephraums, J. J. (ed.) (1990). *Climate Change: The IPCC Scientific Assessment*. Cambridge: Cambridge University Press.

Hazel, J. R. & Prosser, C. L. (1974). Molecular mechanisms of temperature compensation in poikilotherms. *Physiological Reviews*, **54**, 620–77.

Hill, D. K. & Magnuson, J. J. (1990). Potential effects of global climate warming on the growth and prey consumption of Great Lakes fish. *Transactions of the American Fisheries Society*, **119**, 265–75.

Hokanson, K. E. F., Kleiner, C. F. & Thorslund, T. W. (1977). Effects of constant temperatures and diel temperature fluctuations on specific growth and mortality rates and yield of juvenile rainbow trout, *Salmo gairdneri*. *Journal of the Fisheries Research Board of Canada*, **34**, 639–48.

Houlihan, D. F. (1991). Protein turnover in ectotherms and its relationships to energetics. In *Advances in Comparative and Environmental Physiology*, Vol. 7, ed. R. Gilles, pp. 1–43. Berlin: Springer-Verlag.

Houlihan, D. F., Carter, C. G. & McCarthy, I. D. (1995a). Protein synthesis in fish. In *Biochemistry and Molecular Biology of Fishes* Vol. 4, ed. P. Hochachka & T. Mommsen, pp. 191–220. Amsterdam: Elsevier.

Houlihan, D. F., Carter, C. G. & McCarthy, I. D. (1995b). Protein synthesis in animals. In *Nitrogen Metabolism and Excretion*, ed. P. Walsh & P. Wright, pp. 1–32. Boca Raton, Fla: CRC Press.

Houlihan, D. F., Costello, M. J., Secombes, C. J., Stagg, R. & Brechin, J. (1994). Effects of sewage sludge exposure on growth, feeding and protein synthesis of dab, *Limanda limanda* L. *Marine Environmental Research*, **37**, 331–53.

Houlihan, D. F., Hall, S. J. & Gray, C. (1989). Effects of ration on protein turnover in cod. *Aquaculture*, **79**, 103–10.

Houlihan, D. F., McCarthy, I. D., Carter, C. G. & Marttin, F. (1995c). Protein turnover and amino acid flux in fish larvae. *ICES Marine Science Symposium*, **201**, 87–99.

Houlihan, D. F., Mathers, E. M. & Foster, A. R. (1993). Biochemical correlates of growth rate in fish. In *Fish Ecophysiology*, ed. J. C. Rankin & F. B. Jensen, pp. 45–71. London: Chapman & Hall.

Huey, R. B. & Kingsolver, J. G. (1989). Evolution of thermal sensitivity of ectotherm performance. *Trends in Ecology and Evolution*, **4**, 131–5.

Jensen, A. J. (1991). Possible effects of climatic changes on the ecology of Norwegian Atlantic salmon (*Salmo salar* L.). *ICES C.M. 1991/M:34*.

Koban, M. (1986). Can cultured teleost hepatocytes show temperature acclimation? *American Journal of Physiology*, **250**, R211–20.

Laurent, G. J., Sparrow, M. P., Bates, P. C. & Millward, D. J. (1978). Turnover of muscle protein in the fowl (*Gallus domesticus*). Rates of protein synthesis in fast and slow skeletal, cardiac and smooth muscle of the adult fowl. *Biochemical Journal*, **176**, 393–405.

Lobley, G. E., Connell, A., Milne, E., Newman, A. M. & Ewing, T. A. (1994). Protein synthesis in splanchnic tissues of sheep offered two levels of intake. *British Journal of Nutrition*, **71**, 3–12.

Lobley, G. E., Milne, V., Lovie, J. M., Reeds, P. J. & Pennie, K. (1980). Whole body and tissue protein synthesis in cattle. *British Journal of Nutrition*, **43**, 491–502.

Loughna, P. T. & Goldspink, G. (1985). Muscle protein synthesis rates during temperature acclimation in a eurythermal (*Cyprinus carpio*) and a stenothermal (*Salmo gairdneri*) species of teleost. *Journal of Experimental Biology*, **118**, 267–76.

MacCrimmon, H. R. & Gott, B. L. (1979). World distribution of Atlantic salmon, *Salmo salar*. *Journal of the Fisheries Research Board of Canada*, **36**, 422–57.

McCarthy, I. D. (1993). Feeding behaviour and protein turnover in fish. PhD thesis, University of Aberdeen.

McCarthy, I. D., Houlihan, D. F., Carter, C. G. & Moutou, K. (1993). Variation in individual consumption rates of fish and its implications for the study of fish nutrition and physiology. *Proceedings of the Nutrition Society*, **52**, 427–37.

McCarthy, I. D., Houlihan, D. F. & Carter, C. G. (1994). Individual variation in protein turnover and growth efficiency in rainbow trout, *Oncorhynchus mykiss* (Walbaum). *Proceedings of the Royal Society of London, Series B*, **257**, 141–7.

McLain, A. S., Magnuson, J. J. & Hill, D. K. (1994). Latitudinal and longitudinal differences in thermal habitat for fishes influenced by climate warming: expectations from simulations. *Verhandlungen der Internationalen Vereinigung für Theoretische und Angewandte Limnologie*, **25**, 2080–5.

McMillan, D. N. & Houlihan, D. F. (1988). The effect of refeeding on tissue protein synthesis in rainbow trout. *Physiological Zoology*, **61**, 429–41.

Magnuson, J. J., Meisner, D. & Hill, D. K. (1990). Potential changes in the thermal habitat of Great Lakes fish after global climate warming. *Transactions of the American Fisheries Society*, **119**, 254–64.

Mathers, E. M., Houlihan, D. F., McCarthy, I. D. & Burren, L. J. (1993). Rates of growth and protein synthesis correlated with nucleic acid content in fry of rainbow trout, *Oncorhynchus mykiss*: effects of age and temperature. *Journal of Fish Biology*, **43**, 245–63.

Metcalfe, N. B. & Thorpe, J. E. (1990). Determinants of geographical variation in the age of seaward migrating salmon, *Salmo salar*. *Journal of Animal Ecology*, **59**, 135–45.

Millward, D. J., Garlick, P. J., Nnanyelugo, D. O. & Waterlow, J. C. (1976). The relative importance of muscle protein synthesis and breakdown in the regulation of muscle mass. *Biochemical Journal*, **156**, 185–8.

Nielsen, J. B. K., Plant, P. W. & Haschemeyer, A. E. V. (1977). Control of protein synthesis in temperature acclimation. II. Correlation of elongation factor 1 activity with elongation rate *in vivo*. *Physiological Zoology*, **50**, 22–30.

Oñate, S., Amthauer, R. and Krauskopf, M. (1987). Differences in the tRNA population between summer and winter acclimatized carp. *Comparative Biochemistry and Physiology*, **86B**, 663–6.

Pain, V. M. & Clemens, M. J. (1986). Mechanism and regulation of protein synthesis in eukaryotic cells. In *Protein Deposition in Animals*, ed. P. J. Buttery & D. B. Lindsay, pp. 1–20. London: Butterworths.

Pannevis, M. C. & Houlihan, D. F. (1992). The energetic cost of protein synthesis in isolated hepatocytes of rainbow trout (*Oncorhynchus mykiss*). *Journal of Comparative Physiology B*, **162**, 393–400.

Peragón, J., Ortega-García, F., Barroso, J. B., De la Higuera, M. & Lupiáñez, J. A. (1992). Alterations in the fractional rates of protein turnover rates in rainbow trout liver and white muscle caused by an aminoacid-based diet and changes in the feeding frequency. *Toxicological and Environmental Chemistry*, **36**, 217–24.

Peragón, J., Ortega-García, F., Barroso, J. B., De la Higuera, M. & Lupiáñez, J. A. (1993). Effect of frequency of feeding of diets with different protein content on the fractional rate of protein synthesis and degradation in liver and white muscle of rainbow trout (*Oncorhynchus mykiss*). In *Fish Nutrition in Practice*, ed. S. J. Kaushik & P. Luquet, pp. 287–92. Paris: INRA.

Peragón, J., Barroso, J. B., Garcia-salguero, L., De la Higuera, M. & Lupiánez, J. A. (1994). Dietary effects on growth and fractional protein synthesis and degradation rates in liver and white muscle of rainbow trout (*Oncorhynchus mykiss*). *Aquaculture*, **124**, 35–46.

Perera, W. M. K. (1995). Growth performance, nitrogen balance and protein turnover of rainbow trout (*Oncorhynchus mykiss* (Walbaum)) under different dietary regimens. PhD thesis, University of Aberdeen.

Pocrnjic, Z., Mathews, R. W., Rappaport, S. & Haschemeyer, A. E. V. (1983). Quantitative protein synthetic rates in various tissues of a temperate fish *in vivo* by the method of phenylalanine swamping. *Comparative Biochemistry and Physiology*, **74B**, 735–8.

Power, G. (1990). Salmonid communities in Quebec and Labrador; temperature relations and climate change. *Polskie Archiwum Hydrobiologii*, **37**, 13–28.

Preedy, V. R., Paska, L., Sugden, P. H., Schofield, P. S. & Sugden, M. C. (1988). The effects of surgical stress and short-term fasting on protein synthesis *in vivo* in diverse tissues of the mature rat. *Biochemical Journal*, **250**, 179–88.

Prouzet, P. (1990). Stock characteristics of Atlantic salmon (*Salmo salar*) in France: a review. *Aquatic Living Resources*, **3**, 85–97.

Reeds, P. J. & Davies, T. A. (1992). Hormonal regulation of muscle protein synthesis and degradation. In *The Control of Fat and Lean Deposition*, ed. P. J. Buttery, K. N. Boorman & D. B. Lindsay, pp. 1–26. Oxford: Butterworth Heinemann.

Reid, S. D., Dockray, J. J., Linton, T. K., McDonald, D. G. & Wood, C. M. (1995). Effects of a summer temperature regime representative of a global warming scenario on growth and protein synthesis in hardwater- and softwater-acclimated juvenile rainbow trout (*Oncorhynchus mykiss*). *Journal of Thermal Biology* **20**, 231–44.

Saez, L., Goicoechea, O., Amthauer, R., Rodriguez, E. & Krauskopf. M. (1982). Behaviour of RNA and protein synthesis during the acclimatisation of the carp. Studies with isolated hepatocytes. *Comparative Biochemistry and Physiology*, **72B**, 31–8.

Sayegh, J. F. & Lajtha, A. (1989). *In vivo* rates of protein synthesis in brain, muscle and liver of five vertebrate species. *Neurochemical Research*, **14**, 1165–8.

Schindler, D. W., Beaty, K. G., Fee, E. J. *et al.* (1990). Effects of climatic warming on lakes of the central boreal forest. *Science*, **250**, 967–70.

Sidell, B. D. (1983). Cellular acclimatisation to environmental change by quantitative alterations in enzymes and organelles. In *Cellular Acclimatisation to Environmental Change*, ed. A. R Cossins & P. Shelterline, pp. 103–20. Cambridge: Cambridge University Press.

Simon, E. (1987). Effect of acclimation temperature on the elongation step of protein synthesis in different organs of rainbow trout. *Journal of Comparative Physiology B*, **157**, 201–7.

Smith, M. A. K. & Haschemeyer, A. E. V. (1980). Protein metabolism and cold-adaptation in Antarctic fishes. *Physiological Zoology*, **53**, 373–82.

Smith, R. W. & Houlihan, D. F. (1995). Protein synthesis and oxygen consumption in fish cells. *Journal of Comparative Physiology B*, **165**, 93–101.

Smith, R. W., Houlihan, D. F., Nilsson, G. E. & Brechin, J. G. (1996). Tissue-specific changes in protein synthesis rates *in vivo* during anoxia in crucian carp. *American Journal of Physiology* (in press).

Sugden, P. H. & Fuller, S. J. (1991). Regulation of protein turnover in skeletal and cardiac muscle. *Biochemical Journal*, **273**, 21–37.

Tessauraud, S., Larbier, M., Chagneau, A. M. & Geraert, P. A. (1992). Effect of dietary lysine on muscle protein turnover in growing chickens. *Reproduction, Nutrition and Development*, **32**, 163–71.

Thorpe, J. E., Metcalfe, N. B. & Huntingford, F. A. (1992). Behavioural influences on life/history variation in juvenile Atlantic salmon, *Salmo salar*. *Environmental Biology of Fishes*, **33**, 331–40.

Watt, P. W., Marshall, P. A., Heap, S. P., Loughna, P. T. & Goldspink, G. (1988). Protein synthesis in tissues of fed and starved carp, acclimated to different temperatures. *Fish Physiology and Biochemistry*, **4**, 165–73.

Yacoe, M. E. (1983). Protein metabolism in the pectoralis muscle and liver of hibernating bats, *Eptesicus fuscus*. *Journal of Comparative Physiology*, **152**, 137–44.

Zuvic, T., Brito, M., Villaneuva, J. & Krauskopf, M. (1980). *In vivo* levels of aminoacyl-tRNA species during acclimatisation of the carp, *Cyprinus carpio*. *Comparative Biochemistry and Physiology*, **67**B, 167–70.

IAN A. JOHNSTON and DEREK BALL

Thermal stress and muscle function in fish

Introduction

Temperature is one of the most important extrinsic factors determining muscle performance in ectotherms. In considering the effects of temperature on muscle contraction in fish it is important to distinguish between sustainable and maximum effort, since they are supported by different muscle fibre types. A great variety of muscle phenotypes are observed, brought about by the differential expression of different protein isoforms, and by varying the amounts of ion channels, membrane pumps and cellular organelles (see Johnston & Altringham, 1991, for a recent review). Slow swimming activity largely involves the recruitment of slow twitch muscle fibres, which are dependent on aerobic metabolism, and complex respiratory and circulatory support systems to supply oxygen and substrates (Bone, 1966; Johnston, Davison & Goldspink, 1977). Slow twitch fibres, with their high concentrations of myoglobin and mitochondria and well developed capillary supply, are the main fibre type in red muscle (Johnston, 1981; Altringham & Johnston, 1988a).

As swimming speed increases, faster contracting muscle fibres are recruited (Johnston et al., 1977; Rome, Loughna & Goldspink, 1984). Maximum performance is achieved during fast-starts associated with escape responses and predation and involves the recruitment of the entire white muscle mass (Johnston, Franklin & Johnson, 1993). The white muscle is typically composed of a single fibre type expressing fast isotypes of the myofibrillar proteins and containing high concentrations of the cytoplasmic Ca^{2+}-binding protein parvalbumin (Gerday et al., 1979; Rowlerson et al., 1985; Crockford & Johnston, 1993). White fibres have larger average diameters than red fibres and contain higher volume densities of myofibrils, and a more extensive sarcoplasmic reticulum for faster calcium cycling (Akster, Granzier & ter

Society for Experimental Biology Seminar Series 61: *Global Warming: Implications for freshwater and marine fish*, ed. C. M. Wood & D. G. McDonald. © Cambridge University Press 1996, pp. 79–104.

Keurs 1985; Fleming *et al.*, 1990). Slow muscle fibres are orientated parallel to the longitudinal axis of the trunk whereas fast muscle fibres are arranged in complex helical patterns (Alexander, 1969). Rome and Sosnicki (1991) showed that fast fibres have a higher gearing ratio, needing to shorten by only 25% as much as the slow fibres to produce a given change in body curvature.

Muscle performance varies with both the passive properties of the musculoskeletal system and the active properties of the nervous system (Johnston, 1991; Van Leeuwen, 1995). In general, fast muscle is dependent on fewer other physiological systems than is slow muscle since it functions largely independently of the circulation (except for recovery) and utilizes anaerobic metabolism and endogenous fuel supplies (phosphocreatine and glycogen) (Johnston & Altringham, 1991).

Because of these differences the factors limiting muscle performance at the whole animal level may well vary with the type of swimming activity. Thus, although many properties of muscle can best be studied using *in vitro* preparations, it is essential to relate any results to how the muscle functions in the intact animal. It is also important to consider realistic levels of thermal stress. A major problem in considering the likely impact of global warming is that most studies on the effect of temperature on muscle contraction have been directed towards understanding cold-adaptation rather than responses to thermal stress. Similarly, although there is a growing literature on the phenotypic responses to temperature change, the main focus of this research has been on eurythermal species rather than on stenothermal fish which are more likely to be vulnerable to any detrimental effects of global warming.

This chapter starts with a review of the phenotypic and genotypic responses of fish muscle to temperature change and concludes with some speculation on the likely impact of global warming on swimming performance. Experimental paradigms for future research are also discussed.

Phenotypic and genotypic responses to temperature

Whole animal studies

Rome and co-workers recorded electromyograms (EMGs) to investigate the effects of temperature on muscle recruitment in the common carp (*Cyprinus carpio*) and the scup (*Stenotomus chrysops*). The maximum speed the fish could swim using the red muscle alone was found to increase with warming, reflecting its higher power output. The threshold speed for the recruitment of white muscle fibres was correspondingly increased. Indeed in all species studied, warming causes an increase in maximum sustainable swimming speed until some limit is reached, after which performance usually declines (see Fig. 1). The temperature at which swimming performance levels off is a function of habitat temperature. For example, the Antarctic species. *Pagothenia borchgrevinski* reaches its maximum cruising speed (U_{crit}) at -0.8 °C and is unable to swim above 2 °C (Wohlschlag, 1964), whereas in the largemouth bass (*Micropterus salmoides*) from North America U_{crit} reaches a maximum at 25–30 °C (Beamish, 1970). The reason for the drop-off in sustainable performance at temperatures approaching the upper thermal tolerance of a species (Fig. 1) is probably complex, reflecting a failure in one or more of the support systems to the muscle fibres and/or a drop in muscle power output (see Egginton *et al.*, this volume).

There are relatively few data on the effects of thermal stress on escape behaviour in fish. The typical response of a fish to a noxious stimulus or predator is a C-start in which the head and the tail rotate in the same direction away from the centre of mass during the first tailbeat. Batty and Blaxter (1992) studied the maximum speed of fish larvae following C-start escape responses using a high speed video recording. The maximum speed (U_{max}) of newly hatched herring (*Clupea harengus*) larvae increased up to 15 °C, whereas U_{max} in plaice larvae

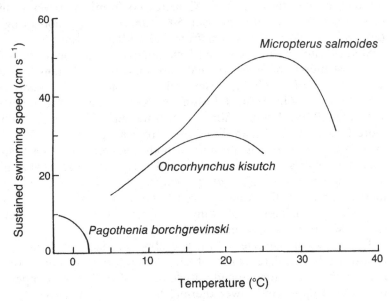

Fig. 1. Influence of temperature on the maximum sustainable swimming speed of fish (adapted from Beamish, 1978).

(*Pleuronectes platessa*) reached a plateau between 8 and 12 °C, due to a reduction in stride length.

Properties of isolated proteins and muscles

The temperature tolerance of extant teleost species ranges from −2 °C in Antarctica to 43–45 °C for fish living in geothermal hot-springs. Studies with isolated proteins and muscle preparations have been used to gain some insight into the limits and mechanisms underlying evolutionary adjustments to different thermal environments.

Antarctic species living permanently between −2 and 1 °C have the lowest upper lethal temperatures reported for fish of around 6 °C (Somero & DeVries, 1967). The species *Notothenia rossii* has been reported to synthesize heat shock proteins in response to thermal stress at temperatures as low as 4 to 5 °C (Maresca *et al.*, 1988). Myosin from coldwater fish has an unstable tertiary structure and it readily aggregates and loses its ATPase activity on purification (Connell, 1960; Johnston *et al.*, 1975a). Johnston and Walesby (1977) showed a strong correlation between the half-life of thermal de-activation of white muscle myofibrillar ATPase and habitat temperature. Under identical conditions of protein concentration, ionic strength and pH, the half-time for thermal deactivation at 37 °C increased from little over a minute in several Antarctic fish to over 500 minutes in the hot-spring fish, *Oreochromis alcalicus grahami* from Lake Magadi, Kenya.

Studies with live and skinned (demembranated) muscle fibres isolated from fast muscle have shown that maximum isometric tensile stress is a function of both experimental and adaptation temperature (Johnston & Brill, 1984; Johnston & Altringham, 1985; Johnson & Johnston, 1991a). Skinned fibres from Antarctic fish produce 5–10 times more tension at 0 °C than fibres from tropical species (Fig. 2a). However, when measured at the normal body temperature of each species maximum tensions are comparable, and may even be somewhat higher for coldwater species (Fig. 2a). At temperatures above 5–10 °C skinned muscle fibres from the Antarctic icefish (*Chaenocephalus aceratus*) failed to relax completely following maximal activations, generating Ca^{2+}-insensitive residual tension (Fig. 3).

The thermal range over which live fibres produce maximum tension is also highly correlated with habitat temperature, but differs from the pattern observed with skinned muscle fibres (Fig. 2b). At temperatures beyond the upper and lower thermal limits for each species live fibre preparations become progressively inexcitable, leading to a reduction in maximum tension. In contrast, maximum tension either rises or stays

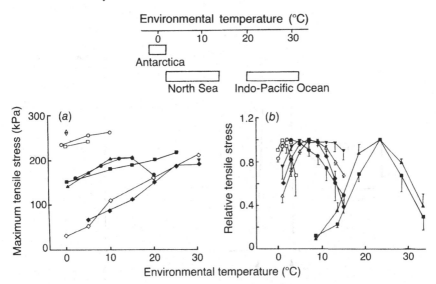

Fig. 2. Effects of temperature on the maximum tensile stress generated by (*a*) skinned and (*b*) live fibres isolated from the fast myotomal (myo) and pectoral fin adductor (m.add.p.) of teleost fish. Symbols refer to the following: (*a*) Antarctic fish: △, *Trematomus hansoni* (myo); ▽, *Notothenia rossii* (myo); ○, *Chaenocephalus aceratus* (myo); □, *Notothenia corriceps* (myo). North Sea fish: ●, *Gadus morhua* (myo); ▲, *Myoxocephalus scorpius* (myo); ■, *Platichthys flesus* (myo). Tropical fish: ◇, *Makaira nigricans* (myo); ◆, *Carangus melampygus* (myo); ▼, *Eurythunnus affinis* (myo). (*b*) Antarctic fish: ○, *Trematomus lepidorhinus* (m.add.p.); □, *Notothenia corriiceps* (m.add.p). North Sea fish: △, *Pollachius virens* (myo); ●, *Limanda limanda* (myo); ◆, *Callionymus lyra* (myo); ▼, *Aggonis cataphractus* (m.add.p.). Tropical fish: ▲, *Abudefduf abdominalis* (m.add.p.); ■, *Thalassoma duperreyi* (myo). Values represent mean ± SE. See Johnston & Altringham (1985) and Johnson & Johnston (1991a) for original data and numbers of fish examined.

relatively constant as temperature is increased in skinned muscle fibres (Fig. 2*a*). The mechanism underlying the loss of excitability of live fibres with thermal stress is likely to be complex, involving a failure of activation processes and/or excitation–contraction coupling (Johnson & Johnston, 1991a). It is also important to bear in mind that in the intact animal the muscle fibres are activated by the motor neurons and the relationship between maximum tensile stress and temperature *in vivo* may well differ from that observed with isolated live fibres.

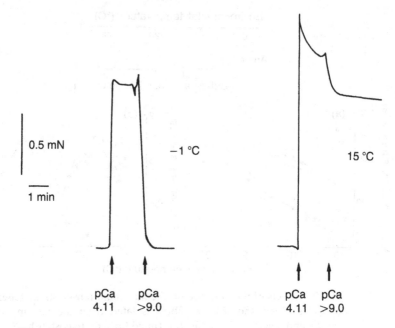

Fig. 3. Skinned fast muscle fibres from the Antarctic icefish (*Chaenocephalus aceratus*) fail to relax completely following activations at high temperatures. The figure shows the tension record from a fibre successively transferred between relaxing ($pCa^{2+} > 9.0$), activating (pCa^{2+} 4.11), and relaxing solutions. (Original data from Johnston, 1985.)

Time-dependent muscle contractile properties are highly temperature-dependent and show much less evidence of evolutionary adjustments to different thermal habitats than does maximum tensile stress (Johnston & Altringham, 1985; Johnson & Johnston, 1991a). For example, the unloaded contraction speed of muscle fibres has a Q_{10} of 1.5 to 2.0, and is relatively independent of the temperature at which the fish is living (Fig. 4). Although the rate of relaxation of isometric twitches is greatly prolonged at low temperatures in muscle fibres from tropical species it does show some limited evidence for temperature compensation (Johnson & Johnston, 1991a). This reflects higher rates of calcium pumping by the sarcoplasmic reticulum at low temperatures in coldwater relative to warmwater fish (McArdle & Johnston, 1980).

The power output of muscle fibres during shortening under constant load can be determined from the force–velocity relationship. The best fit of experimental data is usually obtained if the points are not con-

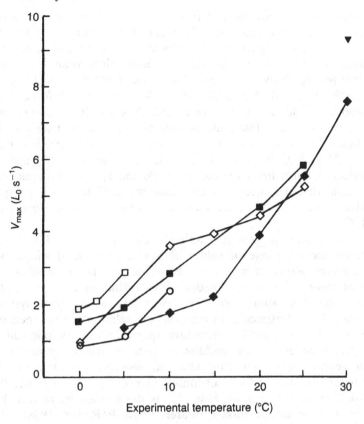

Fig. 4. Effects of temperature on the unloaded contraction velocity (V_{max}) of skinned fast myotomal muscle fibres of teleosts fish. Antarctic: ○, *Chaenocephalus aceratus*; □, *Notothenia corriiceps*. North Sea: ■, *Platichthys flesus*. Tropical: ◇, *Makaira nigricans*; ◆, *Carangus melampygus*; ▼, *Eurythunnus affinis*. (Data from Johnston, 1990.)

strained to pass through the maximum tensile stress (Po) (Altringham & Johnston, 1988b; Rome & Sosnicki, 1990). Rome and Sosnicki (1990) calculated that the power output (W kg⁻¹ wet muscle mass) of red fibres from carp was 60 at 10 °C, rising to 94 at 20 °C (Q_{10} = 1.6). The power output (W kg⁻¹ wet muscle mass) of fast muscle fibres in the short-horned sculpin (*Myoxocephalus scorpius*) increased from 140 at 1 °C to 313 at 8 °C, but was similar (292) at 12 °C (Langfeld, Attringham & Johnston, 1989). The plateau in power output towards the upper temperature limit of this species largely reflected a decrease in the tensile stress generated. Po reached its maximum value of 315 kPa at 8 °C, declining to only 172 kPa at 16 °C (Langfeld *et al.*, 1989). In both the above studies the force–velocity relationship was found to become progressively more curved at higher temperatures (Langfeld *et al.*, 1989; Rome & Sosnicki, 1990). A more curved force–velocity relationship yields a lower velocity and thus a reduced power output for a given load. For fast muscles from the short-horned sculpin, Hill's constant *a*/Po, which is inversely related to the curvature, was 0.27 at 1 °C, 0.24 at 8 °C and 0.17 at 12 °C. Langfeld *et al.* (1989) quantified the effect by normalizing the curves for Po and V_{max} at each temperature and found that the change in curvature was sufficient to decrease the relative power output by 15% on increasing the temperature from 1 to 8 °C.

In vivo studies suggest that animals use their muscle fibres over a relatively narrow range of values of V/V_{max} (0.17–0.36) where power is maximum (Rome *et al.*, 1988), and efficiency is within 90% of the optimal value (Curtin & Woledge, 1991). Thus as the V/V_{max} of red fibres exceeds around 0.36 then faster contracting fibre types are recruited. For interspecific comparisons the relationship between swimming speed and muscle contraction speed (*V*) is complicated. For example, scup use a less undulatory style of swimming and have a lower gearing ratio than carp and can consequently swim at higher speeds for the same velocity of muscle shortening (Rome *et al.*, 1992). It should be noted that whereas V_{max} is determined by myosin heavy and light chain composition (Greaser, Moss & Reiser, 1988), *V* is a function of the rate of activation and relaxation of muscle fibres, tail beat amplitude and the orientation (and hence gearing) of the muscle fibres.

Muscle function during swimming

Isometric contractions cannot be directly related to swimming since both the development of tensile stress and relaxation are speeded up in cyclical contractions (Altringham & Johnston, 1990; Rome & Swank, 1992). Similarly, *in vivo*, muscle fibres seldom shorten under constant load and cross-bridge kinetics are sensitive to the impact of previous contraction cycles, particularly the effects of prestretch (Johnston, 1991; Van Leeuwen *et al.*, 1990; Van Leeuwen, 1995). During steady swimming the myotomal muscle fibres undergo repeated cycles of shortening and lengthening. Under these circumstances the length changes of the myotomal muscles are approximately sinusoidal (Hess & Videler, 1984).

A number of studies have imposed sinusoidal length changes about the *in situ* resting fibre length and stimulated the preparation phasically during each cycle (Altringham & Johnston, 1990; Anderson & Johnston, 1992; Rome & Swank, 1992; Altringham, Wardle & Smith, 1993). The work done by the muscle can be calculated from plots of tensile stress and muscle fibre length (work loops). Power output is calculated from the work per cycle multiplied by the cycle frequency. The work output of a muscle is critically dependent on such factors as its starting length, the number and timing of stimuli, the amount of shortening and the contraction frequency (Johnston, 1991; Van Leeuwen, 1995). In some species, such as the short-horned sculpin, muscle fibres in rostral and caudal myotomes have identical mechanical properties and have similar power outputs under *in vivo* conditions (Johnston *et al.*, 1993; Johnston, Van Leeuwen & Beddow, 1995). However, in saithe (Altringham *et al.*, 1993) and Atlantic cod, *Gadus morhua* (Davies, Johnston & Van der Wal, 1995), muscle fibres in caudal myotomes have longer twitch contraction times and probably function somewhat differently during swimming compared with fibres in rostral myotomes.

The various parameters influencing work can be systematically optimized to determine the maximum power output, taking advantage of any available information on cycle frequencies and the stimulation duty cycle during swimming (Rome & Swank, 1992; Altringham *et al.*, 1993; Johnston *et al.*, 1993). Rome and Swank (1992) used the work loop technique to measure the power output of red muscle fibres from the scup at 10 and 20 °C. They selected values of stimulation duty cycle (0.35–0.5) on the basis of EMG recordings made during steady swimming. The optimal strain for maximum work was 0.05–0.06 of resting fibre length. Net positive work per cycle decreased with increasing oscillation frequency and was significantly higher at 20 °C than at 10 °C, except at 1 Hz. Net power output increased to a maximum value and then decreased as the oscillation frequency increased. The maximum power was obtained at 2.5 Hz at 10 °C (12.8 W kg^{-1} wet muscle mass) and 5 Hz at 20 °C (27.9 W kg^{-1} wet muscle mass). The maximum speed that scup can swim using their red muscle fibres alone was found to increase from 60 cm s^{-1} at 10 °C to 84 cm s^{-1} at 20 °C (for fish 17–20 cm in length) (Rome *et al.*, 1992). Assuming the power required to swim increased with the 2.5 exponent of swimming speed (Webb, 1975), this would require around a 2.7-fold increase in mechanical power. Rome and Swank (1992) showed that the Q_{10} for muscle power production was a function of oscillation frequency. Comparing muscle power at the highest frequencies observed during swimming, of 4.5 Hz at 10 °C (9 W kg^{-1} wet muscle mass) and 7.5 Hz at 20°C (27 W kg^{-1}

wet muscle mass), gave a Q_{10} of around 3, which was in reasonable agreement with the predicted power requirements. Scup use the same tailbeat frequency at any given speed regardless of temperature (Rome *et al.*, 1992). Therefore, in order to swim at their maximum sustainable speed at 20 °C the fish would need to recruit only one-third as much red muscle as at 10 °C, which is the main reason why U_{crit} increases at high temperatures (Rome & Swank, 1992).

Phenotypic plasticity to temperature change

Sustained swimming

Following a period of warm-acclimation lasting several weeks, the maximum sustainable swimming speed of carp (Fig. 5) and goldfish

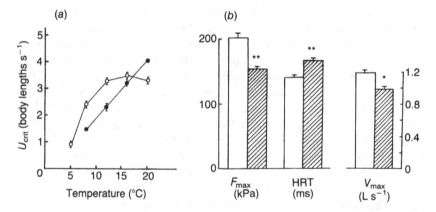

Fig. 5. (*a*) Effects of temperature on the maximum sustainable swimming speed (U_{crit}) over a 60 min period for common carp (*Cyprinus carpio*) acclimated to either 8 °C (open symbols) or 20 °C (closed symbols) for a minimum of 2 months. Error bars represent mean ± SE, $n = 88$ fish at each temperature. Note that the swimming performance of the cold-acclimated fish begins to decline above 15 °C. (Original data from Johnston, 1993.)

   (*b*) Effects of temperature acclimation on the contractile properties of live slow muscle fibre bundles isolated from the pectoral fin muscles of common carp acclimated to either either 8 °C (open bars) or 20 °C (cross-hatched bars) for a minimum of 2 months. The measurements were made at an experimental temperature of 8 °C. F_{max}, maximum tetanic tension (kPa); HRT, time (ms) from last stimuli to 50% maximum tension; V_{max}, unloaded shortening velocity in fibre lengths s^{-1}. Error bars represent mean ± SE, $n = 6$ fish at each temperature. (Original data from Langfeld *et al.*, 1991.)

(Fry & Hart, 1948) decreased at low temperatures and increased at high temperatures, although the tailbeat frequency and amplitude required to swim at any given speed remains the same (Johnston, 1993). In many species the relative proportion of red muscle fibres in the myotome varies with acclimation temperature (Johnston & Lucking, 1978; Sidell, 1980). As acclimation temperature increases, the amount of red muscle decreases, presumably since a smaller volume of red muscle is required to swim at a given speed because of its higher power output.

The contractile properties of slow muscle in common carp vary with acclimation temperature (Johnston, Sidell & Driedzic, 1985). Maximum tensile stress and contraction velocity at 7 °C were found to be 1.5 times higher for skinned fibres from carp acclimated to 7 °C than to 23 °C. Fibres from 7 °C-acclimated carp failed to relax completely following maximal activations at 23 °C. The resulting Ca^{2+}-insensitive force component (50–70% Po) built up with successive activations and resulted in very low contraction velocities, presumably as a result of the formation of abnormal cross-bridge linkages. In contrast, red fibres isolated from 23 °C-acclimated fish relaxed completely following up to 5 maximal activations each of up to 3 min duration (Johnston *et al.*, 1985). The maximum power output was more than 3-fold higher at 23 °C in 23 °C-acclimated than in 7 °C-acclimated fish. However, following acclimation to 7 °C maximum power output increased to values which were similar to that for warm-acclimated fish at 23 °C, indicating perfect thermal compensation.

Similar changes, but of a more modest nature, have been found for live slow muscle fibre bundles isolated from carp pectoral fin *abductor superficialis* muscle (Fig. 5*b*). Following acclimation from 8 to 20 °C, the maximum isometric tensile stress (F_{max}) at 8 °C decreased from 202 kPa to 153 kPa, but was unchanged at 20 °C. Twitch half-relaxation time was significantly longer at all temperatures in 20 °C-acclimated than in 8 °C-acclimated fish. Langfeld *et al.* (1989) studied the force-relationship of slow fibres isolated from the pectoral fin abductor muscle at 8 °C. Maximum shortening speed was 17% higher in 8 °C-acclimated than in 20 °C-acclimated fish at 8 °C (Fig. 5*b*), whereas the curvature of the P–V relationship was independent of acclimation temperature (Langfeld, Crockford & Johnston, 1991). The net result was that slow fibres produced 47% more power at 8 °C following cold-acclimation, largely as a result of increased force production. Since myofibrillar volume density is relatively independent of acclimation, these changes in force production are unlikely to reflect differences in the number of myosin cross-bridges per unit fibre cross-sectional area (Johnston & Maitland, 1980).

The myosin molecule is made up of heavy and light chain sub-units which are present as different isoforms in fast and slow muscle fibres. The globular head of myosin contains one alkali light chain ($LC1_f$ and $LC3_f$ in fast muscles and $LC1_s$ in slow muscles) and one regulatory light chain ($LC2_f$ and $LC2_s$ in fast and slow muscles, respectively). Fibre bundles from 8 °C-acclimated carp were found to contain a higher content of myosin light chain isoforms normally associated with faster contracting muscle fibre types (Langfeld et al., 1991). Since the small (3%) percentage of fast fibres in the preparations did not vary with acclimation temperature, it would appear that $LC1_f$ and $LC2_f$ are co-expressed with $LC1_s$ and $LC2_s$ in the slow fibres of 8 °C-acclimated fish, contributing to the observed increase in shortening speed at low temperatures.

Maximum swimming speeds

Johnston, Davison & Goldspink (1975b) discovered that in goldfish the Mg^{2+}/Ca^{2+}-activated ATPase activity of white muscle myofibrils increased at low temperatures following several weeks cold-acclimation. The ATPase activity in warm-acclimated fish was also less sensitive to thermal denaturation than in cold-acclimated goldfish. Similar changes in ATPase activity with thermal acclimation have been reported for other fish including common carp (Crockford & Johnston, 1990; Hwang, Watabe & Hashimoto, 1990), roach (Rutilus rutilis) and tench (Tinca tinca) (Heap, Watt & Goldspink, 1985). Changes in myofibrillar ATPase activity are reversible, take around 4 weeks to complete, and are inhibited in starved fish in which protein synthesis has been reduced to a very low level (Heap, Watt & Goldspink, 1986).

Fleming et al. (1990) used a nerve–muscle preparation to investigate twitch contraction kinetics following thermal acclimation. At 8 °C, the half-times for both twitch activation and relaxation were 2 to 3 times faster in preparations from 8 °C-acclimated than 20 °C-acclimated fish. The half-time for relaxation at 20 °C was also shorter for warm- than cold-acclimated fish, indicating plasticity of deactivation rates at both low and high temperatures (Fleming et al., 1990). The pCa^{2+}–tensile stress relationship (Johnston, Fleming & Crockford, 1990), parvalbumin content and the surface and volume density of the sarcoplasmic reticulum (SR) (Fleming et al., 1990) were not altered by temperature acclimation in the common carp. The faster relaxation of twitch tensile stress in muscle fibres from cold-acclimated carp has been correlated with an increase in SR Ca^{2+}-ATPase activity (Fleming et al., 1990; Ushio & Watabe, 1993). Studies with skinned fibres have shown that both Po and V_{max} in white muscle fibres of carp are altered by thermal acclimation (Johnston *et al.*, 1985). ATPase activity (Penney & Goldspink, 1979), maximum tension and contraction velocity (Crockford & Johnston, 1990) do not vary continuously with acclimation temperature but reach upper and lower limits. For example, Po for fast myotomal fibres at 0 °C is similar in carp acclimated to 2–11 °C, but declines progressively at higher temperatures (Crockford & Johnston, 1990).

The maximum contraction velocity of isolated single fibres has been correlated with the expression of specific myosin heavy chain isoforms (Lannergren, 1987; Bottinelli *et al.* 1994a) and alkali light chain isoforms (Bottinelli *et al.* 1994b). The molecular mechanisms underlying changes in Po and V_{max} with thermal acclimation in carp include altered expression of myosin heavy chain genes (Gerlach *et al.*, 1990; Hwang *et al.*, 1991) and myosin light chain isoforms (Crockford & Johnston, 1990). Gerlach *et al.* (1990) reported that carp had at least 28 different myosin heavy chain (MHC) genes and found evidence for altered expression with temperature acclimation. The expression of different MHC isoforms with temperature acclimation is supported by peptide mapping studies (Hwang *et al.*, 1991).

Few studies have examined thermal acclimation in marine species that are subject to more limited seasonal temperature changes. The most studied species is the short-horned sculpin which is an essentially coldwater species with a distribution in the eastern Atlantic from the Arctic to the northwest coast of France. Beddow, Van Leeuwen and Johnston (1995) found that the kinematic parameters of fast-starts elicited by predation are modified at high temperatures in summer-acclimated fish (Fig. 6). Fast-starts are used in prey capture and are highly variable depending on the position of the fish at the start of the attack and distance to the prey. The sculpin creep towards their prey using a combination of their pectoral and pelvic fins. For 20 cm fish the mean strike distance was 10 cm for 15 °C-acclimated fish and 5 cm for 5 °C-acclimated fish. Initially the body is bent into an S-shape, the pectoral fins rapidly adducted and the dorsal and ventral fins partially erected. This is followed by a complete tailbeat and an unpowered glide of variable duration. Towards the end of the powered phase of the fast-start the jaws are expanded and protruded, and the fish attempts to suck in the prey, greatly increasing the velocity of the attack. The average velocities reached during the complete tail beat were found to be relatively independent of temperature (Q_{10} 1.2–1.3) (Fig. 6). For 15 °C-acclimated fish average values for maximum acceleration were 16.2 m s^{-2} at 5 °C and 18.0 m s^{-2} at 15 °C. Following 6 to 8 weeks acclimation to 15 °C, the maximum velocity of the fast-start at 15 °C was 33% faster than in 5 °C-acclimated fish acutely exposed to the same temperature (Fig. 6). This increase in performance with warm-acclimation was sufficient to increase the percentage of successful attacks during prey capture from 23.2 to 73.4% (Beddow *et al.*, 1995).

Fast muscle fibres isolated from fish acclimated to 5 °C generated maximum tetanic tensions of 139 kPa at 5 °C, but force generating capabilities declined above 10 °C and Po at 15 °C was only 78 kPa (Fig. 7*a*). In contrast, Po in fibres from 15 °C-acclimated fish increased from 125 kPa at 5 °C to 282 kPa at 15 °C (Fig. 7*a*). The low tensile stresses generated by fibres from winter-acclimated fish at summer temperatures probably reflect a partial failure of excitation–contraction coupling, since peak force at 15 °C was increased 2.2 times following depolarization with a high potassium solution (Beddow & Johnston, 1995).

Ball and Johnston (1996) have also shown that the force–temperature characteristics of skinned muscle fibres are relatively independent of acclimation temperature (Fig. 7*b*). The maximum contraction speeds of fast muscle fibres in the short-horned sculpin are also modified at high (Fig. 8) and low temperatures following thermal acclimation (Beddow & Johnston, 1995). Only relatively minor differences in contractile properties were evident between freshly caught sculpins and fish acclimated to

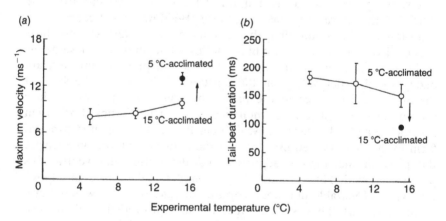

Fig. 6. Influence of temperature and thermal acclimation on (*a*) maximum velocity and (*b*) tailbeat duration in predation fast-starts in the short-horned sculpin (*Myoxocephalus scorpius*). Fish were acclimated to either 5 °C (o) or 15 ° (•) for a minimum of 2 months. Values represent mean ± SE, *n* = 6–8 fish at each temperature. (From Beddow *et al.*, 1995.)

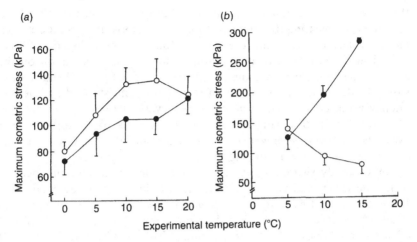

Fig. 7. Maximum isometric tensile stress generated by (*a*) skinned and (*b*) live fibres isolated from the fast myotomal muscle of short-horned sculpin acclimated to either 5 °C (○) or 15 °C (●) for a minimum of 2 months. Values represent mean ± SE, n = 7–11 fish at each temperature. (From Ball & Johnston, 1996 and Beddow & Johnston, 1995, respectively).

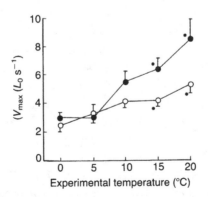

Fig. 8. Influence of temperature and thermal acclimation on the maximum contraction velocity of skinned fibres isolated from the fast myotomal muscle of the short-horned sculpin. Fish were acclimated to either 5 °C (○) or 15 °C (●) for a minimum of 2 months. Values represent mean ± SE, n = 7–11 fish at each temperature. (From Ball & Johnston, 1996.)

the same temperatures under a constant photoperiodic regime (12 h dark: 12 h light), indicating that temperature was the most important environmental variable influencing the results. Johnson and Johnston (1991b) investigated the effects of temperature and thermal acclimation on the ability of fast muscle fibres from the short-horned sculpin to do oscillatory work. Under optimal conditions of stimulation and strain the cycle frequency required for maximum work per cycle increased from 4 Hz at 5 °C to 9 Hz at 15 °C (Fig. 9). At 15 °C, the peak force generated per cycle and the average power output were significantly higher in summer- than winter-acclimated fish (Fig. 9).

For unsteady swimming muscle, length changes deviate significantly from sine waves and vary from tailbeat to tailbeat (Van Leeuwen 1995; Johnston et al., 1995). Johnston et al. (1995) calculated the changes in strain of the fast muscle in short-horned sculpin during the first tailbeat of fast-starts associated with prey capture. Isolated fibres were subjected to the strains calculated for the first tailbeat of the fast-start (abstracted cycle) and stimulated at a range of phases with the in vivo duty cycle determined from EMG recordings. For 5 °C-acclimated fish, the average power per cycle (W kg^{-1} wet muscle mass) was 21.7 at 5 °C, falling to 6.9 at 15 °C. Following acclimation to 15 °C, average power output per cycle increased to 22.8 at 15 °C, indicating near perfect thermal compensation of muscle performance with acclimation (Fig. 10).

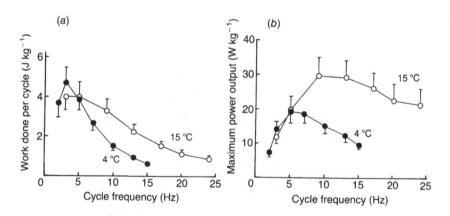

Fig. 9. Influence of thermal acclimation on (a) work and (b) power output during cyclical contractions of fast muscle fibres in the short-horned sculpin. Fish were acclimated to either 4 °C (●) or 15 °C (○) for a minimum of 2 months. (See Johnson & Johnston, 1991b for original data and further details.)

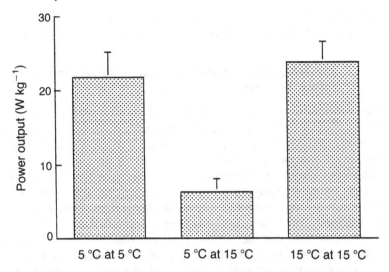

Fig. 10. The effects of temperature and thermal acclimation on the average power output of fast muscle fibres from short-horned sculpin measured under conditions simulating the first tailbeat of a fast-start. Values represent mean $\pm$ SE, $n = 5$ fish at each temperature. (From Johnston *et al.*, 1995.)

Recently, we have investigated the molecular mechanisms underlying changes in V_{max} with temperature acclimation in the short-horned sculpin (Ball & Johnston, 1996). Peptide mapping of myosin heavy chains with four different enzymes revealed no differences between 5 °C-acclimated and 15 °C-acclimated fish. No myofibrillar protein isoforms unique to either acclimation temperature were detected using 2-dimensional polyacrylamide gel electrophoresis. However, using capillary electrophoresis the ratio of myosin light chains $LC3_f$ to $LC1_f$ was significantly higher (1:1.66 $\pm$ 0.60) in muscle from 5 °C-acclimated than from 15 °C-acclimated (1:0.73 $\pm$ 0.20) fish (mean $\pm$ SD, 6 fish per acclimation temperature). The results are consistent with the change in V_{max} with temperature acclimation being due to the altered expression of myosin light chains independently of myosin heavy chain composition. In support of this hypothesis myofibrillar ATPase activity of fast muscle was found to be independent of acclimation temperature (Ball & Johnston, 1996). It is known that myosin heavy chain composition is a major determinant of ATPase activity (Bottinelli *et al.* 1994b), whereas removal of the light chains from myosin decreases contraction velocity without affecting ATPase activity (Lowey, Waller & Trybus, 1993). Thus the molecular mechanisms underlying the plasticity of muscle contractile properties with temperature acclimation are different in the two species (common carp and short-horned sculpin) that have been examined so far.

Implications for global warming and suggestions for future research

The temperatures of the world's oceans and freshwater bodies have changed constantly since their formation. For example, prior to the breakup of the super-continent Gondwanaland around 35 million years ago, the Southern Ocean had a relatively temperate climate (12–15 °C). The early Cenozoic fish fauna contained representatives of cosmopolitan groups found in temperate seas today, e.g. gadoids, salmonids, etc. Following the establishment of present-day oceanic circulation patterns 20 million years ago there has been a gradual cooling to around −1.86 °C at high latitudes. Although the cooling of the Southern Ocean took place over many millions of years there is recent evidence for periods of particularly sharp cooling and intermittent periods of warming (see Fig. 1 in Clarke & Johnston, 1996). The present-day fauna is highly endemic and lacks such groups as gadoids, clupeioids and salmonids, consistent with the extinction of some fishes. It is dominated by a single suborder of bottom-living perciformes, the Notothenioides, which lack swimbladders. Notothenioid fish have undergone a striking adaptive radiation to fill a range of ecological niches since the early Tertiary. This radiation was driven by climate change, the intermittent availability of new habitats during interglacial periods, and perhaps the lack of competition from other groups due to extinctions (see Clarke & Johnston, 1996, for a review). Thus climate change has historically been a major driver of the evolution of fishes. However, even the most rapid periods of cooling/warming in the past are 10 or 100 times slower than predicted by global warming models for the next 100 years. The rate of evolutionary change that is possible is clearly a function of generation time. The time to sexual maturity in fish ranges from a few weeks in the killifish found in seasonal ponds to 5–8 years in many coldwater marine fish. For a rate of temperature change of 2 °C per 100 years the scope for adaptations in the genotype may be limited for the majority of species. Thus, in considering the worst case scenarios for global warming it is the phenotypic responses to temperature change that species are able to make which will be most critical for survival.

    With some important exceptions, such as species living in the deep sea or in polar regions, fish often encounter daily changes in temperature which are far in excess of those expected with global warming. For example, many mesopelagic fishes undergo extensive diurnal vertical migrations from cold deep water during the day to the warmer surface layers at night. It is also common for fish living at the margins of the ocean to undergo extensive heating or cooling with each tidal cycle. An extreme example is provided by the common killifish (*Fundulus heteroclitus*) which inhabits saltmarshes along the Atlantic coast of North America. On warm sunny days in early summer the temperature of isolated pools can rise from 12 °C at high tide to around 30 °C at low tide, without adversely affecting the swimming abilities of the killifish (Sidell *et al.*, 1983). Freshwater fish living in relatively small bodies of water also frequently experience marked diurnal and seasonal changes in temperature, often in excess of 20 °C. Although some species become torpid at low temperature others can remain active over a wide range of temperatures, indicating considerable phenotypic plasticity (Lemons & Crawshaw, 1985). For eurythermal species living in the middle of their thermal tolerance range the relatively small temperature changes expected with global warming are therefore unlikely to have a major impact on swimming abilities.

Many traits contribute to the thermal tolerance of a species. Heat stress in swimming muscles is unlikely to be a common immediate cause of death, since there are other tissues such as the gill epithelium and respiratory neurones which are probably more sensitive and likely to fail (Cossins & Bowler, 1987). However, any impairment of swimming performance may influence fitness through a decreased ability to forage for food or an increased likelihood of predation. Thus thermal effects on muscle function may in the long-term influence the success of a particular species. We know that species differ markedly both in the extent of their phenotypic responses to temperature change and in the cellular and molecular mechanisms underlying the plasticity of particular traits. In general, species from stenothermal environments show less plasticity in muscular performance with temperature change than do eurythermal fish. Species living towards the upper limit of their zone of thermal tolerance are therefore likely to be more vulnerable to a given rise in temperature than species living in the middle of their thermal range. Thus a 2 °C rise in temperature is likely to have a proportionally greater effect on swimming performance in *P. borchgrevinski* living at −1.86 to 0 °C, than in a temperate species such as the largemouth bass which routinely experiences 4 to 30 °C (Fig. 1).

More research is required specifically directed at the impact of thermal stress on muscle function, particularly in stenothermal species. The starting point for future work should be the responses to temperature at the organismal level, in terms of both sustained and maximum swimming performance. Studies on isolated muscle fibres are valuable, but where possible measurements of contractile performance should be carried out under the constraints operating during swimming. The results from organismal and physiological experiments should be used as the appropriate starting point to select the cellular and molecular techniques needed to elucidate underlying mechanisms.

Finally, the great majority of studies in this field have concentrated on adult stages because of their ease of study. However, we should remember that natural selection operates on all stages of the life/cycle and that temperature tolerance generally rises during ontogeny (see Johnston, Vieira & Hill, 1996; Rombough, this volume). The key to understanding the responses of fish to elevated temperatures associated with global warming may well therefore lie with studies of the early life stages.

Acknowledgements

The authors are grateful to the Natural Environment Research Council of the UK for financial support.

References

Akster, H. A., Granzier, H. L. M. & ter Keurs, H. E. D. J. (1985). A comparison of quantitative ultrastructural and contractile characteristics of muscle fibre types in the perch. *Perca fluviatilis* L. *Journal of Comparative Physiology B*, **155**, 685–91.

Alexander, R. McN. (1969). The orientation of muscle fibres in the myomeres of fishes. *Journal of the Marine Biological Association of the United Kingdom*, **49**, 263–90.

Altringham, J. D. & Johnston, I. A. (1988a). Activation of multiply innervated fast and slow myotomal muscle fibres of the teleost *Myoxocephalus scorpius. Journal of Experimental Biology*, **140**, 313–324.

Altringham, J. D. & Johnston, I. A. (1988b). The mechanical properties of polyneuronally innervated, myotomal muscle fibres isolated from a teleost fish *Myoxocephalus scorpius. Pflügers Archives*, **412**, 524–9.

Altringham, J. D. & Johnston, I. A. (1990). Modelling muscle power output in a swimming fish. *Journal of Experimental Biology*, **148**, 395–402.

Altringham, J. D., Wardle, C. S. & Smith, C. I. (1993). Myotomal muscle functions at different locations in the body of a swimming fish. *Journal of Experimental Biology*, **182**, 191–206.

Anderson, M. E. & Johnston, I. A. (1992). Scaling of power output in fast muscle fibres of the Atlantic cod during cyclical contractions. *Journal of Experimental Biology*, **170**, 143–54.

Ball, D. & Johnston, I. A. (1996). Molecular mechanisms underlying the plasticity of muscle contractile properties in fish folllowing temperature accclimation. *Journal of Experimental Biology* **199**, 1363–73.

Batty, R. S. & Blaxter, J. H. S. (1992). The effect of temperature on the burst swimming performance of fish larvae. *Journal of Experimental Biology*, **170**, 187–201.

Beamish, F. W. H. (1970). Oxygen consumption of largemouth bass in relation to swimming speed and temperature. *Canadian Journal of Zoology*, **48**, 1221–8.

Beamish, F. W. H. (1978). Swimming capacity. In *Fish Physiology*, ed. W. S. Hoar & D. J. Randall, pp. 101–87. New York: Academic Press.

Beddow, T. A. & Johnston, I. A. (1995). Plasticity of muscle contractile properties following temperature acclimation in the marine fish *Myoxocephalus scorpius*. *Journal of Experimental Biology*, **198**, 193–201.

Beddow, T. A., Van Leeuwen, J. L. & Johnston, I. A. (1995). Swimming kinematics of fast-starts are altered by temperature acclimation in the marine fish *Myoxocephalus scorpius*. *Journal of Experimental Biology*, **198**, 203–8.

Bone, Q. (1966). On the function of myotomal muscle fibres in elasmobranch fish. *Journal of the Marine Biological Association of the United Kingdom*, **46**, 321–49.

Bottinelli, R., Betto, R., Schiaffino, S. & Reffiani, C. (1994a). Maximum shortening velocity and coexistance of myosin heavy chain isoforms in single skinned fast fibres of rat skeletal muscle. *Journal of Muscle Research and Cell Motility*, **15**, 413–19.

Bottinelli, R., Betto, R., Schiaffino, S. & Reffiani, C. (1994b). Unloaded shortening velocity and myosin heavy chain and alkali light chain isoform composition in rat skeletal muscle fibres. *Journal of Physiology (London)*, **478**, 341–9.

Clarke, A. & Johnston, I. A. (1996). Evolution and adaptive radiation of Antarctic fish. *Trends in Ecology and Evolution*, **11**, 212–18.

Cossins, A. R. & Bowler, K. (1987). *Temperature Biology of Animals*. London: Chapman & Hall.

Connell, J. J. (1960). The relative stabilities of the skeletal muscle myosins of some animals. *Biochemical Journal*, **80**, 503–10.

Crockford, T. & Johnston, I. A. (1990). Temperature acclimation and the expression of contractile protein isoforms in the common carp (*Cyprinus carpio* L.). *Journal of Comparative Physiology B*, **160**, 23–30.

Crockford, T. & Johnston, I. A. (1993). Developmental changes in the composition of myofibrillar proteins in the swimming muscles of Atlantic herring, *Clupea harengus*. *Marine Biology*, **115**, 15–22.

Curtin, N. A. & Woledge, R. C. (1991). Efficiency of energy conversion during shortening of muscle fibres of the dogfish, *Scyliorhinus canicula*. *Journal of Experimental Biology*, **158**, 343–53.

Davies, M. L. F., Johnston, I. A. & Van der Wal, J.-W. (1995). Muscle fibres in rostral and caudal myotomes of the Atlantic cod (*Gadus morhua* L.) have different mechanical properties. *Physiological Zoology*, **68**, 673–97.

Fleming, J. R., Crockford, T., Altringham, J. D. & Johnston, I. A. (1990). Effects of temperature acclimation on muscle relaxation in the carp: a mechanical, biochemical and ultrastructural study. *Journal of Experimental Zoology*, **255**, 286–95.

Fry, F. E. J. & Hart, J. S. (1948). Cruising speed of goldfish in relation to water temperature. *Journal of the Fisheries Research Board of Canada*, **7**, 175–99.

Gerday, C., Joris, B., Gerardin-Otthiers, N., Collin, S. & Hamoir, G. (1979). Parvalbumins from the lungfish, *Protopterus dolloi*. *Biochimie*, **61**, 589–99.

Gerlach, G.-F., Turay, L., Malik, K. T. A., Lida, J., Scutt, A. & Goldspink, G. (1990). Mechanisms of temperature acclimation in the carp: a molecular biology approach. *American Journal of Physiology*, **259**, R237–44.

Greaser, M. L., Moss, R. L. & Reiser, P. J. (1988). Variations in contractile properties of rabbit single muscle fibres in relation to troponin T isoforms and myosin light chains. *Journal of Physiology (London)*, **406**, 85–98.

Heap, S. P., Watt, P. W. & Goldspink, G. (1985). Consequences of thermal change on the myofibrillar ATPase of five freshwater teleosts. *Journal of Fish Biology*, **26**, 733–8.

Heap, S. P., Watt, P. W. & Goldspink, G. (1986). Myofibrillar ATPase activity in the carp (*Cyprinus carpio*): interactions between starvation and environmental temperature. *Journal of Experimental Biology*, **123**, 373–82.

Hess, F. and Videler, J. J. (1984). Fast continuous swimming of saithe (*Pollechius virens*): a dynamic analysis of bending moments and muscle power. *Journal of Experimental Biology*, **109**, 229–51.

Hwang, G. C., Watabe, S. & Hashimoto, K. (1990). Changes in carp myosin ATPase induced by temperature acclimation. *Journal of Comparative Physiology B*, **160**, 233–9.

Hwang, G.-C., Ochiai, Y., Watabe A. S. & Hashimoto, K. (1991). Changes of carp myosin subfragment-1 induced by temperature acclimation. *Journal of Comparative Physiology B*, **161**, 141–6.

Johnson, T. P. & Johnston, I. A. (1991a). Temperature adaptation and the contractile properties of live muscle fibres from teleost fish. *Journal of Comparative Physiology B*, **161**, 27–36.

Johnson, T. P. & Johnston, I. A. (1991b). Power output of fish muscle fibres performing oscillatory work: effects of acute and seasonal temperature change. *Journal of Experimental Biology*, **157**, 409–23.

Johnston, I. A. (1981). Structure and function of fish muscles. *Symposium of the Zoological Society of London*, **48**, 71–113.

Johnston, I. A. (1985). Effects of temperature on force–velocity relationship of skinned fibres isolated from Icefish (*Chaenocephalus aceratus*) skeletal muscle. *Journal of Physiology (London)*, **361**, 40P.

Johnston, I. A. (1990). Cold-adaptation in marine organisms. *Philosophical Transactions of the Royal Society of London, Series B*, **326**, 655–67.

Johnston, I. A. (1991). Muscle action during locomotion: a comparative perspective. *Journal of Experimental Biology*, **160**, 167–185.

Johnston, I. A. (1993). Phenotypic plasticity of fish muscle to temperature change. In *Fish Ecophysiology*, ed. J. C. Rankin & F. B. Jensen, pp. 321–40. London: Chapman & Hall.

Johnston, I. A. & Altringham, J. D. (1985). Evolutionary adaptation of muscle power output to environmental temperature: force–velocity characteristics of skinned muscle fibres isolated from Antarctic, temperate and tropical marine fish. *Pflügers Archives*, **405**, 136–40.

Johnston, I. A. & Altringham, J. D. (1991). Movement in water: constraints and adaptations. In *Biochemistry and Molecular Biology of Fishes*, Vol. 1, ed. P. W. Hochachka & T. Mommsen, pp. 249–68. Amersterdam: Elsevier.

Johnston, I. A. & Brill, I. A. (1984). Thermal dependence of contractile properties of single skinned muscle fibres isolated from Antarctic and various Pacific marine fishes. *Journal of Comparative Physiology B*, **155**, 63–70.

Johnston, I. A., Davison, W. & Goldspink, G. (1975b). Adaptations in Mg^{2+}-activated myofibrillar ATPase activity induced by temperature acclimation. *Federation of European Biochemical Societies Letters*, **50**, 293–5.

Johnston, I. A., Davison, W. & Goldspink, G. (1977). Energy metabolism of carp swimming muscles. *Journal of Comparative Physiology B*, **144**, 203–16.

Johnston, I. A., Fleming, J. D. & Crockford, T. C. (1990). Thermal acclimation and muscle contractile properties in cyprinid fish. *American Journal of Physiology*, **259**, R231–6.

Johnston, I. A., Franklin, C. E., & Johnson, T. P. (1993). Recruitment patterns and contractile properties of fast muscle fibres isolated from rostral and caudal myotomes of the short-horned sculpin. *Journal of Experimental Biology*, **185**, 251–65.

Johnston, I. A. & Lucking, M. (1978). Temperature induced variation in the distribution of different muscle fibre types in the goldfish (*Carassius auratus*). *Journal of Comparative Physiology B*, **124**, 111–16.

Johnston, I. A. & Maitland, B. (1980). Temperature acclimation in crucian carp: a morphometric study of muscle fibre ultrastructure. *Journal of Fish Biology*, **17**, 113–25.

Johnston, I. A., Sidell, B. D. & Driedzic, W. R. (1985). Force-velocity characteristics and metabolism of carp muscle fibres following temperature acclimation. *Journal of Experimental Biology*, **119**, 239–49.

Johnston, I. A., Van Leeuwen, J. & Beddow, T. (1995). How fish power predation fast-starts. *Journal of Experimental Biology*, **198**, 1851–61.

Johnston, I. A., Vieira, V. L. A. & Hill, J. (1996). Temperature and ontogeny in ectotherms: muscle phenotype in fish. In *Phenotypic and Evolutionary Adaptations of Animals to Temperature*, ed. I. A. Johnston & A. F. Bennett. *Society for Experimental Biology Seminar Series*. pp. 153–81. Cambridge: Cambridge University Press.

Johnston, I. A. & Walesby, N. J. (1977). Molecular mechanisms of temperature adaptation in fish myofibrillar adenosine triphosphatases. *Journal of Comparative Physiology B*, **119**, 195–206.

Johnston, I. A., Walesby, N. J., Davison, W. & Goldspink, G. (1975a). Temperature adaptation in myosin of Antarctic fish. *Nature*, **254**, 74–75.

Langfeld, K. S., Altringham, J. D. & Johnston, I. A. (1989). Temperature and the force–velocity relationship of live muscle fibres from the teleost *Myoxocephalus scorpius*. *Journal of Experimental Biology*, **144**, 437–48.

Langfeld, K. S., Crockford, T. & Johnston, I. A. (1991). Temperature acclimation in the common carp: force–velocity characteristics and myosin sub-unit composition of slow muscle fibres. *Journal of Experimental Biology*, **155**, 291–304.

Lannergren, J. (1987). Contractile properties and myosin isoenzymes in various kinds of *Xenopus* twitch muscle fibres. *Journal of Muscle Research and Cell Motility*, **8**, 260–73.

Lemons, D. E. & Crawshaw, L. I. (1985). Behavioural and metabolic adjustments to low temperature in the largemouth bass (*Micropterus salmoides*). *Physiological Zoology*, **58**, 175–80.

Lowey, S., Waller, G. S. & Trybus, K. M. (1993). Skeletal muscle myosin light chains are essential for physiological speeds of shortening. *Nature*, **365**, 454–6.

McArdle, H. J. & Johnston, I. A. (1980). Evolutionary temperature adaptation in fish muscle sarcoplasmic reticulum. *Journal of Comparative Physiology B*, **135**, 157–64.

Maresca, B., Patriarca, E., Goldenberg, C. & Sacco, M. (1988). Heat shock and cold-adaptation in Antarctic fish: a molecular approach. *Comparative Biochemistry and Physiology*, **90B**, 623–9.

Penney, R. K. & Goldspink, G. (1979). Compensation limits of fish muscle myofibrillar ATPase to environmental temperature. *Journal of Thermal Biology*, **4**, 269–72.

Rome, L. C., Choi, I.-H., Lutz, G. & Sosnicki, A. (1992). The influence of temperature on muscle function in the fast swimming scup. I. Shortening velocity and muscle recruitment during swimming. *Journal of Experimental Biology*, **163**, 259–79.

Rome, L. C., Funke, R. P., Alexander, R. McN., Lutz, G., Aldridge, H., Scott, F. & Freadman, M. (1988). Why animals have different muscle fibre types. *Nature*, **335**, 824–7.

Rome, L. C., Loughna, P. T. & Goldspink. G. (1984). Muscle fibre recruitment as a function of swim speed and muscle temperature in carp. *American Journal of Physiology*, **247**, R272–9.

Rome, L. C. & Sosnicki, A. J. (1990). The influence of temperature on mechanics of red muscle in carp. *Journal of Physiology*, **427**, 151–69.

Rome, L. C. & Sosnicki, A. J. (1991). Myofilament overlap in swimming carp. II. Sarcomere length changes during swimming. *American Journal of Physiology*, **260**, C289–96.

Rome, L. C. & Swank, D. (1992). The influence of temperature on power output of scup red muscle during cyclical length changes. *Journal of Experimental Biology*, **171**, 261–81.

Rowlerson, A., Scapolo, P. A., Mascarello, F., Carpene, E. & Veggettii, A. (1985). Comparative study of the myosins present in the lateral muscle of some fish: species variations in myosin isoforms and their distributions in red, pink and white muscle. *Journal of Muscle Research and Cell Motility*, **6**, 601–40.

Somero, G. N. & De Vries, A. L. (1967). Temperature tolerance of some Antarctic fishes. *Science*, **15**, 257–8.

Sidell, B. D. (1980). Response of goldfish (*Carassius auratus*) to temperature acclimation: adaptations in biochemistry and proportions of different fibre types. *Physiological Zoology*, **53**, 948–57.

Sidell, B. D., Johnston, I. A., Moerland, T. S. & Goldspink, G. (1983). The eurythermal myofibrillar complex of the mummichog (*Fundulus heteroclitus*): adaptations to a fluctuating thermal environment. *Journal of Comparative Physiology B*, **153**, 167–73.

Ushio, H. & Watabe, S. (1993). Effects of temperature acclimation on Ca^{2+}-ATPase of the carp sarcoplasmic reticulum. *Journal of Experimental Biology*, **265**, 9–17.

Van Leeuwen, J. L. (1995). The action of muscles in swimming fish. *Experimental Physiology*, **80**, 177–91.

Van Leeuwen, J. L., Lankheet, M. J. M. Akster, H. A. & Osse, J. W. M. (1990). Function of red axial muscles of carp (*Cyprinus carpio*): recruitment and normalized power output during swimming in different modes. *Journal of Zoology (London)*, **220**, 123–40.

Webb, P. W. (1975). Hydrodynamics and energetics of fish propulsion. *Bulletin of the Fisheries Research Board of Canada*, **190**, pp. 159.

Wohlschlag, D. E. (1964) Respiratory metabolism and ecological characteristics of some fishes in McMurdo Sound, Antarctica. *Antarctic Research Series*, **1**, 33–62.

E.W. TAYLOR, S. EGGINTON,
S.E. TAYLOR and P.J. BUTLER

Factors which may limit swimming performance at different temperatures

Introduction

As thermal diffusion is an order of magnitude more rapid than molecular diffusion, it is clear that the same design features that make the gills of fish well suited for respiratory gas exchange from water (large surface area, active convection of water and blood at appropriate ventilation/perfusion ratios across a functional counter-current) also provide for very effective branchial heat exchange. This is reinforced by the relatively high heat capacity of water which is more than 3000 times that of air, so that for most fishes, and indeed all other water-breathing ectotherms, body temperature equilibrates rapidly to any change in environmental temperature. Consequently, in the absence of specific anatomical specialization to maintain thermal gradients, temperature throughout the body of fishes is in equilibrium with the environment to within a fraction of a degree. Thus, large changes in body temperature may be experienced: diurnally, by coastal fish subjected to tidal variations; or by vertically migrating pelagic species, particularly if they cross a thermocline; or seasonally by eurythermal temperate zone fish. Over evolutionary time, speciation of tropical and polar fishes has resulted in species with widely different thermal ranges within the accepted biological temperature range (between the freezing point of water and the temperature for protein denaturation), which do not overlap.

As temperature, which is a measure of kinetic energy, directly affects the kinetics of both physical (for example, diffusion) and chemical (that is, biochemical) processes, changes in body temperature present major metabolic challenges, which require immediate physiological or biochemical compensation if the animals are to maintain relatively constant biological activity. Despite the profound effects of temperature on biochemical reaction rates, which typically have an acute Q_{10} of 2–3, different species have successfully colonized thermal environments

Society for Experimental Biology Seminar Series 61: *Global Warming: Implications for freshwater and marine fish*, ed. C. M. Wood & D. G. McDonald. © Cambridge University Press 1996, pp. 105–133.

ranging from $<-1.5\,°C$ at the Poles to $>40\,°C$ in East African hot springs. More importantly, there appears to be conservation of loco-motory activity (homeokinesis) based on the simple observation that fish swim at similar average speeds across a wide range of global environmental temperatures. The mechanisms whereby homeokinesis is achieved remain little understood and are the subject of this review.

Typical sub-carangiform swimming, as exemplified by carp, generates mechanical power by alternate contractions of the trunk musculature, and swimming speed is directly proportional to tailbeat frequency and seemingly unaffected by tailbeat amplitude or orientation (Rome, Funke & Alexander, 1990). Although temperature will affect the con-traction velocity of muscle fibres, and thus have a direct effect on swimming speed (see the chapter by Johnston & Ball in this volume), fish held at different temperatures show the same relationship between tailbeat frequency and swimming speed (Rome *et al.*, 1990), suggesting that the underlying kinematics are little affected. Neither acclimation to different temperatures nor sudden changes in temperature over a range of $\pm 7\,°C$ from their acclimation temperature had a significant effect on the critical velocity of swimming of 17 species of fish from the McKenzie River (Jones, Kiceniuk & Bamford, 1974). Changes in swimming capacity, therefore, are likely to reflect intrinsic differences in energy supply or demand, rather than in efficiency of locomotion.

What emerges from laboratory studies of the effects of temperature upon the speed of sustained swimming in fish, emanating largely from Brett and co-workers, is that swimming speed is reduced at low tem-peratures, increases to a maximum speed at an optimum temperature (which varies between species and with acclimation temperature) and is reduced at temperatures close to the upper thermal limit for the species. The subsequent bell-shaped curves relating swimming speed to ambient temperature are described by Johnston and Ball in this volume.

It is unclear which characteristics are important in determining the functional plasticity of sustained swimming because, despite decades of extensive research into the factors underlying aerobic muscle perform-ance, the nature of the functional limits to swimming are as yet unclear. Simplistically, we may consider these factors to be evident at several levels, including: the animal–environment interface (varying in nature from branchial O_2 uptake to hydrodynamic constraints); the blood–tissue interface (that is, oxygen transport in the blood and delivery by capillaries); the tissue level (metabolic compensation of muscle fibres); or at the level of control systems (neuronal activity or excitation/contraction coupling). Given that very few integrative studies have been performed, we are unable to make adequate quantitative predictions regarding the relative importance of these various factors in limiting a fish's capacity for sustained activity. This is likely in any event to vary between species, or individuals within a species at different growth stages or final size, or following the physiological impact of direct environmental changes such as temperature.

With regard to the possible effects of any global warming, it is important to bear in mind the scale and rate of change likely to occur. Organisms typically are able to compensate for environmental change when faced with perturbations of low amplitude and periodicity, due to the interactive nature of physiological processes. Perhaps the earliest example to illustrate this point was the experiment of Dallinger (1887), who by very gradual increases in temperature over 7 years was able to produce flagellate protozoa capable of survival and reproduction at 70 °C, well above their normal upper lethal temperature. More modest thermal challenges are likely to be tolerated without difficulty so that deleterious effects become noticeable only if any change is large or rapid, or when physiological limits are approached (i.e. scope for activity is reduced).

Intuitively, a small acute temperature increment would not appear to be a problem unless the prevailing ambient temperature was close to the upper thermal limit for a species. Indeed, increased temperature could be expected to improve swimming performance as it is associated with reduced viscosity of the water and the working muscles (providing an environmental 'warm-up' which could improve mechanical power output of swimming muscles), as well as increased rates of diffusion of gases and metabolites to and from the working muscles and increased rates of metabolism. As swimming speed varies directly with tailbeat frequency in carp (Rome *et al.*, 1990), the increase in maximum velocity of contraction of slow, oxidative (red) muscle with acclimation temperature would also be expected to increase the sustainable (aerobic) speed of swimming, at least in the short term. Further increases in speed are dependent upon recruitment of fast, glycolytic (white) muscle fibres, which fatigue rapidly, so that there is a 'compression of the recruitment order' at low temperatures (Rome, Loughna & Goldspink, 1985), with the inadequate locomotory power of oxidative muscle being supplemented by activation of glycolytic fibres. Examination of the data from carp implies that increasing temperature would cause an expansion of recruitment order, whereby an increasing proportion of locomotory power is derived from oxidative, fatigue resistant muscle and the maximum speed of sustainable swimming is therefore increased. We shall see that other factors tend to negate this simplistic assumption. For example, compression of the recruitment order is not evident in adult brown trout (*Salmo trutta*) acclimated to summer (15 °C) and winter (5 °C) seasonal temperatures (Day & Butler, 1996). In this case white muscle recruitment occurs at approximately one body length per second and continues up to a U_{crit} of just over 2 body lengths s^{-1} at either temperature. Clearly, the responses to progressive seasonal changes in temperature may differ from the responses to changes in laboratory temperature and the factors underlying these differences require further study.

Eurythermal fish show a range of adaptational responses to chronic temperature changes which can improve their performance in the cold (Sidell & Moerland, 1989). These include an increase in red muscle as a proportion of total muscle volume, which may serve to increase the power of contraction and consequent stride length. Increases in capillary and mitochondrial density may shorten effective diffusion distance for the exchange of gases and metabolites in swimming muscles. Diffusion rates may be further increased by increased myoglobin concentrations in red muscle (responsible for facilitated diffusion, rates of which rise by 40% over 10 °C), and by the accumulation of lipid droplets in the tissues which can act both as a store of oxygen and (by increasing oxygen solubility) as a means of increasing diffusivity (Egginton & Sidell, 1989). In addition, the oxidative potential of red muscle may increase in the cold due to the increased mitochondrial density, and hence increased activities of key oxidative enzymes (Sidell, 1980; Johnston & Dunn, 1987). There may also be adequate functional plasticity in the contractile machinery of muscle itself to accommodate quite large temperature changes (see Johnston & Ball, this volume). The combined effects of these responses may compensate completely for the affects of progressive temperature change.

If it can be shown that these integrated changes improve the swimming performance of fish in the cold, what are the likely effects of the reciprocal changes on warming? Could they lead to a reduction in performance which counters the assumed direct effects of increased temperature? It is this rhetorical question that we attempt to answer in this review.

Rates of oxygen uptake

Rates of oxygen uptake are a reliable measure of aerobic metabolism. However, they can vary widely with size, maturation state, activity, and environmental variables such as temperature. When fish are transferred in the laboratory between water of different temperatures it is usually a matter of days, and can be weeks, before they attain metabolic and performance rates characteristic of the new temperature, a process referred to as 'acclimation' to differentiate it from the acclimatization of a species to changes in its natural environment (e.g. seasonal temperature variation). This is important to bear in mind when searching the literature for data on the thermal sensitivity of oxygen uptake, as values from 'acute' (short duration) transfer experiments will be higher than those from acclimated fish, which may in turn be different from those in acclimatized animals. While it is often thought that there is an exponential increase in metabolic rate with increasing temperature, this is an oversimplification, based on acute experiments, and there is often a region in the middle part of the temperature range for a species when metabolic rate becomes relatively independent of temperature (Cossins & Bowler, 1987).

Deviation from this specific thermal optimum or preferendum may result in increased thermal sensitivity, giving rise to Q_{10} values between 1.5 and 3. At the cold end of the spectrum this may reflect entry into torpor, which in itself may not pose a serious problem for the animal concerned and may even protect it from an associated seasonal food shortage. As temperature rises towards the upper limit for the species, however, life becomes increasingly unsustainable. Coincident with the decrease in available dissolved oxygen is an increased demand for oxygen, consequent upon an acute increase in aerobic metabolism and the metabolic cost of maintenance. The cost of ventilation (buccal and branchial pumps) is itself likely to rise because the increased demand for oxygen, exacerbated by the reduced availability of dissolved oxygen, requires that greater volumes of water be presented to the branchial exchange surfaces. Consequently, the respiratory cost of ventilation represents an increasing proportion of total oxygen uptake, and scope for activity inexorably declines (Jones, 1971).

Some data do indeed imply that the metabolic scope for locomotion can vary inversely with temperature. In the brown trout, an increase in temperature from 5 to 15 °C was accompanied by up to a doubling in the routine rate of oxygen uptake with little or no change in sustained swimming speed (Butler, Day & Namba, 1992; Beaumont, Butler & Taylor, 1995a; Fig 1). However, as the active metabolic rate of fish may be as high as nine times their resting rate during intense exercise (Webb, 1971), the increased cost of maintenance metabolism at elevated temperatures is likely to have only a relatively minor influence on their scope for sustained exercise. What can be critically affected by temperature is the anaerobic oxygen debt accumulated during intense exercise. When tilapia were chased to exhaustion at 12, 24 and 34 °C

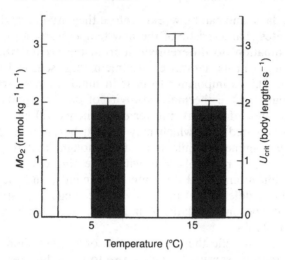

Fig. 1. Changes in routine rate of oxygen uptake (M_{O_2}, open bars) and maximal sustained swimming speed (U_{crit}, filled bars) of brown trout acclimated to 5 and 15 °C, showing thermal sensitivity of the former and the relative insensitivity of the latter (Beaumont *et al.*, 1995a).

the resultant increase in oxygen uptake during recovery increased progressively with temperature (Fig. 2), indicating that a greater oxygen debt was accumulated at higher temperatures. Brett (1964) reported that the extent of oxygen debt accumulated at the time of fatigue doubled between 5 and 15 °C in sockeye salmon, *Oncorhynchus nerka*. This may limit the duration of anaerobic burst swimming at increased temperatures.

Branchial oxygen uptake

The physical structure and physiological function of gills subserve functions other than respiratory gas exchange. These include ionoregulation and the associated functions of excretion, acid–base regulation and water balance. Consequently, as oxygen demand rises in line with increasing temperature, in the face of a falling supply due to decreased solubility, any adjustments likely to increase trans-epithelial oxygen flux could compromise both ionoregulation and water balance, by increasing the functional permeability of the gills.

The maximum oxygen exchange capacity of gills can be expressed as their morphometric diffusing capacity (D_{O_2}), estimated from their structural properties and the permeation coefficent for tissue, on the

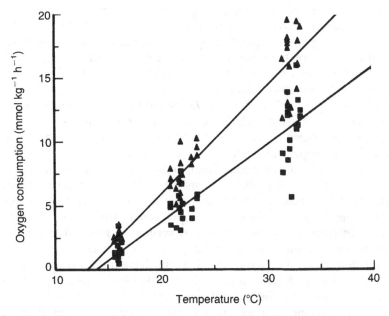

Fig. 2. Oxygen consumption (M_{O_2}) of tilapia, during recovery from being chased to exhaustion (▲), compared to routine rates prior to exercise (■), at 12, 24 and 34 °C.
Mean resting $M_{O_2} = 0.606T(°C) - 8.404$ $r^2 = 0.870$
Mean M_{O_2} in 6 h recovery $= 0.846T(°C) - 11.053$ $r^2 = 0.932$
(McKenzie *et al.*, 1996.)

assumption that oxygen transfer (conductance) is entirely by diffusion, using the formula:

$D_{O_2} = A.\tau_{ht}.K_{O_2}$ (Perry, 1990, based on Hughes & Shelton, 1962) Where A is the respiratory surface area (i.e. total surface area of secondary lamellae), τ_{ht} is the water–blood diffusion distance (usually expressed as the harmonic mean of epithelial thickness), and K_{O_2} is Krogh's permeation coefficient of O_2. The rate of oxygen flux will be directly proportional to the driving force (P_{O_2} gradient) and A, and inversely proportional to τ_{ht}.

If the rate of oxygen flux (i.e. uptake) is to be regulated, then some or all of the factors in this equation must be capable of adjustment. As yet there has been no systematic study of adjustments in anatomical (that is, total measurable) A, and such gross anatomical adaptations may indeed only be feasible in response to sustained selection pressure over many generations. While there are few data available, interspecific differences in whole animal V_{O_2} seem to be reflected in measured differences in A and τ_{ht} values (Perry, 1990). Indeed, the 10-fold range in V_{O_2} among fishes is paralled by a similar range of diffusion distances, and an even greater range for surface area estimates. For example, τ_{ht} in tench (*Tinca tinca*) is 2.47 μm (Hughes, 1972) while in the tropical tilapia (*Oreochromis niloticus*) it is only 1.25 μm (Kisia & Hughes, 1992), giving a 2-fold increase in D_{O_2} should A and K_{O_2} remain constant. As gaseous diffusivity will increase with temperature, so too will K_{O_2} and hence oxygen transfer will be further increased to at least partially offset the temperature-induced increase in V_{O_2}.

Acclimation to increased temperature has also been shown to decrease τ_{ht} in juvenile largemouth bass, *Micropterus salmoides* (Leino & McCormick, 1993), and adult European eel, *Anguilla anguilla* (H. Tuurala and S. Egginton, unpublished data) (Fig. 3). Whether such differences may be developed further at higher temperatures is unknown, but there seems to be no physical reason why not. Tuna, with deep body temperatures of 32 °C, have τ_{ht} values half those of *Oreochromis* (0.6 μm; Hughes, 1972). However, the adaptive benefit may be more obvious at cold temperatures. A thickened respiratory epithelium would decrease oxygen uptake (with a decreased metabolic demand this may not pose a problem) and present a greater barrier to ion loss in fresh water. As the rate of active accumulation of ions over epithelial cells is likely to be depressed in the cold, this may be a major benefit. Given so few data, then, it is difficult to assess whether adjustments in τ_{ht} would most benefit those fish entering overwintering torpor or undergoing thermal hypermetabolism.

Although changes in the anatomical value of A for an individual fish may be relatively immutable, the functional value is under physiological control. Trout perfuse no more than 60% of their gill lamellae at rest (Booth, 1979). The increase in blood flow and pressure associated with exercise opens up previously poorly perfused lamellae, increasing the functional area for gas exchange and improving the diffusion capacity (Farrell, Daxboeck & Randall, 1979). A similar argument may apply to diffusion distance as increased blood flow directed through the branchial circulation may engorge the lamellae stretching their epithelia and reducing water to blood diffusion distances (Farrell, 1980; Soivio & Tuurala, 1981). Similar changes have been reported during hypoxia and they can be interpreted as attempts to compensate for any imbalance between oxygen supply and demand such as that associated with the rise in temperature.

Responses to temperature change are complicated by changes in oxygen availability. A rise in environmental temperature will reduce

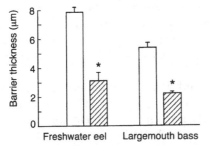

Fig. 3. Thickness of the respiratory epithelium (water–blood barrier thickness), determined from histological sections of the gills, from fish acclimated to different temperatures (open columns, cold; hatched columns, warm). Freshwater eel, 5 vs 25 °C (H. Tuurala & S. Egginton, unpublished data); largemouth bass, 4 vs 20 °C. (From Leino & McCormick, 1993.)

oxygen solubility in water and plasma and typically reduce the oxygen affinity of haemoglobin. In addition, the increased oxygen demand accompanying acute increases in temperature may result in the depletion of the venous oxygen reserve. As arterial chemoreceptors in teleosts are known to be sensitive to oxygen content of blood (Randall, 1982), and there is evidence for venous oxygen receptors in fish (Taylor, 1992), an increase in temperature could stimulate an increase in ventilation and/or perfusion which would serve to supply the greater demand for oxygen. This may in turn increase functional A or τ_{ht} (Taylor, 1985). However, increased ventilation rates may increase the proportion of water shunted past the gills, with a consequent reduction in the effectiveness of oxygen uptake (for example, percentage extraction of oxygen by trout gills falls from 80 to 20% as the rate of gill ventilation increases with exercise; see Hughes & Shelton, 1962). This may limit the effectiveness of compensation for increased temperature. In contrast, patterns of gill perfusion may be actively regulated by sphincters on the vessels controlling the flow of blood through the non-respiratory arteriovenous route in the gill filaments (Nilsson & Pettersson, 1981).

Cardiovascular responses

Our studies on rainbow trout revealed that heart rate increased with temperature by 2.5-fold between 4 and 18 °C (Fig. 4a), with the apparent Q_{10} for heart rate being 1.5 between 4 and 11 °C and 2.1 between 11 and 18 °C. In the brown trout the Q_{10} for heart rate between 5 and 15 °C was 1.7 (Butler *et al.*, 1992). These changes may reflect the intrinsic thermal sensitivity of physiological rate functions, and in rainbow trout may also have somewhat compensated for a reduction in oxygen carrying capacity at 18 °C (Taylor, Egginton & Taylor, 1993). These apparent Q_{10} values for heart rate in trout were within the ranges previously reported for other species between 5 and 20 °C (see Farrell, this volume).

Resting cardiac output increased with water temperature between 5 and 16 °C, with apparent Q_{10} values of 2.0 to 2.7 (winter flounder, Cech *et al.*, 1976; rainbow trout, Barron, Tarr & Hayton, 1987; large-scale sucker, Kolok, Spooner & Farrell, 1993). The last study suggested that 'pockets of thermal insensitivity' occur over a range of seasonal acclimatization temperatures. This was the case for stroke volume in the rainbow trout; the apparent Q_{10} was 1.5 between 4 and 11 °C, but was only 0.3 between 11 and 18 °C (Fig. 4), while Kolok *et al.* (1993) found that thermal insensitivity occurred between the two lower temperatures (5 and 10 °C) where apparent Q_{10} was 0.4.

When acclimated to 5 or 15 °C, *in situ* perfused rainbow trout hearts show an impressive degree of compensation for the thermal sensitivity

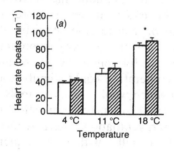

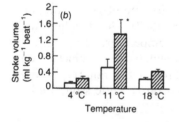
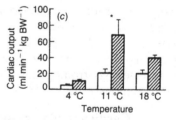

Fig. 4. Cardiovascular response of rainbow trout to sustained exercise at different acclimatization temperatures (i.e. seasonal temperature exposure with appropriate photoperiod): (*a*) heart rate, (*b*) stroke volume (*c*) cardiac output. (From Taylor *et al.*, 1993, 1996a.) Open bars are values determined under resting conditions; shaded bars are for maximal aerobic exercise. Asterisks indicate significant difference from 4 °C.

of heart rate ($Q_{10} = 2.2$) such that maximum cardiac output does not increase to the same extent ($Q_{10} = 1.3$; Graham & Farrell, 1989), which may in part reflect cardiac hypertrophy in the cold (see Farrell, this volume). Any mechanism that compensates for the effects of temperature on biological rate functions will effectively extend the range of thermal tolerance (for example, if cardiac output varied directly with heart rate, then the excessive flow and pressure at high temperatures would further compromise gill function).

Heart rate increased by over 2-fold in brown trout swimming at U_{crit} (Butler *et al.*, 1992). In contrast, swimming at the limit for sustainable (aerobic) exercise did not significantly affect heart rate (Fig. 4*a*) or mean arterial blood pressure in rainbow trout, although stroke volume increased nearly 2-fold at both 4 and 18 °C (Fig. 4*b*). At 11 °C stroke volume increased nearly 3-fold upon exercise, which was significantly greater than the exercise values at 4 and 18 °C. Cardiac output during exercise at 11 °C was therefore significantly greater than at 4 °C, a difference not observed at 18 °C due to the higher resting cardiac output (Fig. 4*c*). The lack of a chronotropic response with exercise in the rainbow trout implies that any increase in cardiac output was met by an inotropic response, and as this did not lead to an increase in blood pressure, it must have been matched by a decrease in total peripheral resistance. Thus at all temperatures examined, cardiac output increased upon maximal sustainable exercise in the rainbow trout, as found for most fish (with the exception of short-finned eels; Davie & Forster, 1980). However, the greatest increase in cardiac performance occurred at the temperature that coincides with the maximum sustainable swimming speed.

While cardiac performance in seasonally acclimatized rainbow trout peaks at 11 °C, it is clear that at 18 °C it is showing signs of deterioration. This may be because the fish were held for long periods at a temperature close to their seasonal maximum, which would normally be experienced for short periods against a varying thermal background. Hence even minor further increases in temperature above this seasonal norm may lead to migratory failure (see Farrell, this volume), and this could be a critical factor in a response to global warming, although it could be offset by compensatory changes in the microcirculation.

Peripheral oxygen transport

Maximum levels of truly sustainable swimming activity (for example, during migration) clearly rely solely on aerobic metabolism, and hence may ultimately be limited by the cardiovascular supply capacity.

Haematocrit generally varies directly with environmental temperature, thereby increasing carrying capacity to offset decreasing solubility. However, this was not the case in brown trout where it was approximately 20% at both seasonal temperatures of 5 and 15 °C (Butler *et al.*, 1992). In fact, some data show the opposite effect, with haematocrit and arterial oxygen content sharply reduced at 18 °C in rainbow trout, possibly reflecting a deterioration in condition of fish acclimatized to a temperature close to their upper limit of thermal tolerance (Taylor *et al.*, 1993).

Regional blood flow

We measured regional blood flow in rainbow trout using the radio-labelled microsphere (marker-dilution) method, in conjunction with electromyography (Wilson & Egginton, 1994; Taylor *et al.*, 1993, 1996b). In resting fish, regional blood flow to all organs except red muscle was higher at 18 °C than at 4 °C. In white muscle and skin, specific blood flow was 0.6 and 0.5 ml.min^{-1}.100 g^{-1} at 4 °C, which increased to 23 and 4 ml.min^{-1}.100 g^{-1} at 18 °C, respectively. These differences reflect changes in blood pressure, as conductance was similar in all tissues examined.

At the limit of sustainable (aerobic) exercise there was a significant increase in specific blood flow to red muscle at 11 °C (14 × resting blood flow), which was significantly greater than the increase at both 4 and 18 °C (4.5 × and 3.5 × resting blood flow, respectively). However, there were 2-fold increases in both specific blood flow and conductance to white muscle at 18 °C which were significantly greater than those at 11 °C. These values were similar to those of other studies using similar techniques (Cameron, 1974; Neumann, Holeton & Heisler, 1983; Wilson & Egginton, 1994).

Barron *et al.* (1987) reported that perfusion of white muscle in rainbow trout increased directly with seasonal acclimatization temperature. They suggested that this maintained blood pressure during temperature-dependent changes in cardiac output, by decreasing peripheral resistance. In contrast, we found no temperature-dependent increase in cardiac output between 11 and 18 °C, and so increases in flow to skin and fast muscle may represent active regulation to accommodate an increased metabolic rate at higher temperatures. However, blood flow to the gonads was also increased at 18 °C, presumably reflecting greater amounts of energy expended on reproductive effort which could limit the scope for activity. Increased blood flow to white muscle at 18 °C would also facilitate mobilization of protein from degenerating muscle fibres to the hypertrophying gonads, as suggested by Weatherly & Gill (1987).

The 14-fold increase in blood flow to red muscle, and decrease in total peripheral resistance at 11 °C during maximal sustainable exercise (see above), was sustained without any major redistribution of blood from other tissues. Increased stroke volume can be achieved by greater cardiac filling (Frank–Starling mechanism), by either increasing the muscle-pump effect via increased tailbeat frequencies, and/or redistribution of blood via sympathetic vasoconstriction to increase venous pressure (Stevens & Randall, 1967; Randall & Daxboeck, 1982). Although the magnitude of these effects would be sensitive to any change in blood volume, there is little evidence of chronic effects of temperature acclimation on blood volume in fish (Nikinmaa, Soivio & Railo, 1981). While acute changes may be marked due to sympathetic effects on splenic contraction and/or vasoconstriction, in rainbow trout there was little evidence for vasoconstriction or splenic contraction as haematocrit was unchanged.

Fish acclimatized to 4 °C had an attenuated slow muscle hyperaemia, reflecting the reduced scope of cardiac output, although the increased volume of slow muscle may offer partial compensation such that restricted blood flow to working muscle is unlikely to be the sole limiting factor for aerobic activity at low temperatures. Fish acclimatized to 18 °C also had a significantly lower blood flow to red muscle during sustained exercise than that at 11 °C. In this case, a reduced volume of red muscle and total arterial oxygen content, coupled with higher metabolism, suggests that limited perfusion may be a significant factor in determining the swimming capacity at higher temperatures. As these changes parallel those seen with cardiac performance, this suggests that regional oxygen delivery by convective transport is determined mainly by changes in cardiac output, with only modest active redistribution in the peripheral circulation.

Structural adaptations

Processes that might limit biological activity at the cellular level, and show significant thermal sensitivity, include (i) catalytic rates of enzymes in pathways of intermediary metabolism, an area which accounts for much of the research effort in this field (Hazel & Prosser, 1974; Johnston & Dunn, 1987; see also Somero & Hofmann, this volume), and (ii) diffusion of metabolites and respiratory gases within tissue and cells, which has received little attention since the studies of Krogh (Krogh, 1919). While molecular diffusion has been considered as a potentially limiting factor for physiological processes in other systems (Jones, 1986), and *in vitro* temperature effects seem to support such a role in fishes (Sidell & Hazel, 1987), it is rarely considered a factor in limiting muscle performance.

A number of studies have shown significant structural changes in skeletal muscle of fish acclimatized to low temperatures (Johnston & Maitland, 1980; Tyler & Sidell, 1984; Egginton & Sidell, 1989), including an increase in relative mitochondrial content and reduction in sarcoplasmic space (Fig. 5). It may be that such cold-induced organelle proliferation ameliorates the effect of temperature on diffusion of small molecules (substrates) between cellular compartments (Egginton & Sidell, 1989). Interestingly, similar changes in mitochondrial content during thermal acclimatization have also been reported in frog muscle (Ballantyne & George, 1978) and fish liver (Campbell & Davies, 1978). However, no analysis of the functional consequences of organelle rarefaction at high temperatures is available.

Recent work suggests that improved muscle performance is actually achieved by cooperative adjustments in structure and/or function at different levels of organization, and hence current models describing only one or two adaptive parameters have limited predictive power

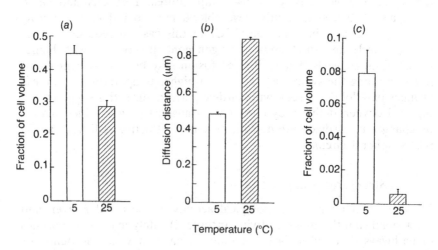

Fig. 5. Adaptation of subcellular structure in red muscle of striped bass, acclimated to winter and summer temperatures. Expansion of the mitochondrial population (*a*) minimizes local diffusion distances (*b*) which, for oxygen, may be accentuated by the presence of lipid droplets in the cold (*c*). (Redrawn from Egginton & Sidell, 1989.)

(Egginton & Ross, 1992). For example, diffusion within mitochondrion-rich and -poor fibres may be compensated by adjustments in subcellular structure, e.g. transcellular oxygen flux may be improved by the presence of lipid deposits, a common adaptation seen in pre-migratory fish (Egginton & Sidell, 1989; Fig. 5). Estimation of intercapillary distances is also an important but neglected consideration in the assessment of factors limiting oxygen supply to fish muscle. Unlike mammals, where capillarization of oxidative muscle is much more homogeneous than glycolytic muscle (and hence raises mean P_{O_2} for a given capillary density), heterogeneity of spacing was found to be similar in both types of muscle in conger eel (Egginton, Ross & Sidell, 1988). Whether this unusual design holds for other species, and following acclimation to different temperatures, is unknown. Likewise, the design of the micro-circulation for delivery of oxygen may well be related to structural modifications affecting post-capillary oxygen consumption. The influence of intracellular lipid has been mentioned, but the relatively poor correlation between capillary density and mitochondrial content of muscle (Egginton, 1990) also indicates that a more integrated analysis of oxygen supply to locomotory muscle is required.

Growth of the capillaries that are essential for delivery of oxygen and substrates to working muscle is normally very restricted. Adaptation of the capillary bed is usually interpreted in terms of oxygen supply, since in slow (red) muscle it often parallels oxidative enzyme activity and mitochondrial content (Egginton, 1990). However, capillary proliferation does not always parallel an increased mitochondrial content in small fibres, nor when oxidative metabolism is increased as a result of training, suggesting that other factors can modulate this basic relationship. Both capillary supply and oxidative capacity would appear to be capable of limiting the endurance of skeletal muscle independently during repetitive work, but little is known about the mechanism(s) by which oxygen delivery may be similarly compensated by these discrete responses. For example, relative changes in capillarity, metabolism and fibre structure of fish skeletal muscle during temperature acclimation may be quite different among species or muscle types (Egginton, 1990, 1992).

Temperature acclimation of fish in the laboratory has shown that capillary density (an important determinant of the driving pressure, ΔP_{O_2}, for O_2 diffusion) may be increased or unchanged in the cold, depending on the degree of muscle fibre hypertrophy (Johnston, 1982; Egginton, 1992). This means that, at the upper end of the thermal range, small diameter fibres (favouring diffusion) will be supplied by a relatively low capillary density (prejudicing diffusion). As mentioned

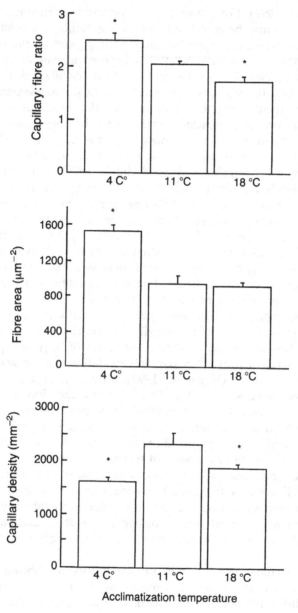

Fig. 6. Growth of capillaries in red muscle of rainbow trout acclimatized to different temperature. In the cold, an increase in C:F is offset by fibre hypertrophy, such that the maximal capillary supply is found at the intermediate temperature. (From Egginton & Cordiner, 1996.)

with respect to the gills, however, the increase in K_{O_2} may offset this disadvantage to some extent, depending on the level of V_{O_2}. In following seasonal changes in capillary supply to the aerobic (red) swimming muscle of trout, we observed a non-linear response to environmental temperature (Egginton & Cordiner, 1996). These data show that fibre hypertrophy is evident only at the coldest time of year, while capillary to fibre ratio is also elevated during spring and autumn, reflecting capillary growth (Fig. 6). The highest capillary density then corresponds to the period of highest muscle blood flow and greatest scope for aerobic swimming, at the intermediate temperature (Taylor *et al.*, 1993). This again points to a positive compensation to the cold, but an inadequate adaptation to the rigours of life in warm water.

Metabolic constraints

Thermal effects on muscle function in fish, within limits, may be viewed as expansion of recruitment order in the warm (see above). Although little studied, the efficiency gains of removing any reliance on fatiguable fast glycolytic fibres for sustained swimming may be considerable. Once again, though, the optimizing of the locomotor system at some intermediate temperature(s) is evident from the limited range of myofibrillar ATPase isoforms found in cyprinids, a response lacking in most other fishes examined (Heap, Watt & Goldspink, 1985; Moerland & Sidell, 1986). To counteract metabolic constraints, fishes may be forced to invest in growth of more red (slow) fibres, which will increase the speed of sustainable swimming at low temperatures. For example, brown trout showed a 30% increase in the mass of red muscle at 5 °C in the winter compared with that at 15 °C in the summer (N. Day and P.J. Butler, unpublished data). In contrast, fishes do not seem to expand this tissue when swimming efficiency declines at higher temperatures, presumably because of the high energetic costs involved. There is little evidence for evolutionary adjustments in the maximal rates of mitochondrial respiration in fish from different thermal environments (Johnston *et al.*, 1994), suggesting that morphometric data on mitochondrial content may be taken to reflect cellular V_{O_2} capacity. Hence reduced red muscle mass combined with decreased mitochondrial content and reliance on carbohydrate, rather than lipid-based metabolism in warm water, is likely to reduce aerobic scope for exercise. For example, striped bass acclimated to 9 °C showed a 50% increase in the speed of initial white muscle recruitment compared with fish acclimated at 25 °C when both were tested at 15 °C (Sisson & Sidell, 1987). In contrast, warm-acclimation may lead to reduction in the proportion of slow muscle, thereby limiting the total power output from this tissue. However, the functional relationship between red and white muscles is complicated by the recent evidence that, during swimming, white muscle can meet some of its energy needs aerobically, by metabolizing lipid (Wang, Heigenhauser & Wood, 1994).

The thermal response of intermediary metabolism depends on both the time-course of adaptation, and the actual acclimation temperature (Johnston & Dunn, 1987; Guderley, 1990). Following cold-acclimation in the goldfish, activity of cytochrome oxidase was significantly higher and lactate dehydrogenase lower in fish acclimated to 5 °C compared with fish at 25 °C (Sidell, 1980). Markers of aerobic function such as succinic dehydrogenase and myoglobin increased markedly in oxidative muscle, while myofibrillar ATPase activity increased dramatically in glycolytic fibres (Fig. 7. However, it may be that laboratory acclimation induces a more pronounced response to altered temperature than does natural acclimatization in the wild (Kleckner & Sidell, 1985). Indeed, there was no significant difference in the activity of cytochrome oxidase in the red muscle of seasonally acclimatized brown trout at 5 °C in the winter compared with those at 15 °C in the summer (N. Day and P. J. Butler, unpublished data). Clearly, we have a reasonable description of the process of cold-acclimation, but as the reasons for a decline in swimming performance above the thermal optimum are undoubtedly different from those below optimum, the mechanisms underlying warm-acclimation are likely to be qualitatively different.

Interactive effects

Global warming is thought to be an overall effect of anthropogenic pollution. This, together with more local thermal pollution of the aquatic environment, may summate with the effects of local toxic pollutant loads to limit fish numbers and their scope for activity. The relative toxicity of a typical pollutant such as the heavy metal copper is strongly temperature dependent. The incipient lethal level for copper to brown trout was 0.47 µmol l^{-1} at 5 °C and 0.08 µmol l^{-1} at 5 °C (Beaumont *et al.*, 1995a). Copper had a number of specific effects

---

Fig. 7. Effect of acclimation temperature on the metabolism of slow oxidative (red) muscle (hatched bars) and fast glycolytic (white) muscle (open bars) in goldfish. (*a*) Myoglobin concentration, (*b*) succinic dehydrogenase activity (*c*) myofibrillar ATPase activity (assayed at 15 °C). (Redrawn from Sidell, 1980.)

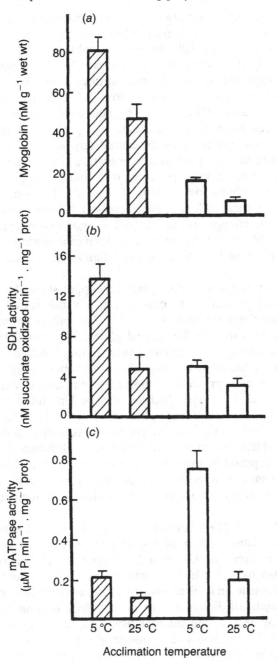

likely to limit aerobic scope. Comprehensive structural damage is caused to gill epithelia by severe copper exposure (Kirk & Lewis, 1993; Taylor *et al.*, 1996b). Such extreme gill damage undoubtedly has a strong effect upon the ability of the gill to function effectively in ionoregulation and as a gas exchange organ, as it is likely to disrupt the perfusion of blood through the gill and even water flow over the gill, and to increase diffusion distances (Hughes, 1976; Tuurala & Soivio, 1982). An expanded mucus layer may exacerbate the problem (Ultsch & Gros, 1979). Sublethal copper exposure caused less severe damage to the gill ultrastructure at all copper and temperature combinations, but the changes in the winter trout exposed to 0.47 μmol l^{-1} copper were quite substantial. Morphometric analysis showed the harmonic mean diffusion distance in these fish to have increased by over threefold. While there was no change in the arterial oxygen partial pressure (Pa_{O_2}) of these trout at rest, there was a significant decline during exercise, indicating an underlying diffusional limitation (Beaumont *et al.*, 1995a).

These winter trout exposed to 0.47 μmol l^{-1} copper showed a sharp decline in swimming performance; only one of six fish tested swam steadily at the lowest test speed of 0.3 m s^{-1}. The others swam only for one or two minutes in brief bursts and glides. However, although they displayed no ability for aerobic exercise they retained some capacity for anaerobic 'burst' swimming. This observation strongly suggests that the effect of copper/acid exposure is on aerobic exercise. Exposure to 0.08 μmol l^{-1} copper had less drastic but still significant effects upon swimming performance, reducing U_{crit} by 25–50% at 15 and 5 °C. Ultrastructural changes at the gill were less severe, and there was no significant difference in the harmonic mean diffusion distance of gills from trout exposed to this level of copper, and that of control fish. Routine oxygen consumption rose in brown trout exposed to sublethal copper concentrations at each acclimation temperature (Fig. 8).

Waiwood and Beamish (1978) also found copper and acid exposure elevated the oxygen consumption of fish at rest. At 12 °C, exposure of fingerling rainbow trout to 0.4 μmol l^{-1} copper at pH 6 increased standard metabolism by 70%. In the same study, U_{crit} was decreased by 40% in comparison to that of control trout in copper-free water at pH 7.8. In the study of Beaumont *et al.* (1995a), routine oxygen consumption rose by 38% in the summer fish. The U_{crit} of these trout was decreased by 26%, in almost the same proportion as the increase in oxygen consumption (cf. Waiwood & Beamish, 1978). However, it is unlikely that these relatively small increases in routine rates of

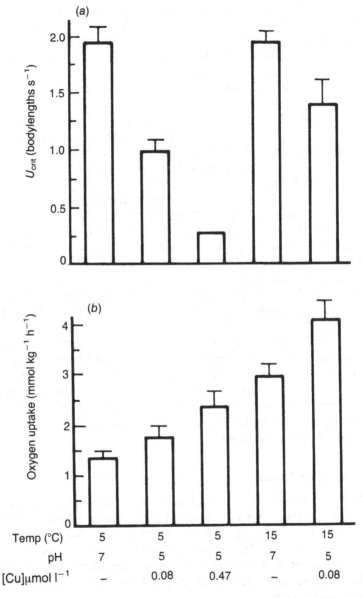

Fig. 8. Maximal sustainable swimming speed and rates of oxygen uptake of brown trout at 5 and 15 °C in water with neutral pH or in acid water plus sublethal levels of dissolved copper. Oxygen uptake increased with temperature and on exposure to copper while U_{crit} was independent of temperature but reduced by exposure to copper. (Redrawn from Beaumont *et al.*, 1995b.)

oxygen uptake would influence scope for exercise to a large extent, when active metabolism can be as much as 12–15 times the resting level (Wood & Perry, 1985). Moreover the winter trout exposed to 0.08 µmol 1^{-1} copper at 5 °C did not have an increased metabolic rate at rest, but displayed a decline in swimming performance similar to that of the brown trout exposed to this level of copper at 15 °C (Fig. 8).

Exposure of freshwater rainbow trout to acutely lethal levels of copper caused a large and progressive increase in plasma ammonia levels. The extent of ammonia accumulation during copper exposure was both dose and temperature dependent (Taylor *et al*. 1996b). Wilson and Taylor (1993) exposed freshwater rainbow trout to 4.9 µmol 1^{-1} copper at 15 °C and found plasma total ammonia concentration, $[T_{Amm}]$, to rise 28-fold in only 19 h. Exposure of brown trout to the sublethal level of 0.47 µmol 1^{-1} copper in acid freshwater (pH 5) at 5 °C for 96 h caused plasma $[T_{Amm}]$ to rise 6-fold (Beaumont, Butler & Taylor, 1995b), whereas exposure of trout to 0.08 µmol 1^{-1} copper at 5 °C increased $[T_{Amm}]$ only 2-fold. This graded accumulation of ammonia showed a clear negative correlation with sustained swimming speed. Ammonia accumulation varied directly with temperature, pH and copper levels, and caused fatigue to occur at lower swimming speeds. Its effect on swimming performance (as with those of temperature alone) may have been on muscle metabolism, or on central or peripheral nervous control of muscle contraction (Taylor *et al*., 1996a). Until we understand the nature of these interactive effects on swimming performance we will remain unable to predict the influences of global warming or low levels of chemical pollution on the survival or distribution of fish populations.

Conclusions

Fishes are sensitive to acute changes in environmental temperature at the cellular level, but have evolved a number of adaptive strategies to minimize the deleterious effects of temperature change on their swimming performance. The majority of data, however, come from studies on cold-adaptation, and extrapolation from these data to the likely effects of the higher temperatures predicted to arise from global warming is problematic.

As metabolic costs vary directly with temperature, whole animal oxygen consumption in the cold is depressed and, in the extreme, the fish enters torpor. As temperature rises, however, there is an increasingly unfavourable balance between cost of ventilation and oxygen uptake. This clearly would limit the duration of sustainable, aerobic exercise and may extend the post-exercise recovery period. Temperature sensitivity over the thermal range of many species is complicated by the existence of a zone of thermal insensitivity around the midpoint of the specific temperature range. Adaptation to a gradual global warming may be possible by extending this zone of viability upwards. In addition to the other systems mentioned in this book that show plasticity, we have highlighted some less well studied candidates for regulating sustained locomotor activity.

1 Oxygen uptake may be regulated by changing the functional respiratory surface area, via recruitment of secondary lamellae, and the effective diffusion distances over the gill lamellae. Few data exist on the structural limits to this process, but evidence is presented of intraspecific changes in the thickness of the physical barrier to oxygen exchange. Given the wide range of interspecific values noted, it is possible that further adaptations may be possible to accompany global warming.

2 Peripheral oxygen delivery is determined by the capacity of the cardiovascular system for convective transport, and the capacity of the microcirculation for diffusional supply of oxygen to respiring cells. Evidence is presented that both components are optimized in the mid-temperature range, and that this is reflected in the measured swimming performance of rainbow trout. While oxygen delivery certainly limits swimming capacity, this is not exclusively determined by cardiac capacity. A contributory factor to cardiac failure at high temperature could be the structural extent of the capillary supply to muscle, which may be inadequate to maintain endurance exercise.

3 The greatest challenges to physiological adaptations come when a system compromised by one variable, such as increased temperature, is then challenged by another. As an example, exposure of trout to a heavy metal pollutant is shown to have a differential effect according to the environmental temperature. Given that thermal and chemical pollution often occur simultaneously the integrative, additive stress needs careful attention.

Our conclusion is that fish have a remarkable resilience to altered temperature, especially with a prolonged time course, and that certain structural limits to gaseous exchange may be capable of adaptive change to accommodate global warming. However, this optimistic analysis is based on woefully inadequate data, and possibly unwarranted extrapolations, and thus cannot be relied on. Clearly, much more work on this topic is required.

References

Ballantyne, J. S. & George, J. C. (1978). An ultrastructural and histological analysis of the effects of cold-acclimation on vertebrate skeletal muscle. *Journal of Thermal Biology*, **3**, 109–16.

Barron, M. G., Tarr, B. D. & Hayton, W. L. (1987). Temperature-dependence of cardiac output and regional blood flow in rainbow trout, *Salmo gairdneri* Richardson. *Journal of Fish Biology*, **31**, 735–44.

Beaumont, M. W., Butler, P. J. & Taylor, E. W. (1995a). Exposure of brown trout, *Salmo trutta*, to sub-lethal copper concentrations in soft acidic water and its effect upon sustained swimming performance. *Aquatic Toxicology*, **33**, 45–63.

Beaumont, M. W., Butler, P. J. & Taylor, E. W. (1995b). Plasma ammonia concentration in brown trout, *Salmo trutta*, exposed to acid soft water and sublethal copper concentrations and its relationship to decreased swimming performance. *Journal of Experimental Biology*, **198**, 2213–20.

Booth, J. H. (1979). Circulation in trout gills: the relationship between branchial perfusion and the width of the lamellar blood space. *Canadian Journal of Zoology*, **57**, 2183–5.

Brett, J. R. (1964). The respiratory metabolism and swimming performance of young sockeye salmon. *Journal of Fisheries Research Board of Canada*, **21**, 1183–226.

Butler, P. J., Day, N. & Namba, K. (1992). Interactive effects of seasonal temperature and low pH on resting oxygen uptake and swimming performance of adult brown trout *Salmo trutta*. *Journal of Experimental Biology*, **165**, 195–212.

Cameron, J. N. (1974). Blood flow distribution as indicated by tracer microspheres in resting and hypoxic Arctic grayling (*Thymallus arcticus*). *Comparative Biochemistry and Physiology*, **52A**, 441–4.

Campbell, C. M. & Davies, P. S. (1978). Temperature acclimation in the teleost, *Blennius pholis*: changes in enzyme activity and cell structure. *Comparative Biochemistry and Physiology*, **61**, 165–7.

Cech, J. J. Jr, Bridges, D. W., Rowell, D. M. & Balzer, P. J. (1976). Cardiovascular responses of winter flounder, *Pseudopleuronectes americanus* (Walbaum), to acute temperature increase. *Canadian Journal of Zoology*, **54**, 1383–8.

Cossins A. R. & Bowler K. (1987). *Temperature Biology of Animals*. London: Chapman & Hall.

Dallinger W. H. (1887). Acclimatization of flagellates to high temperatures. *Journal of the Royal Microscopical Society*, **7**, 185–9.

Davie, P. S. & Forster, M. E. (1980). Cardiovascular responses to swimming in eels. *Comparative Biochemistry and Physiology*, **67A**, 367–73.

Day, N. & Butler, P. J. (1996). Environmental acidity and white muscle recruitment during swimming in the brown trout (*Salmo trutta*). *Journal of Experimental Biology*, **199**, 1947–59.

Egginton, S. (1990). Morphometric analysis of tissue capillary supply. In *Vertebrate Gas Exchange from Environment to Cell*, ed. R. G. Boutilier; *Advances in Comparative Environmental Physiology*, Vol. 6, pp. 73–141. Berlin: Springer-Verlag.

Egginton, S. (1992). Adaptability of the anatomical capillary supply to skeletal muscle of fishes. *Journal of Zoology (London)*, **226**, 691–8.

Egginton, S. & Cordiner, S. (1996). Cold-induced angiogenesis in skeletal muscle of rainbow trout. *International Journal of Microcirculation*, **15**, 209.

Egginton, S. & Ross, H. F. (ed.) (1992). *Oxygen Transport in Biological Systems. Society for Experimental Biology Seminar Series 51.* Cambridge: Cambridge University Press.

Egginton, S., Ross, H. F. & Sidell, B. D. (1988). Morphometric analysis of intracellular diffusion distances. *Acta Stereologica*, **6**, 449–54.

Egginton, S. & Sidell, B. D. (1989). Thermal acclimation induces adaptive changes in subcellular structure of fish skeletal muscle. *American Journal of Physiology*, **256**, R1–9.

Farrell, A. P. (1980). Gill morphometrics, vessel dimensions, and vascular resistance in ling cod (*Ophiodon elongatus*). *Canadian Journal of Zoology*, **58**, 807–18.

Farrell, A. P., Daxboeck, C. & Randall, D. J. (1979). The effect of input pressure and flow on the pattern and resistance to flow in the isolated perfused gill of a teleost fish. *Journal of Comparative Physiology* B, **133**, 233–40.

Graham M. S. & Farrell A. P. (1989). The effect of temperature acclimation and adrenaline on the performance of a perfused trout heart. *Physiological Zoology*, **62**, 38–61.

Guderley, H. (1990). Functional significance of metabolic responses to thermal acclimation in fish muscle. *American Journal of Physiology*, **259**, R245–52.

Hazel, J. R. & Prosser, C. L. (1974). Molecular mechanisms of temperature compensation in poikilotherms. *Physiological Reviews*, **54**, 620–77.

Heap, S. P., Watt, P. W. & Goldspink, G. (1985). Consequences of thermal change on the myofibrillar ATPase of 5 freshwater teleosts. *Journal of Fish Biology*, **26**, 733–8.

Hughes, G. M. (1972). Morphometrics of fish gills. *Respiration Physiology*, **14**, 1–25.

Hughes, G. M. (1976). Polluted fish respiratory physiology. In *Effects of Pollutants on Aquatic Organisms*, ed. A. P. M. Lockwood, pp. 163–83. London: Cambridge University Press.

Hughes, G. M. & Shelton, G. (1962). Respiratory mechanisms and their nervous control in fish. In *Advances in Comparative Physiology and Biochemistry*, Vol. 1, ed. O. E. Lowenstein, pp. 275–364. London: Academic Press.

Johnston, I. A. (1982). Capillarisation, oxygen diffusion distances and mitochondrial content of carp muscles following acclimation from summer to winter temperatures. *Cell and Tissue Research*, **222**, 579–96.

Johnston, I. A. & Dunn, J. (1987). Temperature acclimation and metabolism in ectotherms with particular reference to teleost fish. In *Temperature and Animal Cells*, ed. K. Bowler. *Symposia of the Society for Experimental Biology* 31, pp. 67–93. Cambridge: Cambridge University Press.

Johnston, I. A., Guderley, H., Franklin, C. E., Crockford, T. & Kamunde, C. (1994). Are mitochondria subject to evolutionary temperature adaptation? *Journal of Experimental Biology*, **195**, 293–306.

Johnston, I. A. & Maitland, B. (1980). Temperature acclimation in crucian carp, *Carassius carassius* L., morphometric analyses of muscle fibre ultrastructure. *Journal of Fish Biology*, **17**, 113–23.

Jones, D. R. (1971). Theoretical analysis of factors which may limit the maximum oxygen uptake of fish: the oxygen cost of the cardiac and branchial pumps. *Journal of Theoretical Biology*, **32**, 341–9.

Jones, D. R. (1986). Intracellular diffusion gradients of O_2 and ATP. *American Journal of Physiology*, **250**, C663–75.

Jones, D. R., Kiceniuk, J. W. & Bamford, O. S. (1974). Evaluation of the swimming performance of several fish species from the McKenzie river. *Journal of the Fisheries Research Board of Canada*, **31**, 1641–7.

Kirk, R. S. & Lewis, J. W. (1993). An evaluation of pollutant induced changes in the gills of rainbow trout using scanning electron microscopy. *Environmental Technology*, **14**, 577–85.

Kisia, S. M., & Hughes, G. M. (1992). Estimation of oxygen diffusing capacity in the gills of different sizes of tilapia, *Oreochromis niloticus*. *Journal of Zoology (London)* **227**, 405–15.

Kleckner, N. W. & Sidell, B. D. (1985). Comparisons of maximal activities of enzymes from tissues of thermally-acclimated and naturally-acclimatised chain pickerel (*Esox niger*). *Physiological Zoology*, **58**, 18–28.

Kolok, A. S., Spooner, M. R. & Farrell, A. P. (1993). The effect of exercise on the cardiac output and blood flow distribution of the largescale sucker *Catostomus macrocheilus*. *Journal of Experimental Biology*, **184**, 301–21.

Krogh, A. (1919). The rate of diffusion of gases through animal tissues with some remarks on the coefficient of invasion. *Journal of Physiology (London)*, **52**, 391–415.

Leino, R. L. & McCormick, J. H. (1993). Responses of juvenile largemouth bass to different pH and aluminium levels at overwintering temperatures: effects on gill morphology, electrolyte balance, scale calcium, liver glycogen, and depot fat. *Canadian Journal of Zoology*, **71**, 531–43.

McKenzie, D. J., Serrini, G., Piraccini, G., Bronzi, P. & Bolis, C. L. (1996). Effects of diet on response to exhaustive exercise in Nile tilapia (*Oreochromis nilotica*) acclimated to three different temperatures. *Comparative Biochemistry and Physiology*, **114A**, 43–50.

Moerland, T. S. & Sidell, B. D. (1986). Biochemical responses to temperature in the contractile protein complex of striped bass (*Morone saxatilis*). *Journal of Experimental Biology*, **238**, 287–95.

Neumann, P., Holeton, G. F. & Heisler, N. (1983). Cardiac output and regional blood flow in gills and muscles after exhaustive exercise in rainbow trout (*Salmo gairdneri*). *Journal of Experimental Biology*, **105**, 1–14.

Nikinmaa, M., Soivio, A. & Railo, E. (1981). Blood volume of *Salmo gairdneri*: influence of ambient temperature. *Comparative Biochemistry and Physiology*, **69A**, 767–9.

Nilsson, S. & Pettersson, K. (1981). Sympathetic nervous control of blood flow in the gills of the Atlantic cod, *Gadus morhua*. *Journal of Comparative Physiology*, **144**, 157–63.

Perry, S. F. (1990). Recent advances and trends in the comparative morphometry of vertebrate gas exchange organs. In *Vertebrate Gas Exchange from Environment to Cell*, ed. R. G. Boutilier. *Advances in Comparative Environmental Physiology*, Vol. 6, pp. 45–71. Berlin: Springer-Verlag.

Randall, D. J. (1982). The control of respiration and circulation in fish during exercise and hypoxia. *Journal of Experimental Biology*, **100**, 275–88.

Randall, D. J. & Daxboeck, C. (1982). Cardiovascular changes in the rainbow trout (*Salmo gairdneri* Richardson) during exercise. *Canadian Journal of Zoology*, **60**, 1135–40.

Rome, L. C., Loughna, P. T. & Goldspink, G. (1985). Temperature acclimation: improved sustained swimming performance in carp at low temperatures. *Science*, **228**, 194–6.

Rome, L. C., Funke, R. P. & Alexander, R. M. N. (1990). The influence of temperature on muscle velocity and sustained swimming performance in swimming carp. *Journal of Experimental Biology*, **154**, 163–78.

Sidell, B. D. (1980). Response of the goldfish (*Carassius auratus*, L.) muscle to acclimation temperature: alterations in biochemistry and proportions of different fibre types. *Physiological Zoology*, **53**, 98–107.

Sidell, B. D. & Hazel, J. R. (1987). Temperature affects the diffusion of small molecules through cytosol of fish muscle. *Journal of Experimental Biology*, **129**, 191–203.

Sidell, B. D. & Moerland, T. S. (1989). Effects of temperature on muscular function and locomotory performance in teleost fish. In *Advances in Comparative Environmental Physiology*, Vol. 5, Ch. 5. Berlin: Springer-Verlag.

Sisson, J. E. & Sidell, B. D. (1987). Effect of thermal acclimation on muscle fibre recruitment of swimming striped bass (*Morone saxatilis*). *Physiological Zoology*, **60**, 310–20.

Soivio, A. & Tuurala, H. (1981). Structural and circulatory responses to hypoxia in the secondary lamellae of *Salmo gairdneri* gills at two temperatures. *Journal of Comparative Physiology B*, **145**, 37–43.

Stevens, E. D. & Randall, D. J. (1967). Changes in blood pressure, heart rate and breathing rate during moderate swimming activity in rainbow trout. *Journal of Experimental Biology*, **46**, 307–15.

Taylor, E. W. (1985). Control and co-ordination of gill ventilation and perfusion. *Symposia of the Society of Experimental Biology* 39, pp. 123–61. Cambridge: Cambridge University Press.

Taylor, E. W. (1992). Nervous control of the heart and cardiorespiratory interactions. In *Fish Physiology*, Vol. 12B, ed. S. Hoar, D. J. Randall & A. P. Farrell, pp. 343–87. New York: Academic Press.

Taylor, E. W., Beaumont, M. W., Butler, P. J., Mair, J. & Mujallid, M. S. I. (1996b). Lethal and sub-lethal effects of copper upon fish: a role for ammonia toxicity? In *Toxicology of Aquatic Pollution*, ed. E. W. Taylor. *Society for Experimental Biology Seminar Series* 57, pp. 85–113. Cambridge: Cambridge University Press.

Taylor, S. E., Egginton, S. & Taylor, E. W. (1993). Respiratory and cardiovascular responses in rainbow trout (*Oncorhynchus mykiss*) to aerobic exercise over a range of acclimation temperatures. *Journal of Physiology*, **459**, 19P.

Taylor, S. E., Egginton S. & Taylor, E. W. (1996a). Cardiovascular and morphometric limits to maximal sustainable exercise in the rainbow trout with seasonal temperature acclimatization. *Journal of Experimental Biology*, **199**, 835–45.

Tuurala, H. & Soivio, A. (1982). Structural and circulatory changes in the secondary lamellae of *Salmo gairdneri* gills after sublethal exposures to dehydroabietic acid and zinc. *Aquatic Toxicology*, **2**, 21–9.

Tyler, S. & Sidell, B. D. (1984). Changes in mitochondrial distribution and diffusion distances in muscle of goldfish upon acclimation to warm and cold temperatures. *Journal of Experimental Zoology*, **232**, 1–9.

Ultsch, G. R. & Gros, G. (1979). Mucus as a diffusion barrier to oxygen: possible role in O_2 uptake at low pH in carp (*Cyprinus carpio*) gills. *Comparative Biochemistry and Physiology*, **62A**, 685–9.

Waiwood, K. G. & Beamish, F. W. H. (1978). Effects of copper, pH and hardness on the critical swimming speed of rainbow trout (*Salmo gairdneri* Richardson). *Water Research*, **12**, 611–19.

Wang, Y., Heigenhauser, G. J. F. & Wood, C. M. (1994). Integrated responses to exhaustive exercise and recovery in rainbow trout white muscle: acid–base, phosphogen, carbohydrate, lipid, ammonia, fluid volume and electrolyte metabolism. *Journal of Experimental Biology*, **195**, 227–58.

Weatherly, A. H. & Gill, H. S. (ed.) (1987). *The Biology of Fish Growth*. New York: Academic Press.

Webb, P. W. (1971). The swimming energetics of trout. II. Oxygen consumption and swimming efficiency. *Journal of Experimental Biology*, **55**, 521–40.

Wilson, R. W. & Egginton, S. (1994). Assessment of maximum sustainable swimming performance in rainbow trout (*Oncorhynchus mykiss*). *Journal of Experimental Biology*, **192**, 299–305.

Wilson, R. W. & Taylor, E. W. (1993). The physiological responses of freshwater rainbow trout, *Oncorhynchus mykiss*, during acutely lethal copper exposure. *Journal of Comparative Physiology B*, **163**, 38–47.

Wood, C. M. & Perry, S. F. (1985). Respiratory, circulatory and metabolic adjustments to exercise in fish. In *Circulation, Respiration and Metabolism*, ed. R. Gilles, pp. 1–22. Heidelberg: Springer-Verlag.

A.P. FARRELL

Effects of temperature on cardiovascular performance

Introduction

Two recent events in British Columbia, Canada demonstrate that relatively small temperature shifts can have rapid and pronounced effects on fish populations. The first event involved a warm mass of marine water moving to a higher than normal latitude (an El Niño effect) and bringing with it a large stock of mackerel. These more active mackerel preyed extensively upon juvenile salmon and herring that normally grow in these coastal waters (B. Hargreaves, personal communication).

The second event involved the upstream spawning migration of sockeye salmon in the Fraser River and their normally spectacular but successful negotiation through Hells Gate and other ferocious rapids, the most formidable natural barriers on the river. In 1994 water temperature was several degrees higher than the yearly average for the Fraser River, reaching 20 °C during one week in the migration window. Sockeye salmon, on reaching Hells Gate, were observed to lose their upstream orientation and retreat downstream (S. Hinch, personal communication). Estimates suggest that an unusually large number of sockeye salmon failed to negotiate these warmer than normal rapids, and of those that did reach the spawning grounds, many laid eggs that did not hatch (C. Walters, personal communication).

These two examples underscore the importance of swimming performance (or the lack thereof) for fish survival and in reproductive success. Cardiovascular performance plays a central support role in aerobic swimming and is often regarded as the limiting factor for maximum exercise. Salmonids, for example, appear to exploit their maximum or near maximum cardiac performance during prolonged swimming. In addition, high demands can be placed upon the cardiovascular system during either feeding or recovery from burst exercise. It is important to note, however, that in general all organs cannot be fully perfused

Society for Experimental Biology Seminar Series 61: *Global Warming: Implications for freshwater and marine fish*, ed. C. M. Wood & D. G. McDonald. © Cambridge University Press 1996, pp. 135–158.

at any one time because the circulatory system is a shared resource and the maximum delivery capacity falls short of the summed demands of individual organs.

The operational flexibility of the cardiovascular system is equivalent to the difference between routine and maximum levels, or cardiovascular scope. An important survival issue is whether or not cardiovascular scope is maintained with increased water temperature; does maximum cardiovascular performance have the same Q_{10} relationship as routine cardiovascular performance? The short answer to this question is that it does not. However, to understand why this is so, which is the purpose of this chapter, requires information on (i) the cardiovascular responses to acute increases in temperature, (ii) the degree to which acclimatory and adaptational (over generations) changes occur in cardiovascular tissues, and (iii) the upper temperature limits for key processes underlying cardiovascular activity. This sort of detailed knowledge base is simply unavailable even for one fish species. Instead, what follows is an attempt to pull together disparate information, especially with respect to rainbow trout and other salmonids, and speculate on some general conclusions regarding the extent to which cardiovascular performance changes with temperature. Since this chapter is intended to relate to global warming, emphasis is given to cardiovascular performance at temperatures approaching the upper incipient lethal temperature (UILT). Salmonids are generally considered stenothermal with UILT values in the range 20–26 °C and a preferred temperature of around 15 °C (Brett, 1952; Black, 1953).

Energetic costs associated with increasing internal oxygen convection

For the purpose of the present discussion, oxygen transport is regarded as the primary role for the salmonid circulatory system. (Note: internal convection serves other transport roles which probably should not be completely overlooked since they could take on greater significance at extreme temperatures.) For active fishes, especially when they are swimming, the amount of oxygen delivered to the tissues is essentially equivalent to the total oxygen consumption ($\dot{V}_{O_2}$). Oxygen delivery is given by the product of cardiac output ($\dot{Q}$ = heart rate × stroke volume) and the arteriovenous difference in O_2 content (AV-O_2). It is therefore possible to divide the cardiovascular response to temperature into a cardiac response (a change in $\dot{Q}$) and a haematological response (a change in AV-O_2).

We have little information on the energetic costs of either of these responses to increased temperature. On the basis of cardiac pumping alone, the energetic cost of a cardiac response can be estimated from the linear relationship that is established between cardiac power output ($\dot{Q}$ × ventral aortic blood pressure) and myocardial O_2 consumption (see Farrell & Jones, 1992). Even so, the cost of routine cardiac work in rainbow trout is a small percentage of routine $\dot{V}_{O_2}$ (4.6%) and this relative cost decreases significantly with swimming (to around 1%; Farrell & Steffensen, 1987), primarily because tissue O_2 extraction increases. Haematological responses can also affect cardiac work if blood viscosity changes. For example, an increase in haematocrit (Hct) will increase blood viscosity and arterial blood pressure, given no other cardiovascular changes. However, blood viscosity *per se* will be reduced somewhat at warmer temperatures. Thus, the energetic cost of a haematological response to warm temperature is likely to be much smaller than that associated with a cardiac response. Furthermore, since a substantial decrease in the cost of cardiac work occurs when O_2 extraction increases, the cost of maintaining the resting venous O_2 store is appreciable.

Haematological response to warm temperatures

The haematological response that has the greatest influence on AV-O_2 is a quantitative change in the blood haemoglobin concentration [Hb]. Qualitative and modulatory responses involving Hb-O_2 affinity also need to be considered because they directly affect the degree to which blood binds O_2 at the gills and releases O_2 at the tissues. Houston (1979) provides an excellent review of the haematological response of fishes to environmental temperature.

Temperature and haemoglobin concentration: a quantitative response of limited value?

[Hb] increases through increases in either the numbers of circulating red blood cells or the mean cell Hb concentration (MCHC). Increased Hct can occur rapidly through the release of stored red blood cells into the circulation, and more slowly through erythropoiesis. In addition MCHC, Hb isomorph complement, and the proportion of circulating red blood cells capable of *de novo* Hb synthesis can also vary with temperature (Houston, 1979; Houston & Schrapp, 1994). Of these variables, Hct tends to have the most dramatic effect on [Hb], possibly because [Hb] is frequently close to its solubility maximum in the red blood cells (Riggs, 1976, as quoted by Houston, 1979).

Changes in Hct in response to temperature are neither profound nor consistent. Over a large temperature range the observed changes in Hct are usually small increases or decreases of no more than 10–20% (Houston, 1979). Figure 1 illustrates the situation for salmonids. These selected data clearly show the absence of a major haematological response to temperature between 3 and 22 °C. Figure 1 includes data collected in my laboratory from over 160 cannulated rainbow trout and chinook salmon used in different experiments and at various acclimation temperatures over a period of 4 years (P. Gallaugher and H. Thorarensen, unpublished data). Great care was taken to minimize experimental blood loss during and after surgery. Hct values in other studies with cannulated rainbow trout (Fig. 1) are similar, with only two exceptions (Wood, Pieprzak & Trott, 1979; Taylor, Egginton & Taylor, 1993). In these two studies Hct decreased at warm tem-

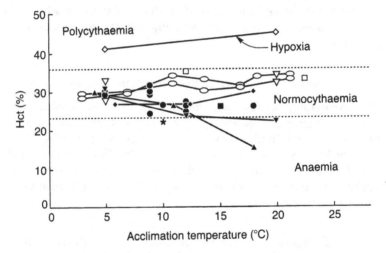

Fig. 1. Effect of temperature on haematocrit (Hct) in salmonids. Selected data are plotted for blood samples taken from cannulated (closed symbols) and non-cannulated fish (open symbols) to illustrate the effect of sampling on the measured Hct. Points connected by lines are from the same study. Also included are data for rainbow trout exposed to chronic hypoxia to illustrate that this environmental stimulus can induce a polycythaemic state independent of the temperature. (Data sources: ○, DeWilde & Houston, 1967; ◇, ▽, Tun & Houston, 1986; □, Jones, 1971; ✳, Kiceniuk & Jones, 1977; ▼, Wood et al., 1979; ■, Bushnell et al., 1984; ▲, Taylor et al., 1993; ●, P. Gallaugher, unpublished data.)

peratures (Fig. 1) to a level of 23% or below, which is probably an anaemic state in salmonids. (The possibility that cannulated fish at warmer temperatures have an increased propensity for blood loss should be seriously considered.) Hct measured in uncannulated rainbow trout (e.g. DeWilde & Houston, 1967; Tun & Houston, 1986) tends to be slightly higher than that in cannulated fish (Fig. 1). This difference probably reflects the small degree of haemoconcentration (red blood cell swelling and red blood cell release from storage tissues) known to occur with sampling uncannulated fish. Regardless, Hct shows a similar insensitivity to temperature.

The lack of an appreciable increase in Hct with temperature has been attributed to an increase in blood viscosity which sets an upper limit to Hct (e.g. Weber & Jensen, 1988). Thus, in theory, increased blood viscosity would compromise maximum cardiac performance and offset any potential gain in terms of internal O_2 convection. However, two lines of evidence indicate that this explanation is untenable for rainbow trout.

One line of evidence shows that, in contrast to the above prediction, blood doping (infusion of red blood cells to increase Hct) significantly increases $\dot{V}_{O_2max}$ in rainbow trout at 6 and 13 °C (Gallaugher, Thorarensen & Farrell, 1995). In fact, it was possible to increase Hct up to 42% (i.e. well above the range for normocythaemia; Hct =27–30%) before $\dot{V}_{O_2max}$ declined (Fig. 2). Furthermore, swimming performance (as measured by the critical swimming speed, U_{crit}) increased up to Hct values of 55% (Gallaugher *et al.*, 1995). These blood doping studies clearly demonstrated that cardiac performance, metabolic performance and swimming performance are not compromised by a significant degree of polycythaemia in rainbow trout.

Rainbow trout clearly have an upper safety margin for their Hct and warm temperature does not appear to be an appropriate environmental stimulus to exploit this safety margin. However, hypoxia can cause polycythaemia (Tun & Houston, 1986), providing further evidence to indicate that blood viscosity is not limiting Hct. Figure 1 illustrates that hypoxia with a short photoperiod stimulates Hct to increase from 28.1 to 41.2% at 5 °C and from 34.1 to 45.9% at 20 °C. Interestingly, these elevated Hct values correspond well with the Hct level determined to be the peak for $\dot{V}_{O_2max}$ (Gallaugher *et al.*, 1995).

Temperature and haemoglobin-O_2 affinity: qualitative and modulatory responses of little benefit?

Acute increases in temperature reduce both O_2 affinity (increased P_{50} value) and O_2 carrying capacity of fish blood, including rainbow trout

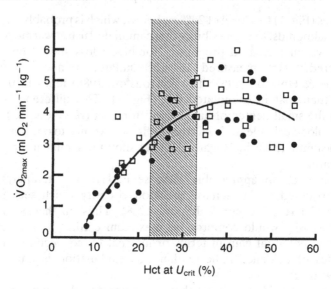

Fig. 2. Effect of haematocrit (Hct) on maximum oxygen consumption ($\dot{V}_{O_2max}$) in rainbow trout. Fish were acclimated to and tested in the summer at 13 °C (●) and in the winter at 6 °C (□). Hct was experimentally adjusted through blood doping (addition or removal of red blood cells) prior to a stepwise swimming challenge which measured the critical swimming speed (U_{crit}) and $\dot{V}_{O_2max}$. The peak $\dot{V}_{O_2max}$ occurred at a Hct of 42%, a value well above the range for normocythaemia (23–32%; shaded area) in this group of fish. (Data adapted from Gallaugher *et al.*, 1995.)

(Eddy, 1971). Furthermore, in the rainbow trout there is little change in the Hb-O$_2$ affinity with temperature acclimation (Eddy, 1971) even though other species may show qualitative shifts in the Hb system and changes to the intracellular modulators that affect the Hb-O$_2$ affinity (Houston, 1979; Weber & Jensen, 1988).

Whereas the decrease in O$_2$ content reduces AV-O$_2$, the increase in P_{50}, by favouring O$_2$ unloading at tissues, could benefit O$_2$ delivery provided the right-shift is not so large as to prevent arterial blood from becoming fully saturated at the gills. However, excessive O$_2$ unloading at the tissues could be detrimental in that venous blood leaving the systemic circulation in fish must still deliver O$_2$ to the working heart because most fish species do not have a coronary circulation (see Davie & Farrell, 1991). Even in salmonids only 30–40% of the ventricular muscle is supplied by a coronary circulation. This means that the venous partial pressure of O_2 ($P_{V_{O_2}}$) and the thickness of the ventricular wall are primary determinants of the maximum rate of venous myocardial O_2 delivery. Consequently, it is not surprising to find that fish that are either very active or routinely experience hypoxia possess a coronary circulation, since both situations reduce $P_{V_{O_2}}$. Whether the role of the coronary circulation varies with temperature is unknown.

In summary, the haematological response of salmonids to warmer temperature is modest at best. Quantitatively, Hct changes are small. Furthermore, an increased P_{50} on one hand improves tissue O_2 unloading but on the other hampers O_2 delivery to the heart.

Cardiac response to warm temperatures

The introduction posed the question: do fish maintain cardiovascular scope with increased temperature? Given the limited haematological response to warm temperature outlined above, it is clear that a cardiac response is the primary means by which salmonids maintain cardiovascular scope. Unfortunately there are very few direct measurements of routine $\dot{Q}$ and maximum cardiac output ($\dot{Q}_{max}$) in fish (see Farrell & Jones, 1992) from which to draw conclusions regarding the nature by which cardiac scope may change with temperature. Nonetheless, these rather limited data do suggest that, although both routine $\dot{Q}$ and $\dot{Q}_{max}$ can increase with temperature, cardiac scope decreases at warm temperatures.

To maintain cardiac scope across a broad temperature range, routine $\dot{Q}$ and $\dot{Q}_{max}$ need to have similar Q_{10} values. However, this is not the case. Whereas Q_{10} values for routine $\dot{Q}$ are typically around 2.0 or higher (e.g. Cech, Rowell & Glasgow, 1976; Barron, Tarr & Hayton, 1987; Kolok, Spooner & Farrell, 1993), Q_{10} values for $\dot{Q}_{max}$ are lower (e.g. 1.2–1.3 in rainbow trout: Graham & Farrell, 1989; Keen & Farrell, 1994; 1.4 in squawfish: Kolok & Farrell, 1994). As result $\dot{Q}$ increases to a larger degree than $\dot{Q}_{max}$ at higher temperatures and cardiac scope is reduced. The implication of this observation is that a temperature-induced increase in $\dot{Q}_{max}$ appears to be constrained in some way.

Using the limited available data, it is possible to illustrate graphically the general trends for $\dot{Q}$ and $\dot{Q}_{max}$ versus acclimation temperature for rainbow trout (Fig. 3). Routine $\dot{Q}$ tends to follow or exceed a line of identity for a Q_{10} value of 2.0. (Author's note: Some of the routine $\dot{Q}$ values could be elevated because of stress). $\dot{Q}_{max}$ also appears to follow a similar line of identity but only up to 15 °C where $\dot{Q}_{max}$ reaches

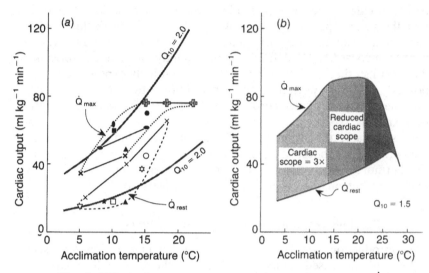

Fig. 3. Effect of temperature on routine cardiac output ($\dot{Q}_{rest}$) and maximum cardiac output ($\dot{Q}_{max}$) in rainbow trout. (*a*) Selected data are plotted for $\dot{Q}_{max}$ and $\dot{Q}_{rest}$. Points from the same study are connected by solid lines. These lines and the broken lines further group the data into two sets for $\dot{Q}_{max}$ and $\dot{Q}_{rest}$. Lines of identity for a Q_{10} value of 2.0 are included for reference. (*b*) An idealized relationship between $\dot{Q}_{rest}$ and $\dot{Q}_{max}$ in fish. This illustrates that $\dot{Q}_{max}$ plateaus at the fish's preferred temperature, 15 °C in this example. As a result cardiac scope is maintained at temperatures below the preferred temperature, but is reduced above the preferred temperature. (Data sources for *in vivo* measurements: ▽, Stevens & Randall, 1967; ✷ Cameron & Davis, 1970; □, ■, Kiceniuk & Jones, 1977; ✿, Wood & Shelton, 1980; ○, ●, Neumann *et al.*, 1983; ×, Barron *et al.*, 1987; ▲, Thorarensen *et al.*, 1996. Data sources for *in vitro* measurements with perfused heart preparations; ✗, Graham & Farrell, 1989; ❶, Farrell *et al.*, 1991; ◗, Keen *et al.*, 1994; ✛, Farrell *et al.*, 1996).

a plateau. Likewise, Brett (1971), in analysing the data for exercising sockeye salmon (Davis, 1968), found a plateau for active $\dot{Q}$ at temperatures above 15 °C. Since $\dot{Q}_{max}$ follows a similar pattern at warm temperature in non-salmonid fish (Yamamitsu & Itazawa, 1990; Kolok *et al.*, 1993), it is possible to generalize in the following manner. Whereas routine $\dot{Q}$ increases in an exponential manner with temperature up to the UILT, $\dot{Q}_{max}$ reaches a plateau at the preferred temperature and then declines just prior to the UILT (Fig. 3*b*). The UILT might be the intersection of the curves for $\dot{Q}_{max}$ and routine $\dot{Q}$. This means that cardiac scope is maintained only up to the preferred temperature (15 °C in this example), and is reduced at warmer temperatures. The preferred temperature is therefore the highest temperature at which cardiac scope is maintained. In the case of the salmonids, the preferred temperature is 7–10 °C below the UILT.

Any reduction in cardiac scope should have a profound effect on swimming activity. In fact, Brett (1971) was able to superimpose the temperature curves for swimming performance and cardiac performance. Both measures of performance peaked at 15 °C and then declined at warmer temperatures. Brett concluded that the debilitating effects of temperatures warmer than 15 °C were multiple and were related to limiting O_2 availability. However, it is equally plausible that the decline in cardiac scope, rather than O_2 availability, is the limiting factor.

To understand why cardiac scope decreases at temperatures above 15 °C in salmonids, it is necessary to consider specifically how heart rate and stroke volume vary with temperature. This will result in two suggestions. First, temperature-induced increases in heart rate are probably not a good predictor of changes in $\dot{Q}_{max}$ at temperatures above the preferred temperature. Second, a decline in cardiac contractility at these high temperatures accounts for the decline in stroke volume, $\dot{Q}_{max}$ and cardiac power output.

Temperature and heart rate: an unreliable relationship to predict $\dot{Q}_{max}$?

Temperature is the single most important environmental determinant of heart rate in fish. Many studies have demonstrated that the Q_{10} for routine heart rate is between 1.3 and 3.0, depending on the species, the temperature range, and the acclimation temperature (Figs. 4 and 5). Furthermore, heart rate continues to increase above the preferred temperature. Therefore, inadequate increases in heart rate clearly are not the reason for the plateau in $\dot{Q}_{max}$. In fact, by using only Q_{10} information for heart rate, one could arrive at erroneous conclusions regarding temperature effects on cardiac scope in fish because, given no other change, the increases in heart rate are such that cardiac scope could be maintained at much warmer temperatures.

To understand fully the effect of temperature on heart rate requires two additional pieces of information. What is the upper temperature limit for pacemaker activity in relation to the UILT? Does the pacemaker have an upper frequency limit?

In answer to the first question, the upper temperature limit for pacemaker activity is species-specific and altered by acclimation tem-

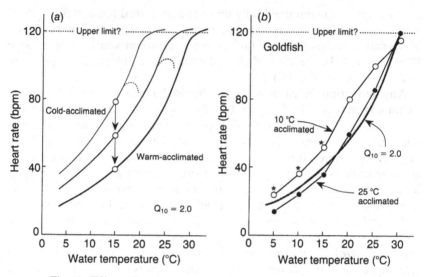

Fig. 4. Effect of temperature on heart rate. (a) An idealized family of curves to illustrate four major points regarding the relationship between pacemaker rate and temperature: (1) pacemaker rate generally follows a Q_{10} relationship; (2) there is an upper frequency limit of around 120 bpm for most fish species. (3) in cold-acclimated fish pacemaker activity can cease at frequencies below the upper frequency limit (broken line), but in warm-acclimated fish pacemaker activity increases up to the UILT; (4) warm acclimation results in a resetting of pacemaker activity to a lower rate and as a result scope for heart rate is maintained. (b) *In vivo* heart rate data from goldfish acclimated to 10 and 25 °C. A line of identity for $Q_{10} = 2.0$ is included for reference. (Data adapted from Morita & Tsukuda, 1994.)

perature. Pacemaker activity typically fails at a temperature well above the acclimation temperature but may not be maintained up to the UILT in cold-acclimated fish (Fig. 4a). However, with warm-acclimation, the temperature at which the pacemaker quits is increased (Tsukuda, Liu & Injii, 1985; Yamamitsu & Itazawa, 1990; Morita & Tsukuda, 1994). Consequently, provided fish are warm-acclimated, heart rate increases with temperature right up to the UILT (Fig. 4b).

In answer to the second question, fish appear to have an upper limit for maximum heart rate of around 120 beats per minute (Fig. 4a: Farrell, 1991). To date, tuna seem to be the only exception to this generalization. The heart rate of warm-acclimated salmonids approaches this upper limit. Maximum heart rate is 114 bpm at 20 °C in rainbow trout and 107 bpm at 22 °C in sockeye salmon (see Fig. 5). As a

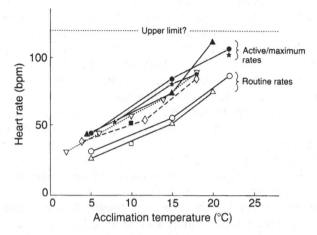

Fig. 5. Heart rates in salmonids as a function of temperature. Because the temperature dependency of routine (open symbols) and maximum/active (closed symbols) heart rates is quite similar, the scope for heart rate is relatively well maintained across a broad temperature range. Points connected by lines are from the same study. The broken line (– –) connecting one set of open symbols is to indicate that this group of fish may have had unusually high routine heart rates. (Data sources for rainbow trout: △, ▲, Wood *et al.*, 1979; □, ■, Kiceniuk & Jones, 1977; ▽, Henry & Houston, 1984; ◇, Taylor *et al.*, 1993; ⋆ Farrell *et al.*, 1996. For sockeye salmon: ○, ●, Davis, 1968.)

result, the scope for increasing heart rate with swimming is reasonably well maintained at warm temperatures in properly acclimated salmonids (Fig. 5) and other fish (Kolok *et al.*, 1993; Kolok & Farrell, 1994).

It is interesting that eurythermal fish, as compared with salmonids, have lower heart rates at equivalent temperatures (compare Figs. 4*b* and 5; e.g. Seibert, 1979; Yamamitsu & Itazawa, 1990; Morita & Tsukuda, 1994). Nonetheless, maximum heart rate in goldfish still approaches 120 bpm, but only at temperatures in excess of the salmonid UILT (Fig. 4*b*). Likewise, the spangled perch, which can survive temperatures up to 44 °C, has a heart rate of around 100 bpm at 35 °C (Gehrke & Fielder, 1988). While the lower heart rates in eurythermic fish could be related to a lower activity level, it is equally plausible that a lower heart rate allows these more thermally tolerant fish to live at higher water temperatures before heart rate reaches the upper limit of 120 bpm. This would mean that there are species-specific 'adjustments' to the pacemaker rate. Thus, either at the species level or through an acclimatory resetting of the pacemaker to a lower rate, fish exposed to warm temperatures are better able to preserve their scope for (and control over) heart rate. If, for example, there was no acclimatory change in heart rate, routine heart rate would be that much closer to the upper frequency limit and scope would be reduced.

Whether or not other cardiovascular responses compensate for the decrease in heart rate with warm-acclimation is not clear at present. It would be interesting to test the hypothesis that the degree of acclimatory change in heart rate is offset by acclimatory changes in [Hb]. The generally larger increase in [Hb] observed in eurythermal fish with warm temperature acclimation (Houston, 1979) might then be accompanied by an equally large decrease in heart rate compared with stenothermal fish.

What is apparent from the above is that, despite the upper frequency limit for heart rate, species differences and acclimatory changes in heart rate are such that they ensure a reasonable scope for heart rate. Furthermore, it certainly is not the absence of an increase in heart rate that limits $\dot{Q}_{max}$ around the preferred temperature. In fact, the opposite is true. $\dot{Q}_{max}$ does not keep pace with the increases in heart rate. The next section addresses why this occurs.

Temperature, stroke volume and cardiac contractility: a qualitative explanation for cardiac failure at high temperature?

In view of the above, it is clear that maximum stroke volume must decrease if $\dot{Q}_{max}$ has a plateau above the preferred temperature. In fact, quantitatively the decrease in maximum stroke volume must be equal to the increase in heart rate. Decreases in maximum stroke volume at warmer temperatures are reported for various fish species *in vivo* and *in vitro* (Fig. 6a: Davis, 1968; Graham & Farrell, 1989; Yamamitsu & Itazawa, 1990; Kolok et al., 1993; Kolok & Farrell, 1994).

In addition to the decrease in maximum stroke volume at high temperatures, Davis (1968) observed a decrease in maximum blood pressure for swimming sockeye salmon. He concluded that maximum cardiac power output decreased above the preferred temperature. This conclusion is supported by a recent *in vitro* study of maximum cardiac performance *in situ* in rainbow trout (Farrell et al., 1996). The maximum stroke volume (Fig. 6a), the maximum pressure generated by the heart

(Fig. 6*b*), and the maximum cardiac power output (Fig. 7) all decreased significantly at 22 °C compared with 15 °C. Thus, in both *in vivo* and *in vitro* studies the relationship between maximum cardiac power output and temperature can be described by an inverted U-shaped curve. As noted above, Brett (1971) found a similar inverted U-shaped curve for maximum prolonged swimming performance in sockeye salmon. These types of agreement are important because they suggest a correlation between the peak performance of the heart *in vivo* and *in vitro* while in isolation from other variables which integrate cardiac performance *in vivo*. Furthermore, additional studies with isolated cardiac tissues suggest that this peak in cardiac performance as a function of temperature occurs because cardiac contractions get weaker when heart rate increases. However, before describing these studies it is important to explain briefly how cardiac contractility is related to stroke volume and arterial blood pressure.

Contractility is the intrinsic property of cardiac muscle which describes the vigour of contraction. Unlike skeletal muscle, every myocardial fibre is activated with each heart beat and the strength of contraction is modulated by alterations in contractility. Cardiac contractility is often measured as the peak isometric force developed by isolated muscle strips. Increased temperature increases both the force generated by attached actinomyosin cross-bridges and the calcium sensitivity of the cardiac muscle (Driedzic & Gesser, 1994). Therefore, isometric force developed by fish hearts should increase with temperature. In contrast to this expectation, spontaneously beating trout and flounder atrial strips show a decrease in maximum isometric tension with increasing temperature (Ask, 1983). Similar findings exist for spontaneously beating carp ventricular muscle (Matikainen & Vornanen, 1992).

Stroke volume is primarily set by the size of the ventricle, by the degree of ventricular filling, and by the force of ventricular contraction which affects the degree of cardiac emptying. Thus, any decrease in contractility with temperature will produce weaker ventricular contractions and poorer ventricular emptying. Such a linkage might explain the observed decrease in maximum stroke volume at high temperatures.

Cardiac contractility is also an important determinant of arterial blood pressure *in vivo*. Therefore, a decrease in cardiac contractility at higher temperatures should adversely affect maximum arterial blood pressures. As shown in Fig. 6*b*, the level to which arterial blood pressure can increase *in vivo* either with swimming activity or after adrenaline injection is reduced at temperatures around 20 °C in salmonids. This

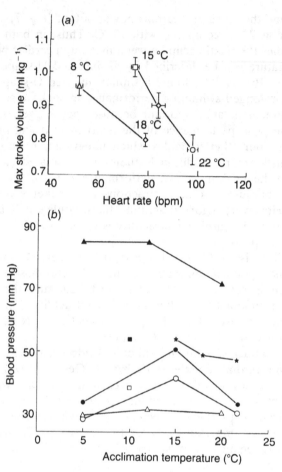

Fig. 6. Effect of temperature on two aspects of cardiac performance (maximum stroke volume, maximum arterial blood pressure) that are directly affected by cardiac contractility. (a) Data from *in situ* perfused heart studies with rainbow trout illustrate that as the heart spontaneously beats faster with increasing temperature, maximum stroke volume decreases significantly. (From Keen & Farrell, 1994; Farrell *et al.*, 1996.) (b) Maximum or active arterial blood pressures in salmonids (closed symbols) decrease at temperatures above the preferred temperature. While this response could be due to decreased vascular resistance, studies with perfused hearts (asterisks) show a similar fall-off in maximum pressure. Routine arterial blood pressures are indicated by same-shaped open symbols. (Data sources: ○, ●, Davis, 1968; □, ■, Kiceniuk & Jones, 1977; △, ▲, Wood *et al.*, 1979; ⋆, Farrell *et al.*, 1996.)

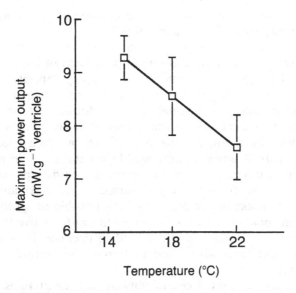

Fig. 7. Effect of temperature on maximum cardiac power output of perfused rainbow trout hearts. Maximum power output decreases at temperatures above 15 °C because of a plateau in maximum cardiac output and a decrease in the maximum pressure that can be developed. (Data adapted from Farrell *et al.*, 1996.)

decrease in the *in vivo* pressure generating ability of the salmonid heart above the preferred temperature could be explained by a decrease in cardiac contractility.

The most plausible explanation for the decrease in cardiac contractility with increasing temperature is that cardiac contractions get weaker with the associated increase in heart rate. It is well known that the interval between cardiac contractions is an important intrinsic determinant of the force of contraction. In fact, isometric force of fish cardiac tissues decreases with increased pacing rate, a response termed a negative force–frequency relationship (Driedzic & Gesser, 1985). Furthermore, Vornanen (1989) suggests that the negative effect of temperature on the contractility of the carp heart is closely correlated with a shortening of the active state. In fact, when carp ventricular strips are paced at a constant frequency, isometric force increased at warmer temperatures, the opposite of what happens when the rate of spontaneous beating increases with warmer temperatures (Matikainen & Vornanen, 1992). (Note: Maximum stroke volume of rainbow trout hearts also decreases at high heart rates without any temperature increase; Farrell, Small & Graham, 1989.)

Matikainen and Vornanen (1992) nicely illustrate these simultaneous negative inotropic and positive chronotropic effects of increasing temperature by deriving a 'maximum tissue pumping capacity' term for cardiac muscle strips (that is, the product of heart rate and maximum isometric force; $g.mg^{-1}$ tissue.min^{-1}). They found that an inverted U-shaped curve describes the 'tissue pumping capacity' as a function of temperature. Furthermore, the peak tissue pumping capacity for carp hearts occurs at a temperature well below the UILT in much the same manner as described above for maximum cardiac power output in rainbow trout, and for swimming performance and maximum cardiac power output in sockeye salmon. Thus, it is possible to speculate that the increase in spontaneous beat frequency (a shorter active state) with warmer temperatures causes a corresponding decrease in contractility which, 'at temperatures above the preferred temperature', leads to cardiac failure.

If, as it seems, decreased contractility at high heart rates leads to the observed decrease in cardiac performance at warmer temperatures, then some aspect of excitation–contraction (E–C) coupling is probably rate-limiting. However, which aspect of E–C coupling is limiting force development is open for speculation. One possibility is that the rate of diffusion of extracellular calcium either into or out of the fish myocyte could limit the force of contraction at high heart rates.

Most of the calcium that binds to myofilaments to initiate cardiac contraction in fish is derived from an extracellular source (see review by Tibbits, Moyes & Hove-Madsen, 1992). This idea is supported by the insensitivity of fish hearts to ryanodine, a known inhibitor of calcium release from the sarcoplasmic reticulum (Driedzic & Gesser, 1988; Keen et al., 1994). With increasing temperature, the time for calcium handling becomes shortened because the time taken to develop peak tension and the time required for 50% relaxation can both decrease significantly. Consequently, either calcium must be handled at a faster rate to maintain contractility, or calcium availability is reduced and contractility decreases, even though the myofibrillar calcium sensitivity is known to be greater in rainbow trout than in mammals (Churcott et al., 1994).

In contrast to the situation in fish, the principal source of activator calcium for mammalian cardiac muscle is the calcium that is released from the sarcoplasmic reticulum (SR). Consequently, calcium handling is faster because this intracellular (vs extracellular in fish) source and sink for activator calcium is considerably closer to the myofibrils.

Furthermore, mammalian cardiac muscles often have positive force–frequency relationships at physiological frequencies and heart rates are typically much higher than in fishes. Similarly, the skipjack tuna heart is sensitive to ryanodine under physiologically relevant conditions and has a positive force–frequency relationship in the frequency range in which salmonids have a negative relationship (Keen *et al.*, 1992). Thus, an involvement of the SR-derived calcium in E–C coupling appears to be beneficial in terms of force development at high contraction frequencies.

When the availability of extracellular calcium is experimentally increased in fish cardiac tissue (for example, by increasing extracellular calcium or adrenaline), the result is a shift in the force–frequency relationship so that force at a given frequency is higher. However, the negative force–frequency effect is not abolished (Driedzic & Gesser, 1985; Keen *et al.*, 1994). In addition, warm-acclimation also decreases the inotropic sensitivity of the rainbow trout heart to adrenaline, in part due to a decrease in the number of sarcolemmal adrenoceptors compared with cold-acclimated trout (Keen *et al.*, 1993). Even so, it is more likely that calcium removal from the fish cardiac cell during relaxation limits contractility at high heart rates. In fish, calcium removal from the cytoplasm is primarily dependent on the sarcolemmal $Na^+–Ca^{2+}$ exchanger (Tibbits *et al.*, 1992). Interestingly, the activity of the rainbow trout $Na^+–Ca^{2+}$ exchanger is relatively insensitive to temperature, unlike the situation in mammals (Tibbits, Philipson & Kashihara, 1992). Warm-acclimation does result in an increase in the maximum rate of contraction and a greater ability to maintain isometric tension at higher heart rates (Bailey & Driedzic, 1990; Bailey, Sephton & Driedzic, 1991), but how these changes relate to calcium handling requires further study.

Summary

From the above account it is reasonable to conclude that the salmonid cardiovascular system does not fare well at warm temperatures. A significant haematological response to elevated temperature is essentially absent (Fig. 8). Maximum cardiac performance begins to plateau at the preferred temperature, that is, a temperature 7–10 °C below the UILT. It is suggested that although heart rate increases with temperature, the high heart rate creates a problem with calcium handling during E–C coupling, thereby decreasing the force of contraction at the tissue level, and decreasing maximum stroke volume and maximum arterial pressure at the organ level. As a result, 'tissue pumping

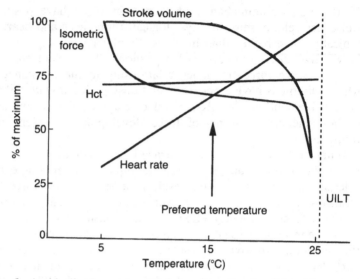

Fig. 8. An idealized summary of the cardiac changes with temperature that lead to a peak in the maximum cardiac performance at the fish's preferred temperature. Heart rate increases with temperature. However, above the preferred temperature maximum stroke volume decreases to such an extent with the increased heart rate that maximum cardiac output plateaus. Underlying the decrease in maximum stroke volume is probably a negative force–frequency effect that causes maximum isometric tension to decrease with the increases in heart rate at warmer temperatures.

capacity' and maximum cardiac power output have similar inverted U-shaped curves as a function of temperature. As noted above, Brett (1971) saw similar relationships between temperature and both maximum metabolic performance and swimming performance in sockeye salmon. He concluded that oxygen availability was rate-limiting. The evidence presented here suggests that a decline in cardiac performance is more likely the primary limitation.

In view of this conclusion, it is not unreasonable to suggest that the failure of sockeye salmon to negotiate Hells Gate in 1994 was related to insufficient cardiac scope at elevated water temperatures. An extension of this line of thought is that one of the important survival factors for salmonids in a global warming situation may be a limitation on cardiac contractility, in particular, and maximum cardiac performance, in general. Using this as a given, it is now possible to explore the question: What cardiovascular features might undergo natural selection if salmonids were to survive the unsettled, pulse changes in temperatures during a gradual trend towards an increased mean global temperature?

A hypothetical scenario for salmonids surviving global warming

Three cardiovascular adaptations are suggested. First, increased ventricular mass would be a simple and effective way to improve force development and solve the problem of reduced cardiac contractility. A larger ventricular mass would necessitate an increased role for the coronary circulation, but this is advantageous because it would reduce the reliance of the fish heart on $P_{v_{O_2}}$. Second, a greater involvement of calcium released from the SR during E–C coupling might reduce the negative effect of increased heart rate on isometric force and, in addition, allow heart rates greater than 120 bpm. Third, routine Hct could be higher, or at least there might be a larger store of red blood cells, capable of being released under a high oxygen demand situation. The advantages of using a larger store are that the routine cost of maintaining a venous O_2 reserve, which is already high, is not increased.

Although these hypothetical cardiovascular features reflect potential 'solutions' to problems faced by modern-day salmonids, the cardiovascular features of skipjack tuna provide empirical evidence to suggest that these are realistic ways in which the fish's cardiovascular system may adapt to warmer water temperatures. Skipjack tuna have a ventricle four times larger than the rainbow trout, have a highly developed coronary circulation, utilize SR-calcium in E–C coupling, have heart rates higher than 120 bpm, and can maintain stroke volume at heart rates and blood pressures that exceed those in rainbow trout. In view of this, perhaps we should not be surprised that a related scombrid, the mackerel, is out-competing juvenile salmon when the offshore waters of British Columbia periodically become uncharacteristically warm!

Acknowledgements

Discussions with Dr David Randall, Dr Kurt Gamperl and Dr Arthur Houston during the preparation of this manuscript were much appreciated. The author's research is support by grants from the Natural Sciences and Engineering Research Council of Canada.

References

Ask, J. A. (1983). Comparative aspects of adrenergic receptors in the hearts of lower vertebrates. *Comparative Biochemistry and Physiology*, **76A**, 543–52.

Bailey, J. R. & Driedzic, W. R. (1990). Enhanced maximum frequency and force development of fish hearts following temperature acclimation. *Journal of Experimental Biology*, **149**, 239–54.

Bailey, J. R., Sephton, D. & Driedzic, W. R. (1991). Impact of an acute temperature change on performance and metabolism of pickerel (*Esox niger*) and eel (*Anguilla rostrata*) hearts. *Physiological Zoology*, **64**, 697–716.

Barron, M. G., Tarr, B. D. & Hayton, W. L. (1987). Temperature-dependence of cardiac output and regional blood flow in rainbow trout, *Salmo gairdneri* Richardson. *Journal of Fish Biology*, **31**, 735–44.

Black, E. C. (1953). Upper lethal temperatures of some British Columbian freshwater fishes. *Journal of the Fisheries Research Board of Canada*, **10**, 196–210.

Brett, J. R. (1952). Temperature tolerance in young Pacific salmon, genus *Oncorhynchus*. *Journal of the Fisheries Research Board of Canada*, **9**, 265–323.

Brett, J. R. (1971). Energetic responses of salmon to temperature. A study of some thermal relations in the physiology and freshwater ecology of sockeye salmon. *American Zoologist*, **11**, 99–113.

Bushnell, P. G., Steffensen, J. F. & Johansen, K. (1984). Oxygen consumption and swimming performance in hypoxia-acclimated rainbow trout, *Salmo gairdneri*. *Journal of Experimental Biology*, **113**, 225–35.

Cameron, J. N. & Davis, J. C. (1970). Gas exchange in rainbow trout *Salmo gairdneri* with varying blood oxygen capacity. *Journal of the Fisheries Research Board of Canada*, **27**, 1069–85.

Cech, J. J., Rowell, D. M. & Glasgow, J. S. (1976). Cardiovascular responses of the winter flounder, *Pseudopleuronectes americanus* to hypoxia. *Comparative Biochemistry and Physiology*, **57A**, 123–5.

Churcott, C. S., Moyes, C. D., Bressler, B. H., Baldwin, K. M. & Tibbits, G. F. (1994). Temperature and pH effects on Ca^{2+} sensitivity of cardiac myofibrils: a comparison of trout with mammals. *American Journal of Physiology*, **267**, R62–70.

Davie, P. S. & Farrell, A. P. (1991). The coronary and luminal circulations of the myocardium of fishes. *Canadian Journal of Zoology* **69**, 1993–2001.

Davis, J. C. (1968). The influence of temperature and activity on certain cardiovascular and respiratory parameters in adult sockeye salmon. MSc thesis, University of British Columbia.

DeWilde, M. A. & Houston, A. H. (1967). Haematological aspects of the thermoacclimatory process in the rainbow trout. *Journal of the Fisheries Research Board of Canada*, **24**, 2267–81.

Driedzic, W. R. & Gesser, H. (1985). Ca^{2+} protection from the negative inotropic effect of contraction frequency on teleost hearts. *Journal of Comparative Physiology B*, **156**, 135–42.

Driedzic, W. R., & Gesser, H. (1988). Differences in force–frequency relationships and calcium dependency between elasmobranch and teleost hearts. *Journal of Experimental Biology*, **140**, 227–42.

Driedzic, W. R. & Gesser, H. (1994). Energy metabolism and contractility in ecotothermic vertebrate hearts: hypoxia, acidosis, and low temperature. *Physiological Reviews*, **74**, 221–58.

Eddy, F. B. (1971). Blood gas relationships in the rainbow trout, *Salmo gairdneri*. *Journal of Experimental Biology*, **55**, 695–711.

Farrell, A. P. (1991). From hagfish to tuna – a perspective on cardiac function. *Physiological Zoology*, **64**, 1137–64.

Farrell, A. P., Gamperl, A. K., Hicks, J. M. T., Shiels, H. A. & Jain, K. E. (1996). Maximum cardiac performance of rainbow trout (*Oncorhynchus mykiss*) at temperatures approaching their upper lethal limit. *Journal of Experimental Biology*, **199**, 663–72.

Farrell, A. P., Johansen, J. A. & Suarez, R. K. (1991). Effects of exercise-training on cardiac performance and muscle enzymes in rainbow trout, *Oncorhynchus mykiss*. *Fish Physiology and Biochemistry*, **9**, 303–12.

Farrell, A. P. & Jones, D. R. (1992). The heart. In *Fish Physiology*, Vol. 12A. *Cardiovascular Systems*, ed. W. S. Hoar, D. J. Randall & A. P. Farrell, pp. 1–88. New York: Academic Press.

Farrell, A. P., Small, S. & Graham, M. S. (1989). Effect of heart rate and hypoxia on the performance of a perfused trout heart. *Canadian Journal of Zoology*, **67**, 274–80.

Farrell, A. P. & Steffensen, J. F. (1987). An analysis of the energetic cost of the branchial and cardiac pumps during sustained swimming in trout. *Fish Physiology and Biochemistry*, **4**, 73–9.

Gallaugher, P., Thorarensen, H. & Farrell, A. P. (1995). The role of hematocrit in oxygen transport and swimming in rainbow trout *Oncorhynchus mykiss*. *Respiratory Physiology*, **102**, 279–92.

Gehrke, P. C. & Fielder, D. R. (1988). Effects of temperature and dissolved oxygen on heart rate, ventilation rate and oxygen consumption of spangled perch, *Leipotherapon unicolor* (Gunther 1985) (Percoidei, Teraponidae). *Journal of Comparative Physiology B*, **157**, 771–82.

Graham, M. S. & Farrell, A. P. (1989). The effect of temperature acclimation and adrenaline on the performance of a perfused trout heart. *Physiological Zoology*, **62**, 38–61.

Henry, J. A. C. & Houston, A. H. (1984). Absence of respiratory acclimation to diurnally-cycling temperature conditions in rainbow trout. *Comparative Biochemistry and Physiology*, **77**A, 727–34.

Houston, A. H. (1979). Components of the hematological response of fishes to environmental temperature change: a review. In *Environmental Physiology of Fishes*, ed. M.A. Ali, pp. 241–98. New York: Plenum Press.

Houston, A. H. & Schrapp, M. P. (1994). Thermoacclimatory hematological response: Have we been using appropriate conditions and assessment methods? *Canadian Journal of Zoology*, **72**, 1238–42.

Jones, D. R. (1971). The effect of hypoxia and anaemia on the swimming performance of rainbow trout. *Journal of Experimental Biology*, **55**, 541–51.

Keen, J. E. & Farrell, A. P. (1994). Maximum prolonged swimming speed and maximum cardiac performance of rainbow trout, *Oncorhynchus mykiss*, acclimated to two different water temperatures. *Comparative Biochemistry and Physiology*, **108**A, 287–95.

Keen, J. E., Farrell, A. P., Tibbits, G. F. & Brill, R. W. (1992). Cardiac physiology in tunas. II. Effects of ryanodine, calcium, and adrenaline on force–frequency relationships in atrial strips from skipjack tuna, *Katsuwonus pelamis*. *Canadian Journal of Zoology*, **70**, 1211–17.

Keen, J. E., Vianzon, D.-M., Farrell, A. P. & Tibbitts, G. F. (1993). Thermal acclimation alters both adrenergic sensitivity and adrenoceptor density in cardiac tissue of rainbow trout. *Journal of Experimental Biology*, **181**, 27–41.

Keen, J. E., Vianzon, D.-M., Farrell, A. P. & Tibbits, G. F. (1994). Effect of acute temperature change and temperature acclimation on excitation–contraction coupling in trout myocardium. *Journal of Comparative Physiology B*, **164**, 438–43.

Kiceniuk, J. W. & Jones, D. R. (1977). The oxygen transport system in trout *Salmo gairdneri* during sustained exercise. *Journal of Experimental Biology*, **69**, 247–60.

Kolok, A. S. & Farrell, A. P. (1994). Individual variation in the swimming performance and cardiac performance of northern squawfish, *Ptytocheilus oregonensis*. *Physiological Zoology*, **67**, 706–22.

Kolok, A. S., Spooner, M. & Farrell, A. P. (1993). The effect of exercise on the cardiac output and blood flow distribution of the largescale sucker, *Catastomus macrocheilus*. *Journal of Experimental Biology*, **183**, 301–21.

Matikainen, N. & Vornanen, M. (1992). Effect of season and temperature acclimation on the function of crucian carp (*Carassius carassius*) heart. *Journal of Experimental Biology*, **167**, 203–20.

Morita, A. & Tsukuda, H. (1994). The effect of thermal acclimation on the electrocardiogram of goldfish. *Journal of Thermal Biology*, **19**, 343–348.

Neumann, P., Holeton, G. F. & Heisler, N. (1983). Cardiac output and regional blood flow in gills and muscles after strenuous exercise in rainbow trout *Salmo gairdneri*. *Journal of Experimental Biology*, **105**: 1–14.

Riggs, A. (1976). Factors in the evolution of hemoglobin function. *Federation Proceedings*, **35**, 2115–18.

Seibert, H. (1979). Thermal adaptation of heart rate and its parasympathetic control in the European eel *Anguilla anguilla* (L.). *Comparative Biochemistry and Physiology*, **64**C, 275–8.

Stevens, E. D. & D. J. Randall (1967). Changes in blood pressure, heart rate, and breathing rate during moderate swimming activity in rainbow trout. *Journal of Experimental Biology*, **46**, 307–16.

Taylor, S. E., Egginton, S. & Taylor, E. W. (1993). Respiratory and cardiovascular responses in rainbow trout (*Oncorhynchus mykiss*) to aerobic exercise over a range of acclimation temperatures. *Journal of Physiology (London)*, **459**, 19P.

Thorarensen, H., Gallaugher, P. & Farrell, A. P. (1996). Cardiac output in swimming rainbow trout, *Oncorhynchus mykiss* acclimated to seawater. *Physiological Zoology*, **69**, 139–53.

Tibbits, G. F., Moyes, C. D. & Hove-Madsen, L. (1992). Excitation–contraction coupling in the teleost heart. In *Fish Physiology*, Vol. 12A, ed. W. S. Hoar, D. J. Randall & A. P. Farrell, pp. 267–304. San Diego: Academic Press.

Tibbits, G. F., Philipson, K. D. & Kashihara, H. (1992). Characterization of myocardial Na^+–Ca^{2+} exchange in rainbow trout. *American Journal of Physiology*, **262**, C411–17.

Tsukuda, H., Liu, B. & Injii, K. I. (1985). Pulsation rate and oxygen consumption of isolated hearts of the goldfish, *Carassius auratus*, acclimated to different temperatures. *Comparative Biochemistry and Physiology*, **82**A, 281–3.

Tun, N. & Houston, A. H. (1986). Temperature, oxygen, photoperiod, and the hemoglobin system of the rainbow trout, *Salmo gairdneri*. *Canadian Journal of Zoology*, **64**, 1883–8.

Vornanen, M. (1989). Regulation of contractility of the fish (*Carassius carassius* L.) heart ventricle. *Comparative Biochemistry and Physiology*, **94**C, 477–83.

Weber, R. E. & Jensen, F. B. (1988). Functional adaptations in hemoglobins from ectothermic vertebrates. *Annual Review of Physiology*, **50**, 161–79.

Wood, C. M., Pieprzak, P. & Trott, J. N. (1979). The influence of temperature and anaemia on the adrenergic and cholinergic mechanisms controlling heart rate in the rainbow trout. *Canadian Journal of Zoology*, **57**, 2440–7.

Wood, C. M. & Shelton, G. (1980). Cardiovascular dynamics and adrenergic responses of the rainbow trout *in vivo*. *Journal of Experimental Biology*, **87**, 247–70.

Yamimitsu, S. & Itazawa, Y. (1990). Effects of preload, afterload and temperature on the cardiac function and ECG in the isolated perfused heart of carp. *Nippon Suisan Gakkaishi*, **56**, 229–37.

GLEN Van Der KRAAK and
NED W. PANKHURST

Temperature effects on the reproductive performance of fish

Introduction

Scientists are increasingly being called upon to predict the outcomes of global temperature changes on fish populations. Both reproduction and early development in fish are particularly sensitive to temperature perturbations. Numerous *in vitro* and *in vivo* studies have shown that reproductive endocrine homeostasis in fish (Fig. 1) is responsive to changes in temperature. This includes alterations in the secretion and actions of hormones associated with all components of the hypothalamic–pituitary–gonadal axis which controls reproductive processes. By comparison, far fewer studies have considered the longer term consequences of altered temperature profiles on the reproductive cycle or larval development. Even fewer studies have considered the long-term ecological consequences of multigenerational exposures to elevated thermal regimes. Given the existing data and the need for a rapid response to questions on the outcomes of temperature change, our best informed judgement will have to be based, in large part, on information from short-term assays. This leads to the question of whether current methods of assessing the effects of temperature on the molecular and cellular events mediating reproductive processes can be used to predict effects at whole animal and population levels (Fig. 2).

This chapter reviews our current understanding of the effects of elevated temperature on reproductive performance in fish. The initial focus is on endocrine homeostasis and the effect of temperature on hormone biosynthesis, metabolism and actions. The second part examines the effect of temperature on the reproductive cycle and subsequent embryonic development of the larvae. The third part reviews the

Society for Experimental Biology Seminar Series 61: *Global Warming: Implications for freshwater and marine fish*, ed. C. M. Wood & D. G. McDonald. © Cambridge University Press 1996, pp. 159–176.

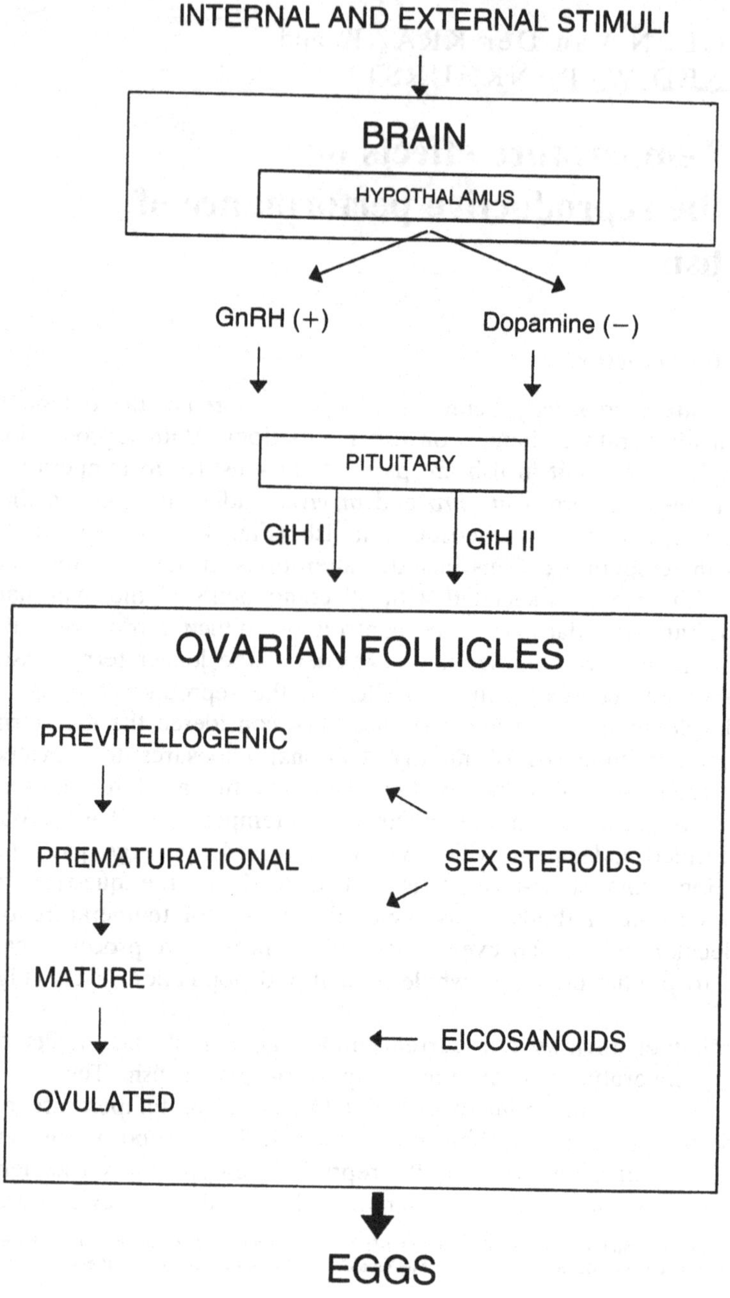

Fig. 1. Simplified schematic diagram of the hypothalamic–pituitary–ovarian axis in female teleosts.

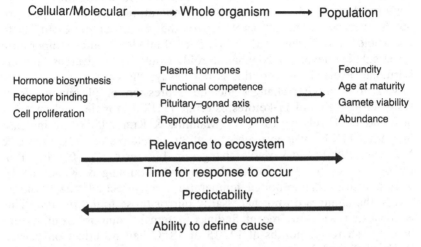

Fig. 2. Description of cellular and molecular markers through to population measures that describe the reproductive competence of fish. The diagram illustrates the inverse relationship between pre-dictability and ecological relevance in describing reproductive measures at the cellular/molecular level that are rapidly measured, whereas longer term population level responses have high ecological relevance but take long time periods to establish and monitor.

limited information on multigenerational exposure to elevated thermal regimes.

Endocrine homeostasis

Hormone biosynthesis and metabolism

Temperature has actions on hormone synthesis, secretion and metab-olism. Consequently, higher than normal temperatures can substantially modify endocrine profiles.

In vitro studies using ovarian or testicular tissue from various species (male rainbow trout, *Oncorhynchus mykiss*: Kime, 1979; Manning & Kime, 1985; male goldfish, *Carassius auratus*: Kime, 1980; female goldfish: Manning & Kime, 1984; male carp, *Cyprinus carpio*: Kime & Manning, 1986; male tilapia, *Oreochromis mossambicus*: Kime & Hyder, 1983) all show that there is an optimum temperature for production of gonadal steroids, above which there is increased *in situ* conversion of free steroid to the glucuronated form. This has generally been interpreted as a mechanism whereby the biologically active forms of the hormone (free steroids) are converted to biologically inactive compounds (steroid glucuronides) at temperatures that are inappropriate for reproduction. Extrapolation of the results of these studies to the whole animal suggests that the effect of an elevation of temperature on circulating levels of gonadal steroids would be to decrease plasma hormone levels. However, this is not always the case.

Measured *in vivo* changes include depression of plasma levels of testosterone (T) and 11-ketotestosterone (11KT) in male rainbow trout at 6 and 17 °C relative to 12 °C (Manning & Kime, 1985), an increase in plasma 11KT levels in male carp at 30 °C relative to 23 °C (Kime & Manning, 1986), and variable changes in 17β-oestradiol (E_2) and T in relation to temperature in female goldfish (Manning & Kime, 1984). This indicates that although temperature can affect plasma hormone levels, the changes do not necessarily mirror those found *in vitro*. This is consistent with our recent findings in female rainbow trout, where holding fish for 2 months at 12, 15 or 18 °C had no effect on plasma levels of T and E_2, whereas the *in vitro* capacity of ovarian follicles to synthesize T and E_2 was impaired at 18 °C (Pankhurst *et al.*, 1996).

All the above studies deal with species which are normally exposed to temperature fluctuation, and might be expected to show some adaptive plasticity in response to moderate shifts in thermal range. Little is known about the effects of temperature change on species that are not normally exposed to any variation in environmental temperature. These include deep-sea forms where the thermal environment is often very stable. For example, orange roughy (*Hoplostethus atlanticus*) is a cosmopolitan deep-slope species with an adult depth distribution of 800–1200 m and in New Zealand waters at least, year round exposure to a temperature of around 5 °C (Pankhurst, McMillan & Tracey, 1987). Similar environmental thermal stability is experienced by near continental Antarctic species, where the sea temperature remains close to −1.8 °C throughout the year (Littlepage, 1965). Species from thermally stable environments such as these might be expected to display little ability to maintain endocrine homeostasis in the face of thermal stress. The limited data available support this. Ovarian follicles of orange roughy incubated at 5 °C produced higher amounts of 17α-hydroxyprogesterone (17P) and T in response to treatment with human chorionic gonadotropin (hCG), salmon pituitary extract or partially purified salmon gonadotropin (GtH), whereas follicles incubated at 14 and 20 °C were unresponsive to GtH preparations (N. Pankhurst, unpublished data).

Temperature also affects the rates of hormone clearance. Intra-arterial injection of female goldfish with ^{125}I-carpGtH (cGtH) showed that metabolic clearance was more rapid at high than at low temperature in sexually regressed but not sexually mature fish. In contrast, the rate of uptake of ^{125}I-cGtH into ovarian tissue was more rapid in mature than regressed fish at high temperature (Cook & Peter, 1980a, b). The disparity between the effects of temperature on hormone uptake and on hormone clearance in fish of different sexual stages emphasizes that the impact of temperature change may vary with the stage of the reproductive cycle at which it is applied.

Hormone action

Temperature apparently modulates hormone action at all levels of the reproductive endocrine cascade. *In vitro*, higher temperatures potentiate the effects of (i) gonadotropin releasing hormone (GnRH) on GtH secretion from the pituitary (J. P. Chang, personal communication), (ii) GtH on ovarian and testicular steroid production (McMaster *et al.*, 1995; G. Van Der Kraak, unpublished data), and (iii) 17β-estradiol on vitellogenin production by isolated goldfish liver cells (Z. Yao and G. Van Der Kraak, unpublished data). Ovarian prostaglandin biosynthesis is also enhanced at higher incubation temperatures (F. Mercure and G. Van Der Kraak, unpublished data). Temperature can also affect hormone action at the level of the oocyte. The maturation inducing steroid (MIS) 17α,20β-dihydroxy-4-pregnen-3-one (17,20β-P) induced final oocyte maturation (FOM) in oocytes incubated *in vitro* at 20 and 25 °C but not at 5 or 10 °C, and was only marginally effective at 15 °C (Epler, Bieiarz & Marosz, 1985). In contrast, exogenous Prostaglandin $F_{2\alpha}$ was able to stimulate ovulation at all test temperatures. These results suggest that the effect of temperature on hormone action may be hormone and tissue specific. The precise mechanism(s) by which temperature affects hormone actions is unknown but effects on receptor–ligand interaction, signal transduction or hormone biosynthesis/release are all possible.

In any case, temperature has an effect on hormone action. For example, increasing the temperature from 12 to 20 °C potentiated the effect of GnRH treatment on GtH release in female goldfish (Sokolowska *et al.*, 1985). Increasing the temperature reduced the response latency between hCG treatment and ovulation in goldfish

(<10 h at 26 °C vs 30–40 h at 12 °C; Stacey, Cook & Peter, 1979). The likely effect of temperature here was to promote the actions of MIS and prostaglandins, which mediate final oocyte maturation and ovulation, respectively.

If, as the limited data suggest, one of the main effects of temperature elevation is to increase the rate at which endocrine responses occur, then temperature may also affect rate-dependent processes such as hormone cycling. It is clear from studies on both fish and other vertebrates that tissue responsiveness to hormones can be maintained only in the presence of fluctuating hormone levels. This is supported by the presence of regular daily cycles of plasma hormone levels in most of the species where the phenomenon has been investigated (reviewed by Hontela, 1984). Modified thermal regimes can interfere with these hormone cycles. Female goldfish shifted from 12 to 20 °C had high and unchanging plasma levels of GtH, a regime of 20 °C by day and 12 °C at night induced a near normal daily cycle of GtH, and fish exposed to 12 °C during the day and 20 °C at night had low and unchanging plasma GtH levels (Hontela & Peter, 1983). Fish which did not show daily cycles of plasma GtH had high levels of ovarian atresia (breakdown and absorption of vitellogenic ovarian follicles), giving further support to the suggestion that cyclic changes in plasma GtH are necessary for normal ovarian maintenance and growth.

Reproductive cycles

Reproductive endocrine homeostasis has as its endpoint the production of viable gametes via the sequential regulation of gametogenesis, maturation, ovulation and spermiation, and finally spawning. Given the temperature-sensitivity of the endocrine processes regulating these events, it is not surprising to find that the events themselves are highly temperature-sensitive.

Gametogenesis and gamete maturation

Gametogenesis begins with the differentiation of gonadal germinal tissue into oogonia or spermatogonia which will subsequently undergo mitotic division to form primary oocytes and spermatocytes in the ovary and testis, respectively. While sexual differentiation is genetically controlled, temperature is one of a host of environmental factors that can also play a role. Thermolabile sex determination is common among Atherinids (silversides), where higher temperatures favour the formation of males and lower temperatures favour the formation of females (Strüssmann & Patino, 1995). As another example, tilapia (*Oreochromis niloticus*) held at 19–20 °C over the period of gonadal differentiation exhibit a 1:1 sex ratio, whereas maintenance at 30–36 °C generates a predominance of males (Baroiller, Clota & Geraz, 1995). It is possible that sex determining mechanisms may be more sensitive to environmental temperature than is generally recognized (Strüssmann & Patino, 1995). There is also circumstantial evidence that sex ratios in species that undergo sex inversion are subject to environmental control. New Zealand snapper (*Pagrus* (= *Chrysophrys*) *auratus*) undergo juvenile sex inversion, with all fish differentiating as females. Later, a proportion of these undergo inversion to males (Francis & Pankhurst, 1988). The adult sex ratio is variable between populations, with females often predominating in populations from more northerly (warmer) areas (Paul & Tarring, 1980), suggesting that the proportion undergoing female to male sex inversion may be temperature-regulated.

Further development of the gonads through gametogenesis may also be influenced by temperature, but the degree to which this occurs varies among species. In salmonids, reproductive development is primarily controlled by photoperiod. The effect of temperature on gametogenesis is thought to be minor provided that the temperature remains within physiological limits (Bye, 1984; Billard, 1985). In contrast, temperature has a much greater effect on gametogenesis in cyprinid fishes. The early stages of gametogenesis require low temperatures, but rising temperatures have a stimulatory effect at later stages of development (Bye, 1984). Temperature also modifies the effects of changing photoperiod in sticklebacks and cyprinids (Lam & Munro, 1987) and some flatfish (Bye, 1987). These modifications may include detrimental effects: for example, unseasonal elevations in temperature result in ovarian atresia in carp (Davies *et al.*, 1986).

There is evidence that the end of spawning and the start of gonadal regression in some spring spawning species is signalled by the rising temperatures of early summer. Gonadal regression in the goby, *Gillichthys mirabilis* (Dodd & Sumpter, 1984), bitterling, *Acheilognathus tabira* (Hanyu *et al.*, 1983), New Zealand snapper (Scott & Pankhurst, 1992), and red gurnard, *Chelidonichthys kumu* (Clearwater & Pankhurst, 1994) occurs when water temperatures rise above some critical value. Collectively these studies show that shifts in environmental temperature modify patterns of gametogenesis. These effects may be on the initiation of gametogenesis or the induction of gonadal regression.

Ovulation and spermiation

Once gonadal growth is complete, the processes of gamete final maturation and ovulation or spermiation must occur before the fish can spawn (the behavioural act of egg and sperm release). Here also, temperature has a potential regulating or modifying role. Most of the information in this regard comes from studies on salmonid fishes. Rainbow trout will ovulate over a temperature range of 9–15 °C given appropriate photoperiod conditions (Scott *et al.*, 1984). FOM and ovulation do not occur at temperatures below 9 °C and although ovulation may occur at temperatures of 15 °C or more, egg quality is usually poor (Billard, 1985). More commonly, rainbow trout held at high temperatures fail to ovulate (Pankhurst *et al.*, 1996). A similar situation has been described for Atlantic salmon (*Salmo salar*), where fish held at 13–14 °C were found to have lower incidence of ovulation than fish held at ambient (10 °C) or lower temperatures (5–8 °C) (Taranger & Hansen, 1993).

The inhibitory effects of elevated temperature appear to act on the processes of FOM and/or ovulation rather than during the preceding period of vitellogenesis. Plasma levels of the gonadal steroids T and E_2 were not different among rainbow trout maintained for several months at temperatures ranging from 12 to 18 °C, nor was there any evidence of gonadal atresia at these temperatures. In contrast, the number of fish subsequently ovulating at 18 °C was very low whereas fish at 12 °C underwent ovulation and produced viable eggs (Pankhurst *et al.*, 1996). High temperature also has inhibitory effects on male salmonids. Rainbow trout held at 18 °C for up to 3 months had lower milt volumes than those held at 10 °C, and a rapid rise from 10 to 18 °C over 2 days led to a decrease in sperm volume within weeks (Billard & Breton, 1977). Temperature shifts also affect ovulation in other species, particularly cyprinids, where rising temperatures along with the presence of spawning substratum stimulate FOM and ovulation (Stacey, 1984).

Spawning

Most fish are external fertilizers and the release of gametes to the environment via the behavioural act of spawning must occur before fertilization can take place. It is not clear what effect temperature change is likely to have on spawning behaviour. However, there is a close association between reproductive behaviour and endocrine state (reviewed in Pankhurst, 1995), and any environmental factor that interferes with normal endocrine function may also disrupt behavioural processes.

Gamete viability

Post-ovulatory egg viability

Following ovulation, gamete viability is dependent on the residence time of the eggs in the oviduct or body cavity before *spawning*. All fish display a window of post-ovulatory egg viability within which spawning must occur if optimum fertility is to be achieved. This window can range from as long as days to weeks in rainbow trout (Billard & Breton, 1977; Springate *et al.*, 1984) and Pacific herring, *Clupea harengus* (Hay, 1986) to as little as a few hours in New Zealand snapper (Scott, Zeldis & Pankhurst, 1993), goldfish (Formacion, Hori & Lam, 1993) and sabalo, *Prochilodus platensis* (Fortuny, Ros & Amutio, 1988). The rate at which post-ovulatory viability declines is temperature dependent. In the loach (*Misgurnus anguillicaudatus*) the period of optimum fertility declined from 6–8 h post-ovulation at 20 °C to 3–4 h at 30 °C (Suzuki, 1975). Similarly, the period of optimum egg viability in rainbow trout (Billard & Breton, 1977) and Pacific herring (Hay, 1986) is shortened at elevated temperatures. The fall in egg viability is thought to be associated with falling intracellular levels of ATP (Boulekbache *et al.*, 1989). At later stages of 'overripening', there is proteolytic breakdown of yolk proteins and loss of small organic molecules or fragments through the egg membranes (Craik & Harvey, 1984). Both processes are likely to be accelerated at high temperature. Collectively, these data indicate that even when oocytes develop and are ovulated normally, exposure of ovulated fish to abnormally high temperature may change the phasing of optimum egg viability relative to the time of spawning.

Embryonic survival

Temperature-compromised gametes, whether by pre- or post-ovulatory events, may still be fertile but fail to undergo normal development and hatching. Fertilized eggs of rainbow trout and Atlantic salmon from females exposed to high temperatures during the last few months of gonad development show low survival to hatching, when incubated at temperatures where high larval survival would normally be expected (Taranger & Hansen, 1993; N. Pankhurst *et al.*, unpublished data). Both studies need to be interpreted with caution, since the eggs were also exposed to varying temperature shock following stripping and transfer to a standard incubation temperature. However, they confirm the prediction that when ovulation and fertilization occur at elevated temperatures, subsequent larval survival will be very low. Similar outcomes have been reported when egg development temperature in the female and subsequent incubation temperature are both elevated above normal. Eggs of Atlantic halibut (*Hippoglossus hippoglossus*) showed an increase in abnormal cleavage patterns and decreased hatching rates when spring water temperatures rose above 8 °C. In contrast, there were no detrimental changes in eggs from females held over the same period at 6 °C (Brown, Bromage & Shields, 1995). Small changes in temperature also have a marked effect on the rate at which fertilized eggs develop. Atlantic salmon eggs incubated at 12 °C had higher mortality of fry and averaged little more than half the weight at hatch of those incubated at 8 or 10 °C (Gunnes, 1979). Development of alevins to the swim-up stage was also abnormal at 12 °C.

The rates of fertilization and survival to hatch of coho salmon (*Oncorhynchus kisutch*) produced from artificial spawning of fish returning to the Fairview hatchery on the American side of Lake Erie were greatly reduced when compared with other stocks including those of similar genetic origin (Flett *et al.*, 1991, 1996). Laboratory cross-fertilization studies showed that the eggs of the Fairview stock were the likely source of the low fertility and this correlated with the presence of overripe eggs and lower levels of testosterone and 17,20β-P but not GtH II in the pre-spawning adults. The underlying basis for the reproductive effects seen in the Fairview stock is not known but did not appear to be related to contaminant burdens of heavy metals, organochlorines or disease aetiologies. One possible explanation is that the Fairview salmon were exposed to temperatures at least 2–4 °C higher than other stocks during the 2 month period encompassing the migration to the spawning grounds. Our studies (Flett *et al.*, 1996) do not conclusively identify temperature as the causative agent in low fertility but illustrate that establishing cause and effect relationships in feral fish populations will be a difficult undertaking.

An empirical model developed by Pauly and Pullin (1988) predicts hatching times of eggs on the basis of their diameter and incubation temperature. The model indicates that temperature (high temperature promotes shorter hatching time) has a four fold greater influence on hatching time than egg size (larger diameter eggs take longer to hatch). If, as the model predicts, exposure of fertilized eggs to elevated temperatures accelerates hatching, then pelagic eggs may hatch in the 'wrong' place. There is evidence that temporal spawning strategies of some nearshore marine species may aid egg survival by maximizing the bulk transport of eggs and hatching larvae away from reef areas, both by removing eggs from areas of high predation intensity, and by placing them in areas suitable for larval feeding (Shapiro, Hensley & Appledoorn, 1988; Robertson, Petersen & Braun, 1990). Accelerated hatching due to elevated temperature may reduce subsequent larval survival if dispersal mechanisms are fine-tuned in terms of the developmental requirements of the eggs and larvae.

A different dispersal problem faces fishes spawning in deep-water. Many deep-water fish spawn buoyant pelagic eggs which undergo development during a slow ascent through the water column (Marshall, 1979). If the rate of egg development is accelerated by increased temperature, then hatching may occur below the shallow productive zones required to support larval feeding, or in a substantially different predation field from normal. For example, based on ascent rates calculated using mean estimates for the density of marine fish eggs and sea water (Robertson, 1978), eggs 2 mm in diameter spawned at 3000 m depth will take 12.8 days to reach the surface waters. Hatching would be predicted to occur after 11.8 days at a mean temperature of 7 °C (at a depth of 225 m), whereas a 2 °C upward shift in mean incubation temperature would shorten the incubation time to 9.4 days, with hatching occurring at a depth of 789 m.

Multigenerational exposure

Few studies have considered the long-term consequences of elevated temperatures on reproductive performance and life/history characteristics in fish. The Biotest Basin, a 1 km^2 reservoir receiving the thermal effluent discharge from the nuclear power plant at Forsmark, Sweden, has provided a unique opportunity to evaluate thermal effects on fish performance (Luksiene & Sandstrom, 1994: Sandstrom, Neuman & Thoresson, 1995). The Biotest Basin has received heated effluent since the early 1980s and water temperatures vary from 4 to 10 °C above ambient. Evaluation of two endemic species, the roach (*Rutilis rutilis*) and the Eurasian perch (*Perca fluviatilis*), revealed marked differences in performance (Table 1). Roach exhibit recruitment failure which has been traced to gonadal malfunctions in the adults, the most serious being oocyte degeneration in more than 50% of the fish (Luksiene & Sandstrom, 1994). Although oocyte development was initiated earlier in the year, these fish exhibited an increased occurrence of asynchronous development later in the year. By comparison, the abundance of perch in the Biotest Basin increased after exposure to the heated water, although the proportion of larger fish has declined. This effect was related to the earlier maturation of the fish coupled with increased adult mortality. Surviving fish often delayed their next spawning by

Table 1. *Summary of reproductive responses in roach and Eurasian perch collected in the Biotest Basin which receives the thermal effluent from the nuclear power plant at Forsmark, Sweden*

Roach (*Rutilus rutilus*)
  Decreased abundance
  Decreased gonadosomatic index
  Accelerated gametogenesis
  Asynchronous ovarian development and increased oocyte
    degeneration

Eurasian perch (*Perca fluviatilis*)
  Increased abundance
  Altered population structure including increased recruitment but with
    increased adult mortality
  Decreased age and size at maturity
  Decreased net reproductive rate

Based on data reported in Luksiene & Sandstrom (1994) and Sandstrom *et al.* (1995).

one or more years with the result that lifetime fecundity was reduced and reproductive performance shifted to younger ages. There was abnormal gonadal development in other species (pike, bream, silver bream) examined in the Biotest Basin, some of which are free to move from the heated areas (Dr O. Sandstrom, personal communication).

In other work, Meffe (1991, 1992) studied a population of eastern mosquitofish (*Gambusia holbrooki*) from a thermally elevated habitat (outflow from a nuclear reactor) in the United States Department of Energy's Savannah River Site. Fish from the heated environment exhibited several distinct life history trait differences compared with fish from a nearby reference location. Thermally exposed fish reproduced at a low rate throughout the winter rather than ceasing reproduction for 6 months, and had a higher reproductive investment (higher clutch sizes and reproductive biomass) but smaller offspring.

Collectively, these studies illustrate the difficulty in predicting the long-term consequences of elevated temperature regimes. Two species inhabiting the same area exhibit vastly different outcomes in terms of life history responses. This work reinforces the need to look at whole animal physiology rather than discrete components of the reproductive axis.

Temperature and stress

Inappropriate temperature regimes or rapid changes in temperature are stressful to fish and may also augment the response to other stressors (reviewed in Barton & Iwama, 1991; Pickering, 1992). The primary effects of stress are to activate the sympathetico-chromaffin tissue, and hypothalamic–pituitary–interrenal pathways, resulting in the release to the bloodstream of catecholamines and corticosteroid hormones, respectively. These hormones collectively enhance metabolic processes that provide a short-term benefit to the animal in terms of responding to the stressor. If, however, the stress is maintained, then detrimental effects often begin to occur as a result of the stress response (Barton & Iwama, 1991). One of these is inhibition of reproductive function, with stress resulting in depression of plasma levels of reproductive hormones, cessation of ovulation, and ovarian atresia (reviewed in Pankhurst, Van Der Kraak & Peter, 1995). The mechanisms whereby stress exerts its effects on reproductive processes are not well understood, but if exposure to elevated temperatures results in sustained activation of the physiological stress response, then one of the probable outcomes will be the inhibition of reproductive function.

Concluding remarks

Temperature affects virtually all aspects of reproduction in fish including gametogenesis and gamete maturation, ovulation/spermiation, spawning and subsequent early development. The endocrine events which mediate reproductive processes are well understood but we have a limited understanding of which endocrine responses best predict whole animal and population level effects. Accordingly, our ability to predict the long-term effects of altered endocrine homeostasis is poor. Perhaps more significantly, the task of establishing cause and effect relationships with various stressors in wild fish populations will be difficult.

It seems that the onset of sexual differentiation, coupled with the initiation of sexual maturation and gonadal growth and the peri-ovulatory periods, represent stages in reproductive development that are particularly sensitive to elevated temperatures. These are areas in which we should be focusing our research efforts. At present there are no surrogates for long-term whole animal testing to predict the effects of elevated thermal regimes on the reproductive and developmental fitness of fish.

References

Baroiller, J. F., Clota, F. & Geraz, E. (1995). Temperature sex determination in two tilapia species, *Oreochromis niloticus* and the red tilapia (Red Florida strain): effect of high or low temperatures. In *Reproductive Physiology of Fish 1995*, ed. F. W. Goetz & P. Thomas, pp. 158–60. Austin: University of Texas at Austin.

Barton, B. A. & Iwama, G. K. (1991). Physiological changes in fish from stress in aquaculture with emphasis on the response and effects of corticosteroids. *Annual Review of Fish Diseases*, 1, 3–26.

Billard, R. (1985). Environmental factors in salmonid culture and the control of reproduction. In *Salmonid Reproduction*, ed. R. N. Iwamoto & S. Sower, pp. 70–87. Seattle: Washington Sea Grant Program, University of Washington

Billard, R. & Breton, B. (1977). Sensibilité à la température des différentes étapes de la réproduction chez la truite arc-en-ciel. *Cahiers Laboratory Montereau*, 5, 5–24.

Boulekbache, H., Bastin, J., Andriamihaja, M., Lefebvre, B. & Joly, C. (1989). Ageing of fish oocytes: effects on adenylic nucleotides content, energy charge and viability of carp embryo. *Comparative Biochemistry and Physiology*, 93B, 471–6.

Brown, N. P., Bromage, N. R. & Shields, R. J. (1995). The effect of spawning temperature on egg viability in the Atlantic halibut (*Hippoglossus hippoglossus*). In *Reproductive Physiology of Fish 1995*, ed. F. W. Goetz & P. Thomas, p. 181. Austin: University of Texas at Austin.

Bye, V. J. (1984). The role of environmental factors in the timing of reproductive cycles. In *Fish Reproduction: Strategies and Tactics*, ed. G. W. Potts & R. J. Wootton, pp. 187–206. London: Academic Press.

Bye, V. J. (1987). Environmental management of marine fish reproduction in Europe. In *Reproductive Physiology of Fish 1987*, ed. D. R. Idler, L. W. Crim & J. M. Walsh, pp. 289–98. St John's: Memorial University of Newfoundland.

Clearwater, S. J. & Pankhurst, N. W. (1994). Reproductive biology and endocrinology of female red gurnard, *Chelidonichthys kumu* (Lesson and Garnot) (Family Triglidae), from the Hauraki Gulf, New Zealand. *Australian Journal of Marine and Freshwater Research*, 45, 131–9.

Cook, A. F. & Peter, R. E. (1980a). Plasma clearance of gonadotropin in goldfish, *Carassius auratus* during the annual reproductive cycle. *General and Comparative Endocrinology*, 42, 76–90.

Cook, A. F. & Peter, R. E. (1980b). The metabolism of gonadotropin in goldfish, *Carassius auratus*: tissue uptake and distribution during the reproductive cycle. *General and Comparative Endocrinology*, 42, 91–100.

Craik, J. C. A. & Harvey, S. M. (1984). Biochemical changes associated with overripening of the eggs of the rainbow trout *Salmo gairdneri* Richardson. *Aquaculture*, **37**, 347–58.

Davies, P. R., Hanyu, I., Furukawa, K. & Nomura, M. (1986). Effect of temperature and photoperiod on sexual maturation and spawning of the common carp. III. Induction of spawning by manipulating photoperiod and temperature. *Aquaculture*, **52**, 137–44.

Dodd, J. M. & Sumpter, J. P. (1984). Fishes. In *Marshall's Physiology of Reproduction*, Vol. 1. *Reproductive Cycles of Vertebrates*, ed. G. E. Lamming, pp. 1–126. Edinburgh: Churchill Livingstone.

Epler, P., Bieiarz, K. & Marosz, E. (1985). Effect of temperature, 17α-hydroxy-20β-dihydroprogesterone and prostagladin $F_{2\alpha}$ on carp oocyte maturation and ovulation *in vitro*. *General and Comparative Endocrinology*, **58**, 192–201.

Flett, P. A., Munkittrick, K. R., Van Der Kraak, G. & Leatherland, J. F. (1991). Reproductive problems in Lake Erie coho salmon. In *Reproductive Physiology of Fish 1991*, ed. A. P. Scott, J. P. Sumpter, D. E. Kime & M. S. Rolfe, pp. 151–3, Sheffield: University of East Anglia.

Flett, P. A., Munkittrick, K. R., Van Der Kraak, G. & Leatherland, J. F. (1996). Over-ripening as the cause of low survival to hatch in Lake Erie coho salmon (*Oncorhynchus kisutch*) embryos. *Canadian Journal of Zoology* **74**, 851–7.

Formacion, M. J., Hori, R. & Lam, T. J. (1993). Overripening of ovulated eggs in goldfish. I. Morphological changes. *Aquaculture*, **114**, 155–68.

Fortuny, A., Espinach Ros, A. & Amutio, V. G. (1988). Hormonal induction of final maturation and ovulation in the sábalo, *Prochilodus platensis* Holmberg: treatments, latency and incubation times and viability of ovules retained in the ovary after ovulation. *Aquaculture*, **73**, 373–81.

Francis, M. P. & Pankhurst, N. W. (1988). Juvenile sex inversion in the New Zealand snapper *Chrysophrys auratus* (Bloch and Schneider, 1801) (Sparidae). *Australian Journal of Marine and Freshwater Research*, **39**, 625–31.

Gunnes, K. (1979). Survival and development of Atlantic salmon eggs and fry at three different temperatures. *Aquaculture*, **16**, 211–18.

Hanyu, I., Asahina, K., Shimizu, A., Razani, H. & Kaneko, T. (1983). Environmental regulation of reproductive cycles in teleosts. In, *Proceedings of the 2nd North Pacific Aquaculture Symposium.* pp. 173–88. Tokyo: Tokyo University.

Hay, D. E. (1986). Effects of delayed spawning on viability of eggs and larvae of Pacific herring. *Transactions of the American Fisheries Society*, **115**, 155–61.

Hontela, A. (1984). Daily cycles of serum gonadotropin hormone in fish. *Transactions of the American Fisheries Society*, **113**, 458–66.

Hontela, A. & Peter, R. E. (1983). Entrainment of daily serum gonadotropin cycles in the goldfish to photoperiod, feeding and daily thermocycles. *Journal of Experimental Zoology*, **228**, 129–34.

Kime, D. E. (1979). The effect of temperature on the steroidogenic enzymes of the rainbow trout. *General and Comparative Endocrinology*, **39**, 290–6.

Kime, D. E. (1980). Androgen biosynthesis by testes of the goldfish *Carassius auratus in vitro*: the effect of temperature on the formation of steroid glucuronides. *General and Comparative Endocrinology*, **41**, 164–72.

Kime, D. E. & Hyder, M. (1983). The effect of temperature and gonadotropin on testicular steroidogenesis in *Sarotherodon* (Tilapia) *mossambicus in vitro*. *General and Comparative Endocrinology*, **50**, 105–15.

Kime, D. E. & Manning, N. J. (1986). Maturational and temperature effects on steroid hormone production by testes of the carp, *Cyprinus carpio*. *Aquaculture*, **54**, 49–55.

Lam, T. J. & Munro, A. D. (1987). Environmental control of reproduction in teleosts: an overview. In *Reproductive Physiology of Fish 1987*, ed. D. R. Idler, L. W. Crim and J. M. Walsh, pp. 279–88. St John's: Memorial University of Newfoundland.

Littlepage, J. L. (1965). Oceanographic investigations in McMurdo Sound Antarctica. *Biology of Antarctic Seas*, **2**, 1–37.

Luksiene, D. & Sandstrom, O. (1994). Reproductive disturbance in a roach (*Rutilis rutilis*) population affected by cooling water discharge. *Journal of Fish Biology*, **45**, 613–25.

McMaster, M. E., Munkittrick, K. R., Jardine, J. J., Robinson R. D. & Van Der Kraak, G. J. (1995). Protocol for measuring *in vitro* steroid production by fish gonadal tissue. *Canadian Technical Report Fisheries and Aquatic Sciences 1961*, 78 pp.

Manning, N. J. & Kime, D. E. (1984). Temperature regulation of ovarian steroid production in the common carp, *Cyprinus carpio* L. *in vivo* and *in vitro*. *General and Comparative Endocrinology*, **56**, 376–88.

Manning, N. J. & Kime, D. E. (1985). The effect of temperature on testicular steroid production in the rainbow trout, *Salmo gairdneri*, *in vivo* and *in vitro*. *General and Comparative Endocrinology*, **57**, 377–82.

Marshall, N. B. (1979). *Developments in Deep-sea Biology*. Poole, UK: Blandford Press.

Meffe, G. (1991). Life history changes in eastern mosquitofish (*Gambusia holbrooki*) induced by thermal elevation. *Canadian Journal of Fisheries and Aquatic Sciences*, **48**, 60–6.

Meffe, G. (1992). Plasticity of life history characters in eastern mosquitofish (*Gambusia holbrooki*: Phoecilidae) in response to thermal stress. *Copeia, 1992*, 94–102.

Pankhurst, N. W. (1995). Hormones and reproductive behaviour in male damselfish. *Bulletin of Marine Science*, **57**, 569–81.

Pankhurst, N. W., McMillan, P. J. & Tracey, D. M. (1987). Seasonal reproductive cycles in three commercially exploited fishes from the slope waters off New Zealand. *Journal of Fish Biology*, **30**, 193–211.

Pankhurst, N. W., Purser, G. J., Van Der Kraak, G., Thomas, P. M. & Forteath, G. N. R. (1996). Effect of holding temperature on ovulation, egg fertility, plasma levels of reproductive hormones and *in vitro* ovarian steroidogenesis in the rainbow trout *Oncorhynchus mykiss*. *Aquaculture* (in press).

Pankhurst, N. W., Van Der Kraak, G. & Peter, R. E. (1995). Evidence that the inhibitory effects of stress on reproduction in teleost fish are not mediated by the action of cortisol on ovarian steroidogenesis. *General and Comparative Endocrinology*, **99**, 249–57.

Paul, L. J. & Tarring, S. C. (1980). Growth rate and population structure of snapper, *Chrysophrys auratus* in the East Cape region, New Zealand. *New Zealand Journal of Marine and Freshwater Research*, **14**, 237–47.

Pauly, D. & Pullin, R. S. V. (1988). Hatching time in spherical, pelagic marine fish eggs in response to temperature and egg size. *Environmental Biology of Fishes*, **22**, 261–71.

Pickering, A. D. (1992). Rainbow trout husbandry: management of the stress response. *Aquaculture*, **100**, 125–39.

Robertson, D. A. (1978). Spawning of tarakihi (Pisces: Cheilodactylidae) in New Zealand waters. *New Zealand Journal of Marine and Freshwater Research*, **12**, 277–86.

Robertson, D. R., Petersen, C. W. & Braun, J. D. (1990). Lunar reproductive cycles of genetic-brooding reef fishes: reflections of larval biology or adult biology? *Ecological Monographs*, **60**, 311–29.

Sandstrom, O., Neuman, E. & Thoresson, G. (1995). Effects of temperature on life history variables in perch, *Perca fluviatilis*. *Journal of Fish Biology*, **47**, 652–70.

Scott, A. P., Baynes, S. M., Skarphédinsson, O. & Bye, V. J. (1984). Control of spawning time in rainbow trout *Salmo gairdneri*, using constant long day lengths. *Aquaculture*, **43**, 225–33.

Scott, S. G. & Pankhurst, N. W. (1992). Interannual variation in the reproductive cycle of the New Zealand snapper *Pagrus auratus* (Bloch and Schneider) (Sparidae). *Journal of Fish Biology*, **41**, 685–96.

Scott, S. G., Zeldis, J. R. & Pankhurst, N. W. (1993). Evidence of daily spawning in natural populations of the New Zealand snapper *Pagrus auratus* (Sparidae). *Environmental Biology of Fishes*, **36**, 149–59.

Shapiro, D. Y., Hensley, D. A. & Appledoorn, R. S. (1988). Pelagic spawning and egg transport in coral reef fishes: a skeptical overview. *Environmental Biology of Fishes*, **22**, 3–14.

Sokolowska, M., Peter, R. E., Nahorniak, C. S. & Chang, J. P. (1985). Seasonal effects of pimozide and des-Gly^{10}[D-Ala^6] LHRH ethylamide on gonadotropin secretion in goldfish. *General and Comparative Endocrinology*, **57**, 472–9.

Springate, J. R. C., Bromage, N. R., Elliott, J. A. K. & Hudson, D. H. (1984). The timing of ovulation and stripping and their effects on the rates of fertilization and survival to eyeing, hatch and swim-up in the rainbow trout *Salmo gairdneri* R. *Aquaculture*, **43**, 313–22.

Stacey, N. E. (1984). Control of the timing of ovulation by exogenous and endogenous factors. In *Fish Reproduction: Strategies and Tactics*, ed. G. W. Potts & R. W. Wootton, pp. 207–22. London: Academic Press.

Stacey, N. E., Cook, A. F. & Peter, R. E. (1979). Spontaneous and gonadotropin-induced ovulation in the goldfish, *Carassius auratus* L.: effects of external factors. *Journal of Fish Biology*, **15**, 349–61.

Strüssmann, C. A. & Patino, R. (1995). Temperature manipulation of sex differentiation in fish. In *Reproductive Physiology of Fish 1995*, ed. F. W. Goetz & P. Thomas, pp. 153–7. Austin: University of Texas at Austin.

Suzuki, R. (1975). Duration of development capacity of eggs after ovulation in the loach, cyprinid fish. *Aquaculture*, **23**, 93–9.

Taranger, G. L. & Hansen, T. (1993). Ovulation and egg survival following exposure of Atlantic salmon, *Salmo salar* L., broodstock to different water temperatures. *Aquaculture and Fisheries Management*, **24**, 151–6.

PETER JOHN ROMBOUGH

The effects of temperature on embryonic and larval development

Introduction

Anthropogenic increases in the levels of carbon dioxide and other greenhouse gases over the next half-century are expected to result in an increase in global mean temperature of 1.5–4.5 °C (Houghton & Woodwell, 1989; Intergovernmental Panel on Climate Change, 1992). We know that temperature increases of this magnitude are certain to have a major impact on fish populations. The historical record indicates that annual temperature anomalies smaller than those projected under most global warming scenarios have had significant short-term impacts on species distribution and abundance (Murawski, 1993). What we do not know are the details concerning which populations are likely to be most affected in the future and how much temperatures will have to rise before effects become obvious. The first step in attempting to answer these questions is to identify those periods in the life cycle that are most sensitive to temperature change. In this regard, attention has focused on the thermal tolerance of fish embryos and larvae (Houde, 1989; Pepin, 1991; Blaxter, 1992). Embryos and larvae are generally assumed to be more sensitive to temperature change than older fish (Brett, 1970) and, thus, would appear to be likely candidates in the search for critical periods.

The aim of this review is to examine the literature critically to see whether embryos and larvae are, in fact, more sensitive than juvenile and adult fish. I begin with an examination of lethal levels. However, based on what is known about older fish, direct lethality may not be the response that ultimately limits species distributions. Upper lethal temperatures for juvenile and adult fish typically are 3–6 °C higher than maximum field temperatures (Eaton et al., 1995). Sublethal effects, particularly a decline in energetic efficiency at higher temperatures (Brett, Clarke & Shelbourn, 1982), have been identified as playing a

Society for Experimental Biology Seminar Series 61: *Global Warming: Implications for freshwater and marine fish*, ed. C. M. Wood & D. G. McDonald. © Cambridge University Press 1996, pp. 177–223.

major role in defining adult distributions. Therefore, I have extended
the review to look at how temperatures within the zone of tolerance
affect various rate functions, specifically the rates of development,
metabolism and growth. These three processes are highly intercon-
nected and disruption of any one is likely to have a major impact on
overall energetic efficiency. I employed two approaches in attempting
to evaluate how temperature affects the various rate functions. The
first was to compile data from multiple sources for a large number of
species. The pooled data were then analysed for general trends. The
second approach was to focus on the response within particular species.
Trends that operate across species do not necessarily apply to lower
taxonomic units and, ultimately, it is the response of the lowest
taxonomic unit, the individual, that determines how populations
respond.

Lethal levels

The zone of thermal tolerance is considerably narrower during embryonic development for most temperate zone fish species than it is later
in life. The width of the zone of tolerance (upper lethal temperature
[ULT] minus lower lethal temperature [LLT]) for larvae and juveniles
of mid-temperate zone species typically is around 20–25 °C while that
for embryos is only about 11.6 °C (Fig. 1). Embryos of temperate
species are most sensitive to temperature change early in development,
particularly during activation, cleavage and gastrulation (Hokanson,
McCormick & Jones, 1973; Hokanson & Kleiner, 1974; Irvin, 1974;
Hassler, 1982; Beacham & Murray, 1990; Cloud, Erdahl & Graham,
1988; Rana, 1990; Buddington et al., 1993; K. Cousins, J. Jensen
and B. Robins, personal communication). The capacity to tolerate
temperature extremes increases as development proceeds, although
species vary in the rate at which this capacity is acquired. Salmonid
embryos become much more tolerant of temperature extremes once
epiboly is complete (i.e. after blastopore closure) so that by the time
they reach the eyed-embryo stage they are only slightly less tolerant
than juveniles (Combs, 1965; Peterson, Spinney & Sreedharan, 1977;
Beacham & Murray, 1990; Murray & Beacham, 1986; Tang, Bryant &
Brannon, 1987; Marten, 1992; K. Cousins et al., personal
communication). Pike (*Esox lucius*), on the other hand, appear to
acquire tolerance more gradually and it is not until they are well into
the larval stage that their ability to tolerate temperature extremes
approaches that of juvenile fish (Hokanson et al., 1973; Hassler, 1982).

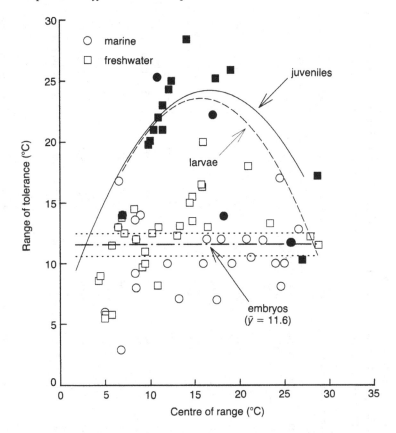

Fig. 1. The range of temperature tolerance (ULT–LLT) of embryonic, larval and juvenile fish as functions of the temperature at the centre of their respective zones of tolerance. Tolerance ranges of larval and juvenile fish varied significantly with central temperature in a parabolic manner with maxima at 15 and 17 °C, respectively. Range of tolerance was independent of temperature for embryos (mean ± SD = 11.6 ± 3.3 °C, $n = 59$). Data for embryos (open symbols) and larvae (closed symbols) are from Table 1; data for juveniles (points not plotted) were obtained from Brett (1970).

Many temperate species also display a shift in the position of the zone of tolerance as development proceeds. Typically, the centre of the zone of tolerance shifts towards higher temperatures (Table 1). This probably reflects the fact that, in nature, most species experience an increasing temperature regime as the water warms in the spring.

Table 1. *Temperature tolerance of fish embryos and larvae.*
In most cases, limits are LT$_{50}$s. LT$_{50}$s were estimated, where possible, from probit plots of corrected per cent mortality vs temperature. Corrections were based on survival rates at the temperature optima.

Species	Family	Stage	Age	Salinity (ppt)	Test duration (d)	Lower (°C)	Upper (°C)	Range (°C)	Centre (°C)	Reference
Marine										
Atherinops affinis (topsmelt)	Atherinidae	embryo	f-h	34	20–50	<12.8	26.8	>14.0	19.8	Hubbs, 1965
Leuresthes tenuis (grunion)	Atherinidae	embryo	f-h	34	20–50	14.8	26.8	12.0	20.8	Hubbs, 1965
		embryo	f-h			15.0	28.5	13.5	21.75	Ehrlich & Farris, 1971
		larva				11.3	25.2	13.9	18.25	Ehrlich & Muszynski, 1981
Fundulus parvipinnis (California killifish)	Cyprinodontidae	embryo	f-h	34	20–50	16.6	28.5	11.9	22.55	Hubbs, 1965
Cyprinodon macularius (desert pupfish)	Cyprinodontidae	embryo	f-h	35	4.6–53	16.0	33.0	17.0	24.5	Kinne & Kinne, 1962
Hypsoblennius sp. (blenny)	Blenniidae	embryo	f-h	34	20–50	<12.0	26.8	>14.8	19.4	Hubbs, 1965
Sardinops caerulea (Pacific sardine)	Clupeidae	embryo	f-h	34		14.0	21.0	7.0	17.5	Lasker, 1962
Brevoortia tyrannus (Atlantic menhaden)	Clupeidae	larva		24	3	3.4–5.2				Lewis, 1965
Clupea harengus (Atlantic herring)	Clupeidae	embryo	f-h	34	7.9–40	2.0	16.0	14.0	9.0	Blaxter, 1956
		larva	yolksac	34	1	−1.8	22.0–23.5	25.3	10.9	Blaxter, 1960
Clupea pallasi (Pacific herring)	Clupediae	embryo	f-h	17	9–34.7	1.6	15.2	13.6	8.4	Alderdice & Velsen, 1971

Species	Family	Stage		Salinity						Reference
Calotomus japonicus	Scaridae	embryo	f-h	34	0.5-2.4	20.5	28.6	8.1	24.6	Seno et al., 1926
Scomber scombrus (Atlantic mackerel)	Scrombridae	embryo	f-h	34	2.0-8.6	11.0	21.0	10.0	16.0	Worley, 1933
Gadus macrocephalus (Pacific cod)	Gadidae	embryo	f-h	26.5	7-20	2.0	8.0	6.0	5.0	Forrester, 1964a
Gadus morhua (Atlantic cod)	Gadidae	embryo	f-h	23-27		-1.0	12.0	13.0	6.5	Johansen & Krogh, 1914
		embryo				<-1.8				Valerio et al., 1992
		larvae				-1.35				
Parophrys vetulus (English sole)	Pleuronectidae	embryo	f-h	25.9	3.5-11.8	4.5	12.5	8.0	8.5	Alderdice & Forrester, 1968
Pleuronectes platessa (plaice)	Pleuronectidae	embryo	f-h	23-27		0	14.0	14.0	7.0	Johansen & Krogh, 1914
Pseudopleuronectes americanus (winter flounder)	Pleuronectidae	embryo	f-h			-1.8	15	16.8	6.6	Williams, 1975
		embryo	f-h			3	14	12	8.5	Rogers, 1976
Eopsetta jordani (petrale sole)	Pleuronectidae	embryo	f-h	28	6.6-13.5	5.3	8.2	2.9	6.8	Alderdice & Forrester, 1971
Hippoglossoides elassodon (flathead sole)	Pleuronectidae	embryo	f-h	29.5	7.2-16.8	3.8	13.0	9.2	8.4	Alderdice & Forrester, 1974
Solea solea (Dover sole)	Pleuronectidae	embryo	f-h	34	4	7	17	10	12	Irvin, 1974
		larva	yolksac	34	4		23			
		larva	1st-feed	34	4		24			
		larva	met.	34	4					
Paralichthys californicus (California halibut)	Pleuronectidae	embryo	f-h	35	1.4-3.1	6-9	28.2	22.2	17.1	Gadomski & Caddell, 1991
						10.4	22.3	12.0	16.4	
Sparus aurata (guilthead seabream)	Sparidae	larva	yolksac	35	12	10.4	>24.1	>14.1		
		embryo	f-h	33	1.0-3.7	14.2	24.2	10.0	19.2	Polo et al., 1991

Table 1. (cont.)

Species	Family	Stage	Age	Salinity (ppt)	Test duration (d)	Lower (°C)	Upper (°C)	Tolerance limits Range (°C)	Centre (°C)	Reference
Dicentrarchus labrax (sea bass)	Serranidae	embryo	f-h	31	3.3-6.7	9.7	16.8	7.1	13.3	Jennings & Pawson, 1991
Bairdiella icistia (bairdiella)	Sciaenidae	embryo	f-h	15-40	0.7-1.3	20.0	30.0	10.0	25.0	May, 1975a
		larva		25-35	1		29-31			May, 1975b
Mugil cephalus (striped mullet)	Mullidae	embryo	f-h	36.5	1.1-2.9	19.8	31.5	11.7	25.7	Walsh et al., 1991
Caranx mate	Carangidae	embryo	f-h	34	0.4-1.5	20.2	33	12.8	26.6	Santerre, 1976
		larva	0 hph	34	1.3	19.9	31.6	11.7	25.8	
		larva	24 hph	34	ILT		31.3			
			72 hph		ILT		33.2			
			120 hph		ILT		34.0			
					ILT		32.0			
Polydactylus sexfilis (moi)	Polynemidae	embryo	f-h	34	0.5-1.0	19	29.0	10.0	24.0	Santerre & May, 1977
Belone belone (garfish)	Synentognathidae	embryo	f-h	10-45	10-45	12.0	24.0	12.0	18.0	Fonds et al. 1974

Freshwater

Species	Family	Stage		0						Reference
Coregonus artedii (lake cisco)	Salmonidae	embryo	f-h	0	38–188	2.2	8.0	5.8	5.1	Colby & Brooke, 1970
		larva		0	1	<2.3	19.8	>17.8 (19.8)	(9.9)	McCormick et al., 1971
Coregonus clupeaformis (lake whitefish)	Salmonidae	embryo	f-h	0	41.7–182	2.2	7.7	5.5	5.0	Brooke, 1975
Coregonus albula (vendace)	Salmonidae	embryo	f-h	0	42–177	2.9	8.7	5.8	5.8	Luczynski & Kirklewskwa, 1984
Oncorhynchus tshawytscha (Chinook salmon)	Salmonidae	larva	21 dph	0	21	<4.8	22.0	>17.2(22)	(11)	Luczynski, 1991
		embryo	f-h	0	30–206	4.0	15.0	11.0	9.5	Velsen, 1987
		embryo	activation	0	1	<1.0	14.0	>13.0		Cousins et al., pers. comm.
			epiboly	0	1	<1.0	>24.0	>23.0		
			eyed	0	1	<1.0	21.3	>20.3		
			posthatch	0	1	<1.0	21.0	>20.0(21)		
		larvae	f-h	0	32–188	0.9	13.4	12.5	(11.5)	
Oncorhynchus kisutch (coho salmon)	Salmonidae	larva	yolksac	0	60–250	<1.0	14.0	>13.0(14)	7.15	Tang et al., 1987
			activation	0	1	8.0	14.0	6.0	(7)	Beacham & Murray, 1990
			epiboly	0	1	<1.0	19.0	>18.0		Cousins et al., pers. comm.
			eyed	0	1	<1.0	19.5	>18.5		
			posthatch	0	1	<1.0	21.0	>20.0(21)	(10.5)	
Oncorhynchus keta (chum salmon)	Salmonidae	embryo	f-h	0		2.5	>16	>13.5	9.3	Beacham & Murray, 1990
Oncorhynchus nerka (sockeye salmon)	Salmonidae	embryo	f-h	0		1.0	15.5	14.5	8.3	Beacham & Murray, 1990

Table 1. (cont.)

Species	Family	Stage	Age	Salinity (ppt)	Test duration (d)	Tolerance limits				Reference
						Lower (°C)	Upper (°C)	Range (°C)	Centre (°C)	
Oncorhynchus gorbuscha (pink salmon)	Salmonidae	embryo	f-h	0		1.0	13.5	12.5	7.3	Beacham & Murray, 1980
Oncorhynchus mykiss (rainbow trout)	Salmonidae	embryo	f-h	0	18–94	4	17	13	10.5	Humpesch, 1985
Salvelinus fontinalis (brook trout)	Salmonidae	embryo	f-h	0			12.7			Hokanson et al., 1973
		larva	posthatch	0	7		20.1	(20.1)	(10.1)	McCormick et al., 1972
			swim-up	0	7		22.2–24.3	(24.3)	(12.2)	
Salvelinus alpinus (Arctic charr)	Salmonidae	embryo	f-h	0	35–241	0	13.8	13.8	6.9	Humpesch, 1985
		embryo	f-h	0	45–95	0?	9.0	9.0	4.5	Swift, 1965
		embryo	f-h	0	53–206	0	8.6	8.6	4.3	Humpesch, 1985
		embryo	f-h	0	45–116	<4	9.0	9.0	4.5	Jungwirth & Winkler, 1984
Salmo salar (Atlantic salmon)	Salmonidae	larva	mid-ya		7		23	(23)	(11.5)	Bishai, 1960
Salmo trutta (brown trout)	Salmonidae	larva	mid-ya		7		23	(23)	(11.5)	Bishai, 1960
		embryo	f-h	0	34–117	0	13	13	6.5	Humpesch, 1985
		embryo	f-h	0		<4	11.5	11.5	(5.8)	Jungwirth & Winkler, 1984
Hucho hucho (Danube salmon)	Salmonidae	embryo	f-h	0	14–134	2.5	14.5	12	8.5	Humpesch, 1985
		embryo	f-h	0	16–77	4.3	14.0	9.7	9.2	Jungwirth & Winkler, 1984
Thymallus thymallus (grayling)	Salmonidae	embryo	f-h	0	9.9–69	4.0	16.5	12.5	10.3	Humpesch, 1985
		embryo	f-h	0	13–63	4.5	14.5	10	9.5	Jungwirth & Winkler, 1984

Species	Family	Stage	Description		Range					Reference
Perca flavenscens (yellow perch)	Percidae	embryo	f-h	0	5-15	6.8	19.9	13.1	13.4	Hokanson & Kleiner, 1974
Perca fluviatilis (Eurasian perch)	Percidae	embryo	f-h	0	6-28	7.0	22.5	15.5	14.8	Wang & Eckmann, 1994
Stizostedion vitreum (walleye)	Percidae	embryo	f-h	0	6-40	<6.0	20.0	>14		Guma'a, 1978
		embryo	f-h	0	5-34	<6	14			Koenst & Smith, 1976
Stizostedion canadense (sauger)	Percidae	larva	h-ya			7.5	>20.9	>13.4		Koenst & Smith, 1976
		embryo	f-h	0	5-37	<6	20			
Etheostoma fonticola (fountain darter)	Percidae	larva	h-ya			7.5	>20.9	>13.4	15.9	Brandt et al., 1993
		embryo	f-h	0		7.7	24.0	16.3		
Esox lucius (pike)	Esocidae	embryo	f-h	0	?-87	6.9	19.2	12.3	13.1	Hokanson et al., 1973
		larva	1 dph	0	7	<3.2	20.6-25.0	>21.8(25)	(12.5)	
		larva	21 dph	0	7	<3.2	23.4-28.4	>25.2 (28.4)	(14.2)	
		embryo	f-h	0	6	4.5	17.5	13.0	11	
		larvae	1 hph	0		<3	>21	>18		Hassler, 1982
Catostomus commersoni (white sucker)	Catostomidae	larva	posthatch	0	7	4.8	28.8-30	25.2	17.4	McCormick et al., 1977
			swim-up	0		6.1	28.1-32	25.9	19.1	
Oreochromis niloticus (Nile tilapia)	Cichlidae	embryo	f-h	0	2.3-6.0	21.8	34.0	12.2	27.9	Rana, 1990
		larva	swim-up	0		21.8	32.1	10.3	27.0	
Oreochromis mossambica (Mozambique tilapia)	Cichlidae	embryo	f-h	0	2-4	18	>40.0	>22.0	29.0	Subasinghe & Sommerville, 1992
Heterobranchus longifilis (African catfish)	Claridae	larva	swim-up	0	8	20.1	37.3	17.2	28.7	Legendre & Teugels, 1991
		embryo	f-h	0	0.7-1.2	23	34.5	11.5	28.8	

Table 1. (*cont.*)

Species	Family	Stage	Age	Salinity (ppt)	Test duration (d)	Lower (°C)	Upper (°C)	Range (°C)	Centre (°C)	Reference
Chalcalburnus chalcoides (Danube bleak)	Cyprinidae	embryo	f-h	0	2.8–22.7	10	23	13	16.5	Herzig, & Winkler, 1986
Vimba vimba (zahrte)	Cyprinidae	embryo	f-h	0	2.2–9.2	10	23	13	16.5	Herzig & Winkler, 1986
Rutilus rutilus (roach)	Cyprinidae	embryo	f-h	0		7.5	24	16.5	15.8	Herzig & Winkler, 1986
Leuciscus idus (ide)	Cyprinidae	embryo	f-h	0		7	22	15	14.5	Herzig & Winkler, 1986
Leuciscus leuciscus (dace)	Cyprinidae	embryo	f-h	0		6.8	15	8.2	10.9	Herzig & Winkler, 1986
Abramis brama (bream)	Cyprinidae	embryo	f-h	0		6	26	20	16	Herzig & Winkler, 1986
Tinca tinca (tench)	Cyprinidae	embryo	f-h	0		16.7	30	13.3	23.4	Herzig & Winkler, 1986
Cyprinus carpio (carp)	Cyprinidae	embryo	f-h	0		12	30	18	21	Herzig & Winkler, 1986
Polyodon spathula (paddlefish)	Chondrostea	larva	0-3 dph	0	30	16.6	23.0	6.4	19.8	Kroll et al., 1992

Abbreviations: f, fertilization; h, hatch; met, metamorphosis; ya, yolk absorption; hph, hours posthatch; dph, days posthatch; ILT, incipient lethal temperature. Values in parentheses are estimates.

In a few cases, for example the California grunion, *Leuresthes tenuis* (Erhlich & Muszynski, 1981), the centre of the zone of tolerance shifts towards lower temperatures. This, too, appears to be a reflection of what occurs in nature. Grunion spawn on beaches that are subject to solar warming but move into cooler inshore waters after hatch.

Unlike temperate species, tropical species show little change in the width of the zone of tolerance as development proceeds (Fig. 1). For example, upper lethal temperatures for embryos, larvae (Subasinghe & Sommerville, 1992) and adults (Allanson & Noble, 1964) of the fresh-water Mozambique tilapia (*Oreochromis mossambica*) are all within about 2 °C of each other. Similarly, there is little difference between the upper lethal temperatures for embryos and larvae of marine species such as the sciaenid *Bairdiella icistia* (May, 1975a, b) or the carangid *Caranx mate* (Santerre, 1976). Seasonal fluctuations in temperature are relatively small in the tropics so it is not all that surprising that the various life stages differ little in terms of their temperature tolerance. The thermal regime also is relatively stable at high latitudes and in the deep oceans but, unfortunately, we have no information on the temperature tolerance of embryos and larvae of species inhabiting these environments.

In nature, fish typically spawn at temperatures within a few degrees of the centre of their embryonic zone of tolerance (Herzig & Winkler, 1986), which can range from < 0 ° to close to 30 °C depending on species. Once spawned, however, the absolute temperature change that eggs can tolerate is roughly the same, ±5.8 °C, for both temperate and tropical species (Fig. 1). One of the major factors that restricts embryos to this narrow temperature range appears to be their inability to temperature-compensate. Embryos are unable to regulate the fluidity of their cell membranes (Buddington *et al.*, 1993), their metabolic rates shows no sign of temperature acclimation (Rombough, 1988a), and upper lethal temperatures appear to be independent of the thermal history of the embryo (McCormick, Hokanson & Jones, 1972; McCormick, Jones & Hokanson, 1977). The thermal history of the parents, on the other hand, appears to have some effect. Silverside (*Menidia audens*) embryos from parents exposed to high temperatures 1–3 days prior to spawning were more resistant to high temperature at gastrulation than were embryos from parents held at low temperatures (Hubbs & Bryan, 1974). Kokurewicz (1981) reported that the upper lethal temperature for tench (*Tinca tinca*) embryos increased by as much as 2 °C when the parents were raised at high temperatures. Buckley *et al.* (1990) reported that winter flounder (*Pseudopleuronectes americanus*) larvae hatched from eggs from females held at high temperatures grew faster at higher temperatures than larvae from females held at low temperature. There is a limit, however, to the extent to which embryonic tolerance, particularly to high temperatures, can be altered by manipulating the temperature regime of the parents. In at least some species, the temperature limits for gametogenesis are as narrow or narrower than those for embryonic development. For example, brook trout (*Salvelinus fontinalis*) embryos will develop normally at temperatures up to 13.8 °C (Humpesch, 1985) but females will not produce viable eggs at temperatures above about 11.7 °C (Hokanson *et al.*, 1973).

Fish begin to acquire the ability to temperature-compensate shortly after hatch. The capacity for homeoviscous adaptation is evident in white sturgeon (*Acipenser transmontanus*) by mid-way through yolk absorption (Buddington *et al.*, 1993). Acclimation temperature begins to have a significant effect on the incipient upper lethal temperature near mid-yolk absorption in Atlantic salmon (*Salmo salar*) and brown trout (*Salmo trutta*) (Bishai, 1960) and around swim-up in brook trout (McCormick *et al.*, 1972) and white sucker (*Catostomus commersoni*) (McCormick *et al.*, 1977). The ability to temperature-compensate appears to develop gradually. Bishai (1960) noted that, while acclimation to high temperatures increased the incipient upper lethal temperatures of Atlantic salmon and brown trout larvae as early as mid-yolk absorption, the magnitude of the response (0.07 °C/Δ °C) was only about half that observed in juvenile fish (0.14 °C/Δ °C).

The effective zone of temperature tolerance is often restricted in nature by other environmental factors. Of abiotic factors, changes in dissolved oxygen level, salinity and the concentrations of various pollutants appear to be particularly important in this regard. Aquatic poikilotherms exposed to rising temperatures are faced with a temperature–oxygen squeeze. Oxygen demand increases because of rising metabolic rates but oxygen supply tends to decrease because of a decline in the solubility of oxygen at higher temperatures. Developing fish are especially sensitive (Rombough, 1988a). Diffusion plays a much more important role in gas exchange during early life than it does in older fish (Rombough, 1989; Rombough & Ure, 1991). The rate of diffusion is relatively insensitive to temperature change whereas metabolic demand for oxygen roughly doubles or triples with each 10 °C rise in temperature. The net result is a downward shift in both optimum and upper lethal temperatures as oxygen levels fall (Brooke & Colby, 1980). The interactive effects of temperature and salinity on embryonic and larval survival have been investigated in a number of marine species (e.g. Kinne & Kinne, 1962; Alderdice & Forrester, 1968; Alderdice &

Velsen, 1971; Fonds, Rosenthal & Alderdice, 1974). Shifts away from the optimum salinity tend to reduce the range of tolerated temperatures. In most cases, the effect of temperature far outweighs the effect of salinity (Hempel, 1984) but under certain conditions the shift in temperature optima may be sufficient to limit larval recruitment. The toxicities of some pollutants are strongly temperature dependent (Kennedy & Walsh, this volume; Reid, McDonald & Wood, this volume). Developing fish are particularly sensitive to point sources because of their limited mobility. Widely distributed lipid-soluble pollutants, such as PCBs and DDT, also are of concern because of their propensity to bioaccumulate in yolk, thereby exposing the developing embryo to concentrations that are much higher than those in the general environment (von Westernhagen, 1988).

Sublethal effects

Rate of development

The rate of embryonic development is strongly temperature dependent in fish. Averaged across species, development rate roughly triples with each 10 °C increase in temperature (i.e. $Q_{10} \cong 3.0$). The overall Q_{10} for the incubation periods of the 93 marine and 78 freshwater species plotted in Fig. 2 is 3.1. Pepin's (1991) analysis of 124 marine species suggests a value of about 2.9. Individual species, on average, appear to be somewhat more sensitive to temperature change than cross-species comparisons suggest. The average Q_{10} is 3.6 for the 37 individual species listed in Table 2. Coldwater species appear particularly responsive. A plot of the data in Table 2 revealed a significant inverse relationship between Q_{10} and central incubation temperature. The magnitude of the change ($\Delta Q_{10} \cong 1.4$ between 25 and 5 °C) is greater than can be explained simply on the basis of the difference between Arrhenius and van't Hoff models ($\Delta Q_{10} \cong 0.6$). Development rate also is highly dependent on egg size, with large eggs taking up to 10 times as long to hatch as small eggs at the same temperature (Ware, 1975; Pauly & Pullin, 1988; Pepin, 1991).

The rate of larval development generally appears to be less sensitive to temperature change than the rate of embryonic development. Cross-species comparisons indicate Q_{10} values of about 2.0 for times to complete yolk absorption (Pepin, 1991), point-of-no-return (Pepin, 1991), and metamorphosis (Houde & Zastrow, 1993) in marine larvae (Table 3). It is not clear what the underlying trend is in freshwater fish. Houde & Zastrow (1993) reported that time to metamorphosis was independent of temperature for freshwater larvae (Table 3). How-

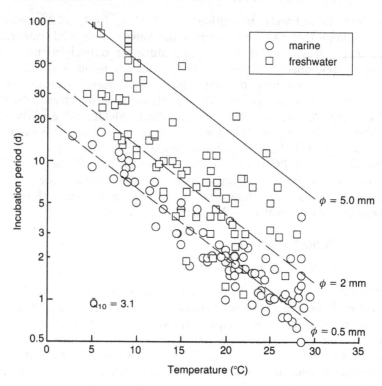

Fig. 2. Average incubation period for eggs of various marine (○) and freshwater (□) species plotted in relation to temperature. Data ($n = 171$) for marine and freshwater species were taken from Pauly & Pullin (1988) and Scott & Crossman (1973), respectively. The regression $\log D = 1.20 - 0.0494T + 0.203\phi$, where D is incubation period (days), T is temperature (°C) and ϕ is egg diameter (mm), accounted for 84.4% of the variability in time to hatch. The diagonal lines illustrate estimated incubation periods for 0.5, 2 and 5 mm diameter eggs.

ever, their sample size was relatively small and it will be interesting to see if this relationship holds up as more species are examined. Reduced thermal sensitivity at the larval stage is evident within species as well as across species. A paired comparison of Q_{10} values for the duration of embryonic (Table 2) and larval (Table 3) stages of the same species indicates that, except for salmonids, thermal sensitivity is significantly lower at the larval stage (mean $Q_{10} \cong 2.1$ vs $\cong 3.6$). Salmonid yolk-sac larvae appear to be as sensitive (Crisp, 1988) and,

Table 2. *Incubation periods of fish embryos as a function of temperature*

Where possible, data relating times from fertilization to hatch were fitted to the exponential model $D = ae^{bT}$, where D is number of days to hatch, T is temperature (°C) and a and b are constants. Mean Q_{10} values were estimated as e^{10b}.

Species	Egg diameter (mm)	Incubation temperature (°C)		Incubation period (days)		Slope (b)	Q_{10}	Reference
		Range	Centre	at centre	at 0°C (a)			
Freshwater								
Oncorhynchus tshawytscha	6.5	4.0–15.0	9.5	61.7	174.0	−0.109	2.98	Velsen, 1987
		1.6–18.1	9.9	58.6	199	−0.124	3.44	Crisp, 1981
Oncorhynchus keta	7.8	2.5–16.0	9.3	59.7	163.0	−0.108	2.95	Velsen, 1987
Oncorhynchus kisutch	5.3	1.0–13.5	7.3	65.1	165.0	−0.128	3.58	Velsen, 1987
Oncorhynchus gorbuscha	6.0	1.0–13.5	7.3	84.8	151.0	−0.079	2.21	Velsen, 1987
Oncorhynchus nerka	4.8	1.0–15.5	8.3	86.8	201.0	−0.102	2.76	Velsen, 1987
Oncorhynchus mykiss	4.0	1.0–16.0	8.5	42.3	128.0	−0.130	3.66	Velsen, 1987
		5–15	10.0	33.0			3.48	Humpesch, 1985
		3.2–15.5	9.4	36.6	135	−0.138	3.99	Crisp, 1981
Salmo trutta	4.5	5–15	10.0	40.0			2.48	Humpesch, 1985
		4–12	8.0	54.2			4.74	Jungwirth & Winkler, 1984
		1.9–11.2	6.6	71.2	207	−0.161	5.02	Crisp, 1981
Salmo salar	6.0	2.4–12	7.2	72.9	211	−0.148	4.38	Crisp, 1981
Salvelinus fontinalis	4.3	2.5–10.0	6.3	77.1	176.0	−0.131	3.71	Garside, 1966
		5–10	7.5	55.5			2.83	Humpesch, 1985
		1.6–14.8	8.2	62.3	183	−0.132	3.73	Crisp, 1981
Salvelinus alpinus	4.5	5–10	7.5	57.9			2.38	Humpesch, 1985
		4–10	7.0	61.4			5.10	Jungwirth & Winkler, 1984
		4–12	8.0	56.5	160	−0.130	3.66	Swift, 1965
Hucho hucho		5–15	10.0	23.0			3.92	Humpesch, 1985
		4–14	9.0	30.2			4.72	Jungwirth & Winkler, 1984

Table 2. (cont.)

Species	Egg diameter (mm)	Incubation temperature (°C)		Incubation period (days)		Slope (b)	Q₁₀	Reference
		Range	Centre	at centre	at 0°C (a)			
Thymallus thymallus		5–15	10.0	20.0			4.64	Humpesch, 1985
		4–14	9.0	25.1			4.72	Jungwirth & Winkler, 1984
Coregonus albula	2.0	1.1–9.9	5.5	84.4	202.0	−0.159	4.90	Luczynski & Kiridewskwa, 1984
Coregonus artedii	2.0	0.5–10.0	5.3	97.2	267.0	−0.193	6.89	Colby & Brooke, 1973
Coregonus clupeaformis	3.0	0.5–10.0	5.3	89.2	204.0	−0.158	4.86	Brooke, 1975
Perca fluviatilis	3.5	6.0–22.0	14.0	11.8	57.4	−0.108	2.95	Wang & Eckermann, 1994
		6–20	13.0	13.9	89.7	−0.144	4.21	Guma'a, 1978
Gobio albipinnatus	1.5	12.0–24.0	18.0	6.7	40.0	−0.099	2.69	Wanzenbock & Wanzenbock, 1993
Clarias gariepinus		22.0–28.0	25.0	1.3	40.4	−0.137	3.93	Kamler et al., 1994
Heterobranchus longifilis		25–33	29.0	0.9	7.69	−0.074	2.09	Legendre & Teugels, 1991
Acipenser transmontanus	4.0	11–20	15.5	6.9	49.4	−0.127	3.56	Wang et al., 1985
Acipenser fulvescens	3.5	10–20	15.0	8.9	73.7	−0.141	4.10	Wang et al., 1985
Fundulus parvipinnis		17–28	22.5	20.7	178	−0.096	2.60	Hubbs, 1965
Marine								
Cyprinodon macularis	2.0	13.0–35.8	24.4	9.7	110.0	−0.101	2.75	Kinne & Kinne 1962
Sardina pilchardus	1.5	13.0–20.0	16.5	2.7	16.0	−0.108	2.95	Miranda et al., 1990
Dicentrachus labrax	1.3	10.6–16.8	13.7	4.6	24.8	−0.123	3.42	Jennings & Pawson 1991
Sparus aurata		16.0–22.0	19.0	2.1	13.9	−0.099	2.70	Polo et al., 1991
Hippoglossus hippolossus	3.1	3.0–9.0	6.0	14.5	41.3	−0.174	5.70	Pittman et al., 1990
Paralichthys californicus		12.4–20.8	16.6	2.1	9.43	−0.092	2.51	Gadomski & Caddell 1991
Pseudopleuronectes americanus		3–14	8.5	12.7	27.7	−0.092	2.51	Rogers 1976

Mugil cephalus		20.0–32.0	26.0	1.6	13.9	−0.083	2.29	Walsh et al., 1991
Polydactylus sexfilis	0.8	21.4–29.8	25.6	0.7	6.8	−0.089	2.42	Santerre & May 1977
Caranx mate	0.7	17.2–25.6	21.4	0.7	16.8	−0.147	4.35	Santerre 1976
Belone belone	3.0	13.2–24.0	18.6	17.5	155.0	−0.117	3.22	Fonds et al., 1973
Leuresthes tenuis		14–24	19.0			−0.151	4.52	Ehrlich & Farris, 1971
Theragra chalcagramma		2–11	6.5	11.8	28.7	−0.137	3.94	Haynes & Ignell, 1983
		3.8–7.7	5.8	12.5	27.1	−0.133	3.78	Blood et al., 1994
Gadus macrocephalus	1.5	5–11	8.0	11.8	28.8	−0.111	3.04	Forrester, 1964a
Lepidopsetta bilineata	0.9	5–11	8.0	9.7	30.8	−0.144	4.22	Forrester, 1964b

Table 3. *Effect of temperature on the rate of development of fish larvae*
Experimental data relating times from hatch to various events were fitted to the exponential model $D = ae^{bT}$, where D is number of days posthatch, T is temperature (°C) and a and b are constants. * indicates the regression was significant at $p < 0.05$.
MAWW, maximum alevin wet weight; PNR, point-of-no-return.

Species	Event	Temperature		Duration (days)		Slope (b)	Q$_{10}$	Reference
		Range	Centre	at centre	at 0 °C (a)			
Freshwater								
Oncorhynchus tshawytscha	MAWW	5–12.5	8.8	48.1	290	−0.218 *	8.90	Rombough, 1985
Salmo salar	swim-up	6.3–12.3	9.3	39.6	214	−0.181 *	6.10	Brannas, 1988
Coregonus albula	1st feeding	4.5–19.0	11.8	3.3	15.1	−0.128 *	3.61	Luczynski & Kirklewskwa, 1984
Clarias gariepinus	yolk absorption	22.1–28.1	25.1	4.9	96.6	−0.119	3.27	Kamler et al., 1994
Morone saxatilis	starvation	15–24	19.5	21.0	59.5	−0.0534*	1.71	Rogers & Westin, 1981
Perca flavescens	starvation	5.2–19.9	12.6	14.3	21.9	−0.0341	1.41	Hokanson & Kleiner, 1974
15 species combined	metamorphosis	9.5–24	16.8	27.0	32.1	−0.0120	1.13	Houde & Zastrow, 1993

Marine

Sardina pilchardus	yolk absorption	13.0–20.0	16.5	6.0	11.4	−0.0389	1.48	Miranda et al., 1990
	starvation			9.7	12.0	−0.0128	1.14	
Sparus aurata	yolk absorption	16.0–22.0	19.0	4.0	38.3	−0.119	3.29	Polo et al., 1991
	starvation			10.3	39.4	−0.0705	2.02	
Polydactylus sexfilis	yolk absorption	21.4–29.8	25.6	1.5	6.61	−0.0584*	1.79	Santerre & May, 1977
Caranx mate	yolk absorption	17.2–25.6	21.4		6.05	−0.0434*	1.54	Santerre, 1976
41 species combined	yolk absorption	5–29	17.0	6.0	19.9	−0.071	2.03	Pepin, 1991
29 species combined	PNR	7–29	18.0	9.8	42.1	−0.081	2.25	Pepin, 1991
79 species combined	metamorphosis	4.5–29.8	17.2	50.4	178	−0.0734*	2.08	Houde & Zastrow, 1993

in some cases, more sensitive (Rombough, 1985; Brannas, 1988) to temperature change than embryos. Here again, however, the sample size is relatively small and what appears to be a trend may not hold up under scrutiny.

Metabolism

Temperature has a more pronounced impact on aerobic metabolism during embryonic and early larval development than it does later in life. Q_{10} values for the rate of routine metabolism of individual species average about 3.0 for late-stage embryos and yolk-sac larvae but only about 2.0 for juveniles and adults (Rombough, 1988a). As noted in the section on lethal limits, embryos display little or no ability to compensate metabolically for changes in temperature. The ability of individual species to limit increases in metabolic rate when faced with rising temperatures appears to arise gradually during the course of larval development. For example, Walsh et al. (1989) recorded about an 80% increase in metabolic rate, equivalent to a Q_{10} of 3.1, when they transferred 31 h old striped mullet (*Mugil cephalus*) larvae from 24 to 29 °C (Fig. 3). If they delayed the transfer until the larvae were 578 h old, however, metabolic rate increased by only about 17%, equivalent to a Q_{10} of 1.4. The response was similar in milkfish (*Chanos chanos*) larvae where the Q_{10} based on acute transfer from 28 to 33 °C dropped from $Q_{10} \cong 3.0$ at 2 h posthatch to $Q_{10} \cong 2.2$ at 24 h post-hatch (Walsh, Swanson & Lee, 1991). Metabolic compensation is evident across species as well as within species by the early larval stage. When looked at across species, the overall response of routine metabolism to increases in temperature ($Q_{10} \cong 2.0$; Fig. 4) is less than one would predict on the basis of the response of individual species ($Q_{10} \cong 3.0$).

Temperature appears to have a relatively minor effect on aerobic scope. Wieser & Forstner (1986) reported little difference between the temperature sensitivities of routine (mean $Q_{10} = 2.65$) and active metabolism (mean $Q_{10} = 2.42$) over the range 12–24 °C for larvae of three species of cyprinids (Fig. 5). Temperature appears to have a slightly greater impact on standard ($Q_{10} = 2.28$) and routine ($Q_{10} = 2.58$) metabolism than on active metabolism ($Q_{10} = 1.92$) in larvae of the Danube bleak, *Chalcalburnus chalcoides* (Kaufmann & Wieser, 1992). However, Q_{10} values estimated on the basis of only two temperatures are often unreliable (errors are multiplicative with ratios) and, in any event, the net impact on the relative scope of the bleak was

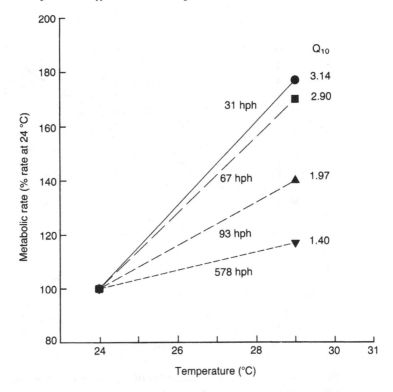

Fig. 3. Ontogenetic changes in the metabolic response of striped mullet (*Mugil cephalus*) larvae acutely transferred from 24 to 29 °C. Rates of oxygen consumption measured 1 hour after transfer to the higher temperature are expressed as a percentage of the pre-transfer rate; hph indicates the age of the larvae in hours posthatch. (Data from Walsh *et al.*, 1989.)

relatively minor (relative scopes values of 2.7 and 2.5, respectively, at 15 and 20 °C).

Growth

Temperature appears to have roughly the same magnitude of effect on growth rate as it does on metabolic rate during embryonic and larval development. Q_{10} values for specific growth rate average about 3.0 at the species level (the grand mean $\pm$ SE for the 19 studies listed in Table 4 is 3.3 ± 0.5) and about 2.0 in cross-species comparisons

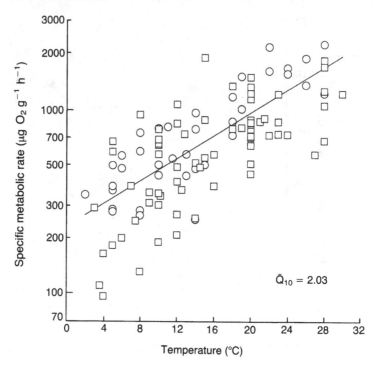

Fig. 4. Mass-specific metabolic rates of late stage embryos and young larvae of marine (○) and freshwater (□) fishes as a function of temperature. (Data from Rombough, 1988a.)

(Fig. 6). There is a tendency for growth rates as well as metabolic rates to begin to level off, as shown by a decline in interval Q_{10}, at higher temperatures within a given species (Table 4). However, unlike the situation in juvenile and adult fish, high temperatures usually do not constrain growth to the extent that growth rates actually begin to decline. With juvenile and adult fish, there is typically an optimum temperature near the centre of the zone of tolerance above which growth rates decline (Brett, 1979). The situation appears to be similar in young sockeye salmon, *Oncorhynchus nerka* (Shelbourn, Brett & Shirahata, 1973), brook trout, *Salvelinus fontinalis* (McCormick *et al.*, 1972; Fig. 7) and, perhaps, Arctic charr, *Salvelinus alpinus* (Wallace & Aasjord, 1984). It is not the case, though, for the other 14 species listed in Table 4. In these species, growth rate typically does not begin to decline until near the upper lethal threshold and, in the case of cisco, *Coregonus artedii* (McCormick, Jones & Syrett, 1971), growth

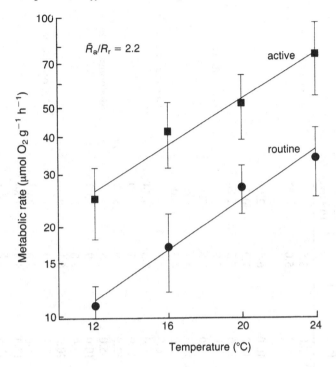

Fig. 5. The effect of temperature on active and routine metabolic rates of larval cyprinids (*Rutilus rutilus*, *Leuciscus cephalus* and *Scardinius erythrophthalmus*). R_a, active metabolic rate; R_r, routine metabolic rate; R_a/R_r, relative scope. Bars indicate SD. (Pooled data from Wieser & Forstner, 1986.)

rates continue to increase up to close to the upper LT_{50} (Fig. 7). The reason sockeye salmon and brook trout appear to respond atypically may reflect the fact that in both cases tests were started using late-stage larvae and continued well into the fry stage. The measured response, thus, may be more typical of juvenile than larval fish.

Growth efficiency during early life appears to vary relatively little with temperature either across species (Houde, 1989; Houde & Zastrow, 1993) or within species. The response at the species level, in particular, is quite different from that seen in juvenile fish. In juvenile fish, growth efficiency typically is maximal at some intermediate temperature near the centre of the zone of tolerance (the optimum temperature for growth efficiency is usually slightly lower than the optimum temperature for growth rate; Brett, 1979). Studies of fish embryos and

Table 4. *Mean specific growth rates of fish embryos and larvae*

Species	Stage	Ration	Temperature (°C)	Growth rate (% day^{-1})	Q_{10} (interval)	Q_{10} (mean)	Reference
Oncorhynchus tshawytscha (chinook salmon)	E–L (bc–mtw)	end	5.0	3.5			Rombough, 1994
			7.3	5.0		4.72	
Oncorhynchus tshawytscha	E–L (f–mtw)	end	10.0	6.5	2.64		Heming, 1982
			10.2	7.0	40.7		
			12.5	7.3	1.20	2.77	
Oncorhynchus tshawytscha	L–F (1st feed– 1634 ATU pf)	ex	6.0	2.0	3.19		Heming et al., 1982
			8.0	2.5	4.63		
			10.0	3.4	4.00		
			12.0	4.5		3.96	
		high	6.0	1.9	7.14		
			8.0	2.8	1.50		
			10.0	3.1	3.80		
			12.0	4.0		3.17	
Oncorhynchus nerka (sockeye salmon)	L–F (post swim-up, 12 d mean)	ex	5.0	1.5	7.09		Shelbourn et al., 1973
		high	10.0	3.9	3.96		
			15.0	7.8	0.61		
			20.0	6.1		2.69	
Salvelinus fontinalis (brook trout)	L–F (56 d)	ex	7.1	2.4	5.39		McCormick et al., 1972
		high	9.8	3.8	1.84		
			12.4	4.5	1.04		
			15.4	4.5	0.61		
			17.9	4.0	0.16		
			19.5	3.0		1.18	

Species	Rearing	Feeding	Ration					Reference
Salvelinus alpinus (Arctic charr)	L (h–100% ya)	mixed	high	3	1.4	6.50		Wallace & Aasjord, 1984
				6	2.4	2.93		
				8	3	0.51		
				12	2.3		1.72	
Coregonus artedii (lake cisco)	L–F (h–28 dph)	ex	ad libitum	3.4	2.4	9.19		McCormick et al., 1971
				7.0	5.4	4.22		
				9.8	8.1	2.02		
				12.8	10.0	2.23		
				15.6	12.5	1.34		
				18.1	13.46	1.74		
				20.8	15.62		2.70	
Coregonus albula (vendace)	L (h–21 dph)	ex	ad libitum	4.8	2.3	32.89		Luczynski, 1991
				7.3	5.5	4.31		
				10.0	8.1	3.49		
				12.3	10.8	2.38		
				15.3	14.0	1.70		
				17.1	15.4	1.21		
				19.8	16.2	1.30		
				22.1	17.2		2.87	
Coregonus albula	L (h–26 dph)	ex	maximum	8.0	5.0	2.19		Koho et al., 1991
				13.5	7.7			
Rutilus rutilus (roach)	L–F (h–35–40 dph)	ex		15.0	6.9	4.24		Wieser et al., 1988
				20.0	14.2	1.32		
				25.0	16.3		2.36	

Table 4. (cont.)

Species	Stage	Ration	Temperature (°C)	Growth rate (% day⁻¹)	Q₁₀ (interval)	Q₁₀ (mean)	Reference
Leuciscus cephalus (chub)	L–F (h–35–40 dph)	ex	20.0	10.8			Wieser *et al.*, 1988
			25.0	14.2	1.73		
Alburnus alburnus (bleak)	L–F (h–35–40 dph)	ex	20.0	13.7			
			25.0	16.9	1.52		
Catostomus commersoni (white sucker)	L (h–28 dph)	ex	10.0	0.9			McCormick *et al.*, 1977
			15.7	7.4	40.29		
			17.9	9.7	3.42		
			20.8	12.0	2.08		
			23.9	13.2	1.36		
			26.9	14.8	1.46		
			29.7	9.2	0.18	2.95	
Morone americana (white perch)	L (8 dph– 1st feed)	mixed high	13.0	4.9			Marguiles, 1989
			17.0	15.2	16.95		
			21.0	21.8	2.46	6.46	
Oreochromis niloticus (Nile tilapia)	L (3–6 dph)	end	17.0	8.4			Rana, 1990
			20.0	14.2	5.75		
			24.0	15.3	1.21		
			28.0	37.8	9.59		
			30.0	39.1	1.18	3.31	
Oreochromis mossambicus (Mozambique tilapia)	L (h–9 dph)	end	20.1	7.9			Subasinghe & Sommerville, 1992
			24.3	13.5	3.53		
			29.8	20.4	2.13		
			34.5	31.3	2.46	2.53	
Clarias gariepinus (African catfish)	L	end	22.1	89.0			Kamler *et al.*, 1994
			25.0	150.0	6.05		
			28.1	183.0	1.90	3.30	

Species		Feeding					Reference
Pseudopleuronectes americanus (white flounder)	L (ya–met)	ex	2.0	2.6			Laurence, 1975
			5.0	5.8	14.50	9.60	
			8.0	10.1	6.35		
Clupea pallasi (Pacific herring)	L h–4 dph	mixed	6.0	3.8			McGurk, 1984
			8.0	4.0	1.29		
			10.0	4.4	1.61	1.44	
Leiostomus xanthrus (spot)	L (?–4–6 dp?)	ex ad libitum	12.0	1.4			Hoss et al., 1988
			16.0	2.3	3.46		
			18.0	2.6	1.85	2.89	

Abbreviations: end, endogenous; ex, exogenous; E, embryo; L, larva, F, fry; bp, blastopore closure; dpf, days post fertilization; dph, days posthatch; met, metamorphosis; mtw, maximum tissue weight; ya, yolk absorption; ATU, accumulated thermal units.

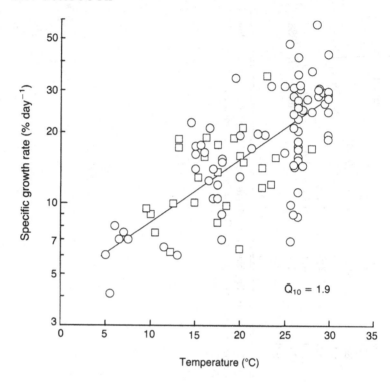

Fig. 6. Specific growth rates for larvae of various freshwater (□) and marine (○) fish plotted as a function of temperature. (Data compiled by Houde & Zastrow, 1993.)

larvae, to date, have revealed no such consistent trend. Of 25 studies of cumulative growth efficiency variation with temperature listed in Table 5, six found essentially no change over the temperature range tested, eight reported an increase in efficiency with temperature, five reported a decrease, and only six reported maximal efficiency at some intermediate temperature. Much of this scatter appears to reflect differences in assessment procedures: in particular, the choice of the interval over which to compare cumulative growth efficiency. Endogenous yolk supplies tend to become limiting earlier in development at higher temperatures (Heming, 1982; Rombough, 1988b). This creates a problem if the interval chosen for comparison extends into the period when yolk is limiting at some temperatures but not at others. For example, if cumulative growth efficiencies of steelhead (*Oncorhynchus mykiss*)

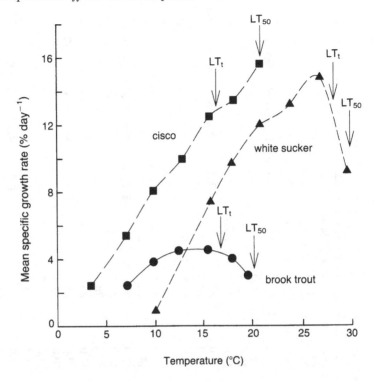

Fig. 7. Mean specific growth rates as a function of temperature for three species of freshwater fish. Data are from McCormick *et al.* 1971 (cisco), McCormick *et al.* 1972 (brook trout) and McCormick *et al.* 1977 (white sucker). LT_t, upper lethal threshold; LT_{50}, temperature corresponding to 50% mortality.

are compared over the interval from fertilization to 80% yolk absorption, fish raised at 15 °C appear to be significantly less efficient than those raised at lower temperatures (Fig. 8). However, if the comparison is made earlier at, say, 60% yolk absorption, temperature appears to have no significant effect. The discrepancy arises from the fact that at 15 °C larvae began to become food-limited at about 70% yolk absorption while at lower temperatures food remained in excess throughout the period of comparison. In this regard it is interesting that the four studies listed in Table 5 that were terminated shortly after hatch (From & Rasmussen, 1991; Kamler & Kato, 1983; Howell, 1980; Laurence, 1973), well before yolk supplies would normally become limiting even at high temperatures, all reported that conversion

Table 5. *Growth efficiency of fish embryos and larvae*

Freshwater

Species	Stage	Ration	Temperature (°C)	Efficiency (%)	Mean	Measure	Reference
Oncorhynchus mykiss (rainbow trout)	E-L (f-90% ya)	yolk	6.0	70.5		P/A d	Rombough, 1988b
			9.0	68.0			
			12.0	68.5			
			15.0	66.2	69.0		
Oncorhynchus mykiss	E (f-h)	yolk	5.0	85.2		P/A e	From & Rasmussen, 1991
			10.0	94.2			
			15.0	96.8	92.1		
	L (h-100% ?ya)	yolk	5.0	64.0		P/A e	
			10.0	74.5			
			15.0	68.1	69.3		
Oncorhynchus mykiss	E (f-h)	yolk	9.0	67.0		P/A+R e	Kamler & Kato, 1983
			10.0	72.7			
			12.0	78.2			
			14.0	86.0	76.0		
	L (f-100% ?ya)		9.0	54.0		P/A+R e	Kamler & Kato, 1983
			10.0	48.5			
			12.0	57.7			
			14.0	73.0	58.3		
Oncorhynchus tshawytscha (chinook salmon)	E-L (f-mtw)		5.0	57.0		P/A e	Rombough, 1994
			7.3	55.6			
			10.0	54.0			
			10.2	53.7			
			12.5	44.6	53.0		

Species	Stage	Yolk	d	P/A	Mean	Reference
Oncorhynchus tshawytscha	E-L (f–mtw)	yolk	6.0	67.9		P/A d Heming, 1982
			8.0	63.0		
			10.0	60.5		
			12.0	48.8	63.8	
Salmo salar (Atlantic salmon)	L (15–80% ya)	yolk	7.6	65.0		P/A d Marr, 1966
			10.0	70.0		
			14.3	64.0	66.3	
Salmo salar	L (early–mid-ya)	yolk	0.2	42.3		P/A d Hayes & Pelluet, 1945 (values recalculated)
			2.8	47.5		
			4.0	35.9		
			6.3	40.7		
			8.4	50.0		
			9.3	75.0		
			10.6	50.8		
			12.2	40.3		
			13.3	56.2		
			14.3	41.0		
			15.2	57.3		
			16.0	38.6	48.0	
Salvelinus alpinus (Arctic charr)	L (h–100% ya)	yolk	3.0	48.6		P/A d Wallace & Aasjord, 1984
			6.0	50.6		
			8.0	47.0		
			12.0	32.1	44.6	
Salmo trutta (brown trout)	L (56 dph–no vis. yolk)	yolk	3.0	48.2		P/A d Wood, 1932
			7.0	60.9		
			12.0	60.2	56.4	

Table 5. (cont.)

Species	Stage	Ration	Temperature (°C)	Efficiency (%)	Mean	Measure	Reference
Clarias gariepinus (African catfish)	E-L (f-no vis. yolk)	yolk	22.0	64.0		P/A e	Kamler et al., 1994
			25.0	71.0			
Clarias gariepinus	L mid-ya	yolk	28.0	68.0	67.7	P/A (max) d (estimated)	Conceicao et al., 1993
			20.0	62.0			
			25.0	65.0	65.3		
			30.0	69.0			
Oreochromis niloticus (Nile tilapia)	L (h–mtw)	yolk	24.0	55.4		P/A d	Rana, 1990
			28.0	57.2			
			30.0	61.7	58.1		
Channa striatus	E-L (f-100% ya)	yolk	20.0	79.0		P/A e	Arul, 1991
			24.0	75.0			
			28.0	69.5			
			32.0	65.0	72.1		
Rutilus rutilus (roach)	L (feeding)	Artemia ad lib.	15.0	70.3		P/P+R e	Wieser et al., 1988
			20.0	73.9	72.1		
Marine							
Paralichthys dentatus (summer flounder)	E-L (f-100% ya)	yolk	11.0	65.0		P/A e	Johns et al., 1981
			16.0	63.0			
			21.0	63.0	63.7		

Species	Stage						Reference
Paralichthys dentatus	E-L (f-100% ya)	yolk	16.0	62.0			Johns & Howell, 1980
			21.0	62.0	62.0	P/A e	
Limanda ferruginea (yellowtail flounder)	E-L (f-100% ya)	yolk	4.0	29.8		P/A o	Howell, 1980
			8.0	43.8			
			10.0	42.2			
			12.0	47.1	40.7		
	E (f-2 hph)	yolk	4.0	46.3		P/A o	
			8.0	73.6			
			10.0	76.1			
			12.0	86.4	70.6		
Tautoga onitis (tautog)	E-L (f-100% ya)	yolk	16.0	36.3		P/A e	Laurence, 1973
			19.0	25.5			
			22.0	25.8	29.2		
Bairdiella icistia (bairdiella)	E-L (f-100% ya)	yolk	21.0	70.6		P/A d	May, 1974
			24.0	72.5			
			27.0	71.7	71.6		
Leuresthes tenuis (grunion)	L (h-100% ya?)	yolk	12.0	55.5		P/A d	Ehrlich & Muszynski, 1982
			15.0	61.5			
			18.0	67.0			
			22.0	66.3			
			25.0	54.5	61.0		
Clupea harengus (herring)	E (f-h)	yolk	8.0	60.2		P/A d	Blaxter & Hempel, 1966
			12.0	60.8	60.5		
	E-L (f-mtw)	yolk	8.0	57.2		P/A d	
			12.0	56.8	57.0		

Abbreviations: P, production; R, respiration; A, assimilated. Subscripts: e, energy basis; d, dry weight basis; o, organic matter basis. Other abbreviations as in Table 4.

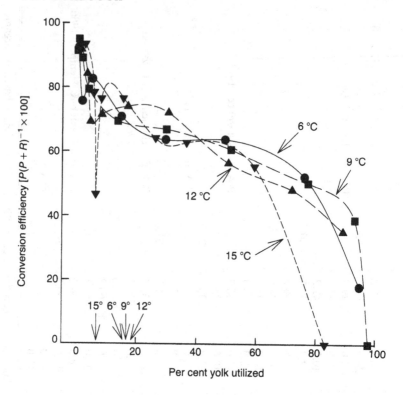

Fig. 8. Ontogenetic changes in energy conversion efficiencies in developing steelhead trout (*Oncorhynchus mykiss*) at four temperatures. *P*, production; *R*, respiration. Arrows indicate the percentage yolk consumed at hatch. (Data from Rombough, 1988b.)

efficiency increased with temperature. It may well be that the underlying trend is for there to be a slight increase in growth efficiency at higher temperatures because of energy savings accruing either directly or indirectly from higher rates of growth. Evidence is beginning to accumulate that the incremental cost of growth may begin to decline in young fish at high rates of growth (Wieser & Medgyesy, 1990a, b; Houlihan *et al.*, 1992; Mathers *et al.*, 1993).

Discussion

Temperate zone fishes do, indeed, appear to be more sensitive to temperature change during early life than they are as juveniles and adults. The tolerance range (ULT–LLT) for embryos of temperate species ($\cong$ 11.6 °C) typically is only about half that of juvenile and adult fish (20–25 °C). Tropical species, on the other hand, appear to display roughly the same degree of sensitivity at all life stages. Direct lethality resulting from disruption of developmental processes appears to be the primary mode of action of temperature during early life. This contrasts with the situation later in life when sub-lethal effects, particularly changes in energetic efficiency, appear to be more important. While temperature generally has a greater impact on metabolism and growth during embryonic and larval development than it does later in life, the net effect on growth efficiency appears to be negligible provided food is not limiting and temperatures are within the zone of tolerance. Cross-species comparisons are useful in predicting overall trends but generally underestimate the effects of temperature on individual species.

Global warming models predict that temperature increases will be more pronounced towards the poles. Fish populations living at mid- to high latitudes are most likely to be directly affected. In general, the impact of global warming is likely to be greater on freshwater than marine species. Highly endemic species such as the fountain darter (*Etheostoma fonticola*), which already spawns at temperatures close to its upper lethal limit (Brandt *et al.*, 1993), appear most vulnerable. Most studies of the effect of global warming on fish have focused on summer temperatures. Temperature increases at higher mid-latitudes, however, are projected to be greatest during the winter months (Coutant, 1990) when many fish spawn. Scott and Poynter (1991) noted that the current distribution of trout in New Zealand appears to be determined by high winter rather than summer temperatures. They speculated that global warming would further restrict suitable spawning habitat and result in a poleward shift in the northern limit of both brown trout and rainbow trout. A rise in winter temperatures in the mid-latitudes also would tend to accentuate the temperature–oxygen squeeze for demersal spawners such as whitefish (*Coregonus* spp.) that inhabit deepwater lakes, particularly if surface temperatures remain high enough to prevent seasonal turnover of the bottom water (Trippel, Eckmann & Hartmann, 1991).

With global warming, there will be winners as well as losers. The poleward range of many freshwater species currently appears to be restricted by the inability of young fish to grow large enough during their first year to survive winter starvation (Shuter & Post, 1990). Temperature greatly accelerates larval growth-rates ($Q_{10} \cong 3.0$) and has little effect on growth efficiency. Provided there is sufficient food, global warming should permit species constrained by overwintering size to extend their ranges poleward. Enhanced larval growth, similarly, is likely to result in increased abundance, at least during the early stages of global warming, of endemic species such as the Lake Constance whitefish (*Coregonus lavaretus*) that are able to tolerate temperatures several degrees higher than those they now experience during the spring bloom when food is most abundant (Trippel *et al.*, 1991).

This review has focused on the direct (autecological) effects of temperature. It has been suggested that indirect (synecological) effects involving interactions with other species, particularly predator–prey relations, may be just as important in the context of global warming (Blaxter, 1992). Unfortunately, we have little information on how temperature affects biotic interactions at the species level. When viewed across species, temperature appears to have surprisingly little effect on larval recruitment. Instantaneous mortality tends to increase at higher temperatures but so does growth rate (Houde, 1989; Pepin, 1991). This led Pepin (1991) to speculate that greater instantaneous mortality at higher temperatures may be largely offset by more rapid development (i.e. a shorter period of vulnerability) and that the net impact of temperature on recruitment may be negligible. However, given that cross-species comparisons tend to underestimate temperature effects at the species level for abiotic interactions, it would be premature to extend this hypothesis to include lower taxonomic units until we have a better understanding of species interactions.

References

Alderdice, D. F. &. Forrester, C. R. (1968). Some effects of salinity and temperature on early development and survival of the English sole (*Parophrys vetulus*) *Journal of the Fisheries Research Board of Canada*, **15**, 229–49.

Alderdice, D. F. & Forrester, C. R. (1971). Effects of salinity and temperature on embryonic development of the Petrale sole (*Eopsetta jordani*). *Journal of the Fisheries Research Board of Canada*, **28**, 727–44.

Alderdice, D. F. & Forrester, C. R. (1974). Early development and distribution of the Flathead sole (*Hippoglossoides elassodon*). *Journal of the Fisheries Research Board of Canada*, **31**, 1899–1918.

Alderdice, D. F. & Velsen, F. P. J. (1971). Some effects of salinity and temperature on early development of Pacific herring (*Clupea pallasi*). *Journal of the Fisheries Research Board of Canada*, **28**, 1545–62.

Allanson, B. R. & Noble, R. G. (1964). The tolerance of *Tilapia mossambica* (Peters) to high temperature. *Transactions of the American Fisheries Society*, **93**, 323–32.

Arul, V. (1991). Effect of temperature on yolk utilization of *Channa striatus*. *Journal of Thermal Biology*, **16**, 1–5.

Beacham, T. D. & Murray, C. B. (1990). Temperature, egg size, and development of embryos and alevins of 5 species of Pacific salmon – a comparative analysis. *Transactions of the American Fisheries Society*, **119**, 927–45.

Bishai, H. M. (1960). Upper lethal temperatures for larval salmonids. *Journal of the Society for the International Exploration of the Sea*, **25**, 129–33.

Blaxter, J. H. S. (1956). Herring rearing – II. The effect of temperature and other factors on development. *Marine Resources of Scotland*, No. 5, 19pp.

Blaxter, J. H. S. (1960). The effect of extremes of temperature on herring larvae. *Journal of the Marine Biological Association of the United Kingdon*, **39**, 605–8.

Blaxter, J. H. S. (1992). The effect of temperature on larval fish. *Netherlands Journal of Zoology*, **42**, 336–57.

Blaxter, J. H. S. & Hempel, G. (1966). Utilization of yolk by herring larvae. *Journal of the Marine Biological Association of the United Kingdom*, **46**, 219–34.

Blood, D. M., Matarese, A. C. & Yoklavich, M. M. (1994). Embryonic development of walleye pollock, *Theragra chalcogramma*, from Shelikof Strait, Gulf of Alaska *Fisheries Bulletin U.S.*, **92**, 207–22.

Brandt, T. M., Graves, K. G., Berkhouse, C. S., Simon, T. P. & Whiteside, B. G. (1993). Laboratory spawning and rearing of the endangered fountain darter. *Progressive Fish-Culturist*, **55**, 149–56.

Brannas, E. (1988). Emergence of Baltic salmon, *Salmo salar L.*, in relation to temperature: a laboratory study. *Journal of Fish Biology*, **33**, 589–600.

Brett, J. R. (1970). Temperature – fishes. In *Marine Ecoloy*, Vol I. *Environmental Factors*, Part 1, ed. O. Kinne, pp. 515–60. London: Willey-Interscience.

Brett, J. R. (1979). Environmental factors and growth. In *Fish Physiology*, Vol. VIII, ed. W. S. Hoar, D. J. Randall & J. R. Brett, pp. 599–675. San Diego: Academic Press.

Brett, J. R., Clarke, W. C. & Shelbourn, J. E. (1982). Experiments on thermal requirements for growth and food conversion efficiency of juvenile chinook salmon *Oncorhynchus tshawytscha*. *Canadian Technical Report of Fisheries and Aquatic Sciences*, No. 1127, 29 pp. Ottawa: Government of Canada.

Brooke, L. T. (1975). Effect of different constant incubation temperatures on egg survival and embryonic development in lake whitefish (*Coregonus clupeaformis*). *Transactions of the American Fisheries Society*, **104**, 555–9.

Brooke, L. T. & Colby, P. J. (1980). Development and survival of embryos of lake herring *Coregonus artedii* at different constant oxygen concentrations and temperatures. *Progressive Fish-Culturist*, **42**, 3–9.

Buckley, L. J., Smigielski, A. S., Halavik, T. A. & Laurence, G. C. (1990). Effects of water temperature on size and biochemical composition of winter flounder *Pseudopleuronectes americanus* at hatching and feeding initiation. *Fisheries Bulletin U.S.*, **88**, 419–28.

Buddington, R. K., Hazel, J. R., Poroshov, S. I. & Vaneehenhaam, J. (1993). Ontogeny of the capacity for homeoviscous adaptation in white sturgeon (*Acipenser transmontanus*). *Journal of Experimental Zoology*, **265**, 18–28.

Cloud, J. G., Erdahl, A. L. & Graham, E. F. (1988). Survival and continued normal development of fish embryos after incubation at reduced temperatures. *Transactions of the American Fisheries Society*, **117**, 503–6.

Colby, P. J. & Brooke, L. T. (1970). Survival and development of lake herring (*Coregonus artedii*) eggs at various incubation temperatures. In *Biology of Coregonid Fishes*, ed. C. C. Lindsey & C. S. Woods, pp. 417–28. Winnipeg: University of Manitoba Press.

Colby, P. J. & Brooke, L. T. (1973). Effects of temperature on embryonic development of lake herring (*Coregonus artedii*). *Journal of the Fisheries Research Board of Canada*, **30**, 799–810.

Combs, B. D. (1965). Effect of temperature on the development of salmon eggs. *Progressive Fish-Culturist*, **27**, 134–7.

Conceicao, L., Verreth, J., Scheltema T. & Machiels, M. (1993). A simulation model for the metabolism of yolk-sac larvae of the African catfish, *Clarias gariepinus* (*Bruchell*). *Aquaculture and Fisheries Management*, **24**, 431–43.

Coutant, C. C. (1990). Temperature–oxygen habitat for freshwater and coastal striped bass in a changing climate. *Transactions of the American Fisheries Society*, **119**, 240–53.

Crisp, D. T. (1981). A desk study of the relationship between temperature and hatching time for the eggs of five species of salmonid fishes. *Freshwater Biology*, **11**, 361–8.

Crisp, D. T. (1988). Prediction, from temperature, of eyeing, hatching and 'swim-up' times for salmonid embryos. *Freshwater Biology*, **19**, 41–8.

Eaton, J. G., McCormick, J. H., Goodno, B. E., O'Brien, D. G., Stefany, H. G., Hondzo, M. & Scheller, R. M. (1995). A field information-based system for estimating fish temperature tolerances. *Fisheries*, **20**(4), 10–18.

Ehrlich, K. F. & Muszynski, G. (1981). The relationship between temperature-specific yolk utilization and temperature selection of larval grunion. *Rapports et Proces-Verbaux des Réunions du Conseil International pour l'Exploration de la Mer*, **178**, 312–13.

Ehrlich, K. F. & Muszynski, G. (1982). Effect of temperature on interactions of physiological and behavioral capacities of larval California grunion: adaptations to the planktonic environment. *Journal of Experimental Marine Biology and Ecology*, **60**, 223–44.

Ehrlich, K. F. &. Farris, F. A. (1971). Some influences of temperature on the development of the grunion *Leuresthes tenuis* (*Ayres*). *California Fish and Game*, **57**, 58–68.

Fonds, M., Rosenthal, H. & Alderdice, D. F. (1974). Influence of temperature and salinity on embryonic development, larval growth and number of vertebrae of the garfish, *Belone belone*. In *The Early Life History of Fish*, ed. J. H. S. Blaxter, pp. 509–25. Heidelberg: Springer-Verlag.

Forrester, C. R. (1964a). Laboratory observations on embryonic development and larvae of the Pacific cod (*Gadus macrocephalus* (*Tilesius*)). *Journal of the Fisheries Research Board of Canada*, **21**, 9–16.

Forrester, C. R. (1964b). Rate of development of eggs of rock sole (*Lepidopsetta bilineata* (Ayres)). *Journal of the Fisheries Research Board of Canada*, **21**, 1533–34.

From, J. & Rasmussen, G. (1991). Growth of rainbow trout, *Oncorhynchus mykiss* (Walbaum, 1792) related to egg size and temperature. *Dana*, **9**, 31–8.

Gadomski, D. M. & Caddell, S. M. (1991). Effects of temperature on early life history stages of California halibut, *Paralichthys californicus*. *Fisheries Bulletin U.S.*, **89**, 567–76.

Garside, E. T. (1966). Effects of oxygen in relation to temperature on the development of embryos of brook trout and rainbow trout. *Journal of the Fisheries Research Board of Canada*, **23**, 1121–34.

Guma'a, S. A. (1978). The effect of temperature on the development and mortality of eggs of perch, *Perca fluviatilis*. *Freshwater Biology*, **8**, 221–7.

Hassler, T. J. (1982). Effect of temperature on survival of northern pike embryos and yolksac larvae. *Progressive Fish-Culturist*, **44**, 174–8.

Hayes, F. R. & Pelluet, D. (1945). The effect of temperature on the growth and efficiency of yolk conversion in the salmon embryo. *Canadian Journal of Research*, **23**, 7–15.

Haynes, E. B. & Ignell, S. E. (1983). Effect of temperature on rate of embryonic development of walleye pollock, *Theragra chalcogramma*. *Fisheries Bulletin U.S.*, **81**, 890–4.

Heming, T. A. (1982). Effects of temperature on utilization of yolk by chinook salmon (*Oncorhynchus tshawytscha*) eggs and alevins. *Canadian Journal of Fisheries and Aquatic Sciences*, **39**, 184–90.

Heming, T. A., McInerney, J. E. & Alderdice, D. F. (1982). Effect of temperature on initial feeding in alevins of chinook salmon (*Oncorhynchus tshawytscha*). *Canadian Journal of Fisheries and Aquatic Sciences*, **39**, 1554–62.

Hempel, G. (1984). *Early Life History of Marine Fish.* Seattle: University of Washington Press.

Herzig, A. & Winkler, H. (1986). The influence of temperature on the embryonic development of three cyprinid fishes, *Abramis brama, Chalcalburnus chalcoides monto* and *Vimba vimba. Journal of Fish Biology*, **28**, 171–81.

Hokanson, K. E. F., McCormick, J. G. & Jones, B. R. (1973). Temperature requirements for embryos and larvae of the northern pike, *Esox lucius* (Linnaeus). *Transactions of the American Fisheries Society*, **102**, 89–100.

Hokanson, K. E., McCormick, J. H., Jones, B. R. & Tucker, H. (1973). Thermal requirements for maturation, spawning and embryo survival of the brook trout, *Salvelinus fontinalis. Journal of the Fisheries Research Board of Canada*, **30**, 975–84.

Hokanson, K. E. F. & Kleiner, Ch. F. (1974). Effects of constant and rising temperatures on survival and developmental rates of embryonic and larval yellow perch, *Perca flavescens* (Mitchill). In *The Early Life History of Fish*, ed. J. H. S. Blaxter, pp. 437–48. Heidelberg: Springer-Verlag.

Hoss, D. E., Coston-Clements, L., Peters, D. S. & Tester, P. A. (1988). Metabolic responses of spot, *Leiostomus xanthurus*, and Atlantic croaker, *Micropogonias undulatus*, larvae to cold temperatures encountered following recruitment to estuaries. *Fisheries Bulletin U.S.*, **86**, 483–9.

Houde, E. D. (1989). Comparative growth, mortality, and energetics of marine fish larvae: temperature and implied latitudinal effects. *Fisheries Bulletin U.S.*, **87**, 471–95.

Houde, E. D. & Zastrow, C. E. (1993). Ecosystem and taxon-specific dynamic and energetic properties of larval fish assemblages. *Bulletin of Marine Science*, **53**, 290–335.

Houghton, R. A. & Woodwell, G. M. (1989). Global climatic change. *Scientific American*, **260**(4), 36–44.

Houlihan, D. F., Wieser, W., Foster, A. & Brechin, J. (1992). *In vivo* protein synthesis rates in larval nase (*Chondrostoma nasus* L.). *Canadian Journal of Zoology*, **70**, 2436–40.

Howell, W. H. (1980). Temperature effects on growth and yolk utilization in yellowtail flounder, *Limanda ferruginea*, yolk-sac larvae. *Fisheries Bulletin U.S.*, **78**, 731–9.

Hubbs, C. & Bryan, C. (1974). Effect of parental temperature experience on thermal tolerance of eggs of *Menidia audens*. In *The Early Life History of Fish*, ed. J. H. S. Blaxter, pp. 431–5. Heidelberg: Springer-Verlag.

Hubbs, C. L. (1965). Developmental temperature tolerance and rates of four southern California fishes, *Fundulus parvipinnis, Atherinops affinis, Leuresthes tenuis* and *Hypsoblennius* sp. *California Fish and Game*, **51**, 113–22.

Humpesch, V. H. (1985). Inter- and intra-specific variation in hatching success and embryonic development of five species of salmonids and *Thymallus thymallus*. *Archives für Hydrobiologie*, **104**, 129–44.

Intergovernmental Panel on Climate Change (IPCC) (1992). *Climate Change 1992: The Supplementary Report to the IPCC Assessment*, ed. J. T. Houghton, B. A. Callander & S. K. Varney. Cambridge: Cambridge University Press.

Irvin, D. N. (1974). Temperature tolerance of early developmental stages of Dover sole, *Solea solea* (L.). In *The Early Life History of Fish*, ed. J. H. S. Blaxter, pp. 449–63. Heidelberg: Springer-Verlag.

Jennings S. & Pawson, M. G. (1991). The development of bass, *Dicentrarchus labrax*, eggs in relation to temperature. *Journal of the Marine Biological Association of the United Kingdom*, **71**, 107–16.

Johansen, A. C. & Krogh, A. (1914). The influence of temperature and certain other factors upon the rate of development of the eggs of fishes. *Publications Circonstantial du Conseil Permanent internationale pour l'Exploration de la Mer*, **68**, 1–44.

Johns, D. M. & Howell, W. H. (1980). Yolk utilization in summer flounder (*Paralichthys dentatus*) embryos and larvae reared at two temperatures. *Marine Ecology*, **2**, 1–8.

Johns, D. M., Howell, W. H. & Klein-MacPhee, G. (1981). Yolk utilization and growth to yolk-sac absorption in summer flounder (*Paralichthys dentatus*) larvae at constant and cyclic temperatures. *Marine Biology*, **63**, 301–8.

Jungwirth, M. & Winkler, H. (1984). The temperature dependence of embryonic development of grayling (*Thymallus thymallus*), Danube salmon (*Hucho hucho*), Arctic char (*Salvelinus alpinus*) and brown trout (*Salmo trutta fario*). *Aquaculture*, **38**, 315–27.

Kamler, E. & Kato, T. (1983). Efficiency of yolk utilization by *Salmo gairdneri* in relation to incubation temperature and egg size. *Polskie Archiwum Hydrobiologii*, **30**, 271–306.

Kamler, E., Szlaminska, M., Kuczynski, M., Hamackova, J., Kouril, J. & Dabrowski, R. (1994). Temperature-induced changes of early development and yolk utilization in African catfish *Clarius gariepinus*. *Journal of Fish Biology*, **44**, 311–26.

Kaufmann, R. & Wieser, W. (1992). Influence of temperature and ambient oxygen on the swimming energetics of cyprinid larvae and juveniles. *Environmental Biology of Fishes*, **33**, 87–95.

Kinne, O. & Kinne, E. M. (1962). Rates of development in embryos of a cyprinodont fish exposed to different temperature–salinity–oxygen combinations. *Canadian Journal of Zoology*, **40**, 231–53.

Koenst, W. M. & Smith L. L., Jr. (1976). Thermal requirements of the early life stages of walleye, *Stizostedion vitreum vitreum*, and sauger, *S. canadense*. *Journal of the Fisheries Research Board of Canada*, **33**, 1130–8.

218   P.J. ROMBOUGH

Koho, J., Karjalainen, J. & Viljanen, M. (1991). Effects of temperature, food density and time of hatching on growth, survival and feeding of vendace (*Coregonus albula* L.) larvae. *Aqua Fennica*, **21**, 63–73.

Kokurewicz, B. (1981). Effect of different thermal regimes on reproductive cycles of tench *Tinca tinca* (L.). Part VII. Embryonal development of progeny. *Polskie Archiwum Hydrobiologii*, **28**, 243–56.

Kroll, K. J., Van Eenennaam, J. P., Doroshov, S. I., Hamilton, J. E. & Russell, T. R. (1992). Effect of water temperature and formulated diets on growth and survival of larval paddlefish. *Transactions of the American Fisheries Society*, **121**, 538–43.

Lasker, R. (1962). Efficiency and rate of yolk utilization by developing embryos and larvae of the Pacific sardine *Sardinops caerulea* (Girard). *Journal of the Fisheries Research Board of Canada*, **19**, 867–75.

Laurence, G. C. (1973). Influence of temperature on energy utilization of embryonic and prolarval tautog, *Tautoga onitis*. *Journal of the Fisheries Research Board of Canada*, **30**, 435–42.

Laurence, G. C. (1975). Laboratory growth and metabolism of the winter flounder *Pseudopleuronectes americanus* from hatching through metamorphosis at three temperatures. *Marine Biology*, **32**, 223–9.

Legendre, M. & Teugels, G. G. (1991). Development and thermal tolerance of eggs in *Heterobranchus longifilis*, and comparison of larval developments of *Heterobranchus longifilis* and *Clarias gariepinus* (Teleostei, Clariidae). *Aquatic Living Resources*, **4**, 227–40.

Lewis, R. M. (1965). The effect of minimum temperature on the survival of larval Atlantic menhaden *Brevoortia tyrannus*. *Transactions of the American Fisheries Society*, **94**, 409–12.

Luczynski, M. (1991). Temperature requirements for growth and survival of larval vendace, *Coregonus albula* L. *Journal of Fish Biology*, **38**, 29–36.

Luczynski, M. & Kirklewskwa, A. (1984). Dependence of *Coregonus albula* embryogenesis rate on the incubation temperature. *Aquaculture*, **42**, 43–55.

McCormick, J. H., Jones, B. R. & Syrett, R. F. (1971). Temperature requirements for growth and survival of larval ciscos (*Coregonus artedii*). *Journal of the Fisheries Research Board of Canada*, **28**, 924–7.

McCormick, J. H., Hokanson, K. E. F. & Jones, B. R. (1972). Effects of temperature on growth and survival of young brook trout, *Salvelinus fontinalis*. *Journal of the Fisheries Research Board of Canada*, **29**, 1107–12.

McCormick, J. H., Jones, B. R. & Hokanson, K. E. F. (1977). White sucker (*Catostomus commersoni*) embryo development, early growth and survival at different temperatures. *Journal of the Fisheries Research Board of Canada*, **34**, 1019–25.

McGurk, M. D. (1984). Effects of delayed feeding and temperature on the age of irreversible starvation and on the rates of growth and mortality of Pacific herring larvae. *Marine Biology*, **84**, 13–26.

Margulies, D. (1989). Effects of food concentration and temperature on development, growth, and survival of white perch, *Morone americana*, eggs and larvae. *Fisheries Bulletin U.S.*, **87**, 63–72.

Marr, D. H. A. (1966). Influence of temperature on the efficiency of growth of salmonid embryos. *Nature (London)*, **212**, 957–9.

Marten, P. S. (1992). Effect of temperature variation on the incubation and development of brook trout eggs. *Progressive Fish-Culturist*, **54**, 1–6.

Mathers, E. M., Houlihan, D. F., McCarthy, I. D. & Burren, L. J. (1993). Rates of growth and protein synthesis correlated with nucleic acid content in fry of rainbow trout, *Oncorhynchus mykiss* – effects of age and temperature. *Journal of Fish Biology*, **43**, 245–63.

May, R. C. (1974). Effects of temperature and salinity on yolk utilization in *Bairdiella icistia* (Jordon and Gilbert) (Pisces: Sciaenidae). *Journal of Experimental Marine Biology and Ecology*, **16**, 213–25.

May, R. C. (1975a). Effects of temperature and salinity on fertilization, embryonic development, and hatching in *Bairdiella icistia* (Pisces: Sciaenidae), and the effect of parental salinity acclimation on embryonic and larval salinity tolerance. *Fisheries Bulletin U.S.*, **73**, 1–22.

May, R. C. (1975b). Effects of acclimation on the temperature and salinity tolerance of the yolk-sac larvae of *Bairdiella icistia* (Pisces: Sciaenidae). *Fisheries Bulletin U.S.*, **73**, 249–55.

Miranda, C., Cal, R. M. & Iglesias, J. (1990). Effect of temperature on the development of eggs and larvae of sardine *Sardina pilchardus* Walbaum in captivity. *Journal of Experimental Marine Biology and Ecology*, **140**, 69–77.

Murawski, S. A. (1993). Climate change and marine fish distributions: forecasting from historical analogy. *Transactions of the American Fisheries Society*, **122**, 647–58.

Murray, C. B. & Beacham, T. D. (1986). Effect of varying temperature regimes on the development of pink salmon (*Oncorhynchus gorbuscha*) eggs and alevins. *Canadian Journal of Zoology*, **64**, 670–6.

Pauly, D. & Pullin, R. S. V. (1988). Hatching time in spherical, pelagic, marine fish eggs in response to temperature and egg size. *Environmental Biology of Fishes*, **22**, 261–71.

Pepin, P. (1991). Effect of temperature and size on development, mortality, and survival rates of the pelagic early life stages of marine fish. *Canadian Journal of Fisheries and Aquatic Sciences*, **48**, 503–18.

Peterson, R. H., Spinney, H. C. E. & Sreedharan, A. (1977). Development of Atlantic salmon (*Salmo salar*) eggs and alevins under varied temperature regimes. *Journal of the Fisheries Research Board of Canada*, **34**, 31–43.

Pittman, K., Bergh, O., Opstad, I., Skiftesvik, A. B., Skjolddal, L. & Strahd, H. (1990). Development of eggs and yolk-sac larvae of halibut (*Hippoglossus hippoglossus* L.). *Journal of Applied Ichthyology*, **6**(3), 142–60.

Polo, A., Yofera, M. & Pascuale, E. (1991). Effects of temperature on egg and larval development of *Sparus auratus* L. *Aquaculture*, **92**, 367–76.

Rana, K. J. (1990). Influence of incubation temperature on *Oreochromis niloticus* (L.) eggs and fry. I. Gross embryology, temperature tolerance and rates of embryonic development. *Aquaculture*, **87**, 165–82.

Rogers, C. A. (1976). Effects of temperature and salinity on the survival of winter flounder embryos. *Fisheries Bulletin U.S.*, **74**, 52–8.

Rogers, B. A. & Westin, D. T. (1981). Laboratory studies on effects of temperature and delayed initial feeding on development of striped bass larvae. *Transactions of the American Fisheries Society*, **110**, 100–10.

Rombough, P. J. (1985). Initial egg weight, time to maximum alevin wet weight, and optimal ponding times for chinook salmon (*Oncorhynchus tshawytscha*). *Canadian Journal of Fisheries and Aquatic Sciences*, **42**, 287–91.

Rombough, P. J. (1988a). Respiratory gas exchange, aerobic metabolism, and effects of hypoxia during early life. In *Fish Physiology*, Vol. XIA, ed. W. S. Hoar & D. J. Randall, pp. 59–161. San Diego: Academic Press.

Rombough, P. J. (1988b). Growth, aerobic metabolism, and dissolved oxygen requirements of embryos and alevins of steelhead, *Salmo gairdneri*. *Canadian Journal of Zoology*, **66**, 651–60.

Rombough, P. J. (1989). Oxygen conductance values and structural characteristics of the egg capsules of Pacific salmonids. *Comparative Biochemistry and Physiology*, **92A**, 279–83.

Rombough, P. J. (1994). Energy partitioning during fish development: additive or compensatory allocation of energy to support growth? *Functional Ecology*, **8**, 178–86.

Rombough, P. J. & Ure, D. (1991). Partitioning of oxygen uptake between cutaneous and branchial surfaces in larval and young juvenile chinook salmon *Oncorhynchus tshawytscha*. *Physiological Zoology*, **64**, 717–27.

Santerre, M. T. (1976). Effects of temperature and salinity on the eggs and early larvae of *Caranx mate* (Cuv. and Valenc.) (Pisces:

Carangidae). *Journal of Experimental Marine Biology and Ecology*, **21**, 51–68.

Santerre, M. T. & May, R. C. (1977). Some effects of temperature and salinity on laboratory-reared eggs and larvae of *Polydactylus sexfilis* (Pisces: Polynemidae). *Aquaculture*, **10**, 341–51.

Scott, D. & Poynter, M. (1991). Upper temperature limits for trout in New Zealand and climate change. *Hydrobiologia*, **222**, 147–52.

Scott, W. B. & Crossman, E. J. (1973). *Freshwater Fishes of Canada*. Fisheries Research Board of Canada Bulletin 184. Ottawa.

Seno, H., Ebina, K. & Okada, T. (1926). Effects of temperature and salinity on the development of the ova of a marine fish (*Calotomus japonicus*). *Journal of the Imperial Fisheries Institute, Tokyo*, **21**, 41–7.

Shelbourn, J. E., Brett, J. R. & Shirahata, S. (1973). Effect of temperature and feeding regime on the specific growth-rate of sockeye salmon fry (*Oncorhynchus nerka*), with a consideration of size effect. *Journal of the Fisheries Research Board of Canada*, **30**, 1191–4.

Shuter, B. J. & Post, J. R. (1990). Climate, population viability, and the zoogeography of temperate fishes. *Transactions of the American Fisheries Society*, **119**, 314–36.

Subasinghe, R. P. & Sommerville, C. (1992). Effects of temperature on hatchability, development and growth of eggs and yolk-sac fry of *Oreochromis mossambicus* (Peters) under artificial incubation. *Aquaculture and Fisheries Management*, **23**, 31–9.

Swift, D. R. (1965). Effect of temperature on mortality and rate of development of the Windermere charr (*Salvelinus alpinus*). *Journal of the Fisheries Research Board of Canada*, **22**, 913–17.

Tang, J., Bryant, M. D. & Brannon, E. L. (1987). Effect of temperature extremes on the mortality and development rates of coho salmon embryos and alevins. *Progressive Fish-Culturist*, **49**, 167–74.

Trippel, E. A., Eckmann, R. & Hartmann, J. (1991). Potential effects of global warming on whitefish in Lake Constance, Germany. *Ambio*, **20**, 226–31.

Valerio, P. F., Goddard, S. V., Kao, M. H. & Fletcher, G. L. (1992). Survival of Northern Atlantic cod (*Gadus morhua*) eggs and larvae when exposed to ice and low temperature. *Canadian Journal of Fisheries and Aquatic Sciences*, **49**, 2588–95.

Velsen, F. P. J. (1987). Temperature and incubation in Pacific salmon and rainbow trout: compilation of data on median hatching time, mortality and embryonic staging. *Canadian Data Report of Fisheries and Aquatic Sciences No. 626, 58pp. Ottawa: Fisheries and Oceans Canada*.

von Westernhagen, H. (1988). Sublethal effects of pollutants on fish eggs and larvae. In *Fish Physiology*, Vol. XIA, ed. W. S. Hoar & D. J. Randall, pp. 253–346. San Diego: Academic Press.

Wallace, J. C. & Aasjord, D. (1984). The initial feeding of Arctic charr (*Salvelinus alpinus*) alevins at different temperatures and under different feeding regimes. *Aquaculture*, **38**, 19–33.

Walsh, W. A., Swanson, C. & Lee, C.-S. (1991). Effects of development, temperature and salinity on metabolism in eggs and yolk-sac larvae of milkfish, *Chanos chanos* (Forsdkal). *Journal of Fish Biology*, **39**, 115–25.

Walsh, W. A., Swanson, C., Lee, C.-S., Banno, J. E. & Eda, H. (1989). Oxygen consumption by eggs and larvae of striped mullet, *Mugil cephalus*, in relation to development, salinity and temperature. *Journal of Fish Biology*, **35**, 347–58.

Walsh, W. A., Swanson, C. & Lee, C.-S. (1991). Combined effects of temperature and salinity on embryonic development and hatching of striped mullet, *Mugil cephalus*. *Aquaculture*, **97**, 281–90.

Wang, N. & Eckmann, R. R. (1994). Effects of temperature and food density on egg development, larval survival and growth of perch (*Perca fluviatilus* L.). *Aquaculture*, **122**, 323–33.

Wang, Y. L., Binkowski, F. P. & Doroshov, S. I. (1985). Effect of temperature on early development of white and lake sturgeon, *Acipenser transmontanus* and *A. fulvescens*. *Environmental Biology of Fishes*, **14**, 43–50.

Wanzenbock, J. & Wanzenbock, S. (1993). Temperature effects on incubation time and growth of juvenile whitefin gudgeon, *Gobio albipinnatus* Lukasch. *Journal of Fish Biology*, **42**, 35–46.

Ware, D. M. (1975). Relation between egg size, growth, and natural mortality of larval fish. *Journal of the Fisheries Research Board of Canada*, **32**, 2503–12.

Wieser, W. & Forstner, H. (1986). Effects of temperature and size on the routine rate of oxygen consumption and on the relative scope for activity in larval cyprinids. *Journal of Comparative Physiology B*, **156**, 791–6.

Wieser, W., Forstner, H., Schiemer, F. & Mark, W. (1988). Growth rates and growth efficiencies in larvae and juveniles of *Rutilus rutilus* and other cyprinid species: effects of temperature and food in the laboratory and in the field. *Canadian Journal of Fisheries and Aquatic Science*, **45**, 943–50.

Wieser, W. & Medgyesy, N. (1990a). Aerobic maximum for growth in the larvae and juveniles of a cyprinid fish, *Rutilus rutilus* L.: implications for energy budgeting in small poikilotherms. *Functional Ecology*, **4**, 233–42.

Wieser, W. & Medgyesy, N. (1990b). Cost and efficiency of growth in the larvae of two species of fish with widely differing metabolic rates. *Proceedings of the Royal Society of London, Series B*, **242**, 51–6.

Williams, G. C. (1975). Viable embryogenesis of the winter flounder, *Pleuronectes americanus*, from −1.8 to 15 °C. *Marine Biology*, **33**, 71–74.

Wood, A. H. (1932). The effect of temperature on the growth and respiration of fish embryos (*Salmo fario*). *Journal of Experimental Biology*, **9**, 271–6.

Worley, L. G. (1933). Development of the egg of the mackerel at different constant temperatures. *Journal of General Physiology*, **16**, 841–57.

MALCOLM JOBLING

Temperature and growth: modulation of growth rate via temperature change

Introduction

Temperature is the most pervasive environmental factor influencing aquatic organisms. Nearly all fish species are, for practical purposes, thermal conformers. In other words, fish of most species are not able to maintain body temperature by physiological means, and their body temperatures fluctuate in close accord with the temperature of the surrounding water. Intimate contact between body fluids and water at the gills, and the high specific heat of water assure this near-identity of internal and external temperatures (Brill, Dewar & Graham, 1994). Consequently, fish species are largely dependent on behavioural control of their body temperatures (Beitinger & Fitzpatrick, 1979; Magnuson, Crowder & Medvick, 1979; Coutant, 1987). Behavioural thermoregulation is widespread among fish species, and in laboratory studies it has often been found that there is good agreement between preferred temperatures and temperatures at which the fish can grow well and perform efficiently (Brett, 1971, 1979, 1995; Beitinger & Fitzpatrick, 1979; Magnuson et al., 1979; Jobling, 1981b; Kellogg & Gift, 1983).

Natural water bodies will seldom, if ever, provide the fish with conditions under which maximum rates of growth can be achieved. Such water bodies will, however, present varied thermal environments offering a wide range of feeding opportunities. It is within such habitats that the ability to thermoregulate by behavioural means may be particularly beneficial to a fish, enabling it to reduce the potentially harmful effects of unfavourable thermal regimes on physiological performance.

The main focus of this chapter is the examination of the effects of temperature upon fish growth under various feeding conditions. The interactions between feeding and growth are discussed in relation to behavioural thermoregulation, temperature selection and temperature compensatory mechanisms.

Society for Experimental Biology Seminar Series 61: *Global Warming: Implications for freshwater and marine fish*, ed. C. M. Wood & D. G. McDonald. © Cambridge University Press 1996, pp. 225–253.

Temperature and growth – maximum feeding

The simplest view of energy partitioning and growth is based upon the concept that 'any change in body weight results from the difference between what enters the body and what leaves it' (Jobling, 1994). This can be represented algebraically as

$$\delta W/\delta t = pR - M$$

where $\delta W/\delta t$ is the change in weight per unit time (somatic and reproductive growth), R is the amount of food consumed (ration), p is a coefficient indicating the availability of nutrients or food energy, and M represents catabolic losses (metabolism). Both R and M are expressed in terms of units per unit time, and therefore represent feeding (or ingestion) and metabolic rates, respectively. The metabolic rate term M incorporates fasting metabolism, activity metabolism, and feeding metabolism, the last term being defined as the metabolism linked to the processing of nutrients and the elaboration of tissues. The coefficient p can be considered to account for faecal and excretory losses.

Numerous workers have investigated the effects of temperature upon the metabolic rate of fishes but most studies have been carried out on fish in a post-absorptive state. Within the range of temperatures under which growth is possible, metabolic rate tends to increase with increasing temperature (Table 1), and the data usually fit well to an exponential equation (Brett, 1979, 1995; Jobling, 1994). The metabolic rate–temperature curve for feeding fish appears to be more complex, and the metabolic rates of fish feeding maximally may plateau, or even decline, at high temperatures (Table 1). Data relating to the effects of temperature on the metabolic rates of feeding fish are, however, scarce, and further studies are sorely needed.

Ingestion rate also increases with temperature (Fig. 1), but there are few detailed studies of this relationship and the fitting of ingestion rate data to mathematical functions is problematic (Elliott, 1982, 1994; Hogendoorn et al., 1983; Binkowski & Rudstam, 1994). Most workers have not attempted a mathematical fit, a few have fitted the data to polynomials and others have fitted logarithmic equations to the ascending limb of the rate–temperature curve (Cox & Coutant, 1981; Elliott, 1982, 1994; Wurtsbaugh & Cech, 1983; Woiwode & Adelman, 1991). Occasionally, more complex functions have been fitted to the data using non-linear regression procedures (Binkowski & Rudstam, 1994). Hogendoorn et al. (1983) suggested that a useful empirical approach was to use Hoerls function to describe the effects of temperature on

Table 1. *The influences of temperature and feeding upon rates of oxygen consumption (metabolic rate) and ventilatory requirements of juvenile sockeye salmon,* Oncorhynchus nerka.

Data relating to metabolic rates are calculated from equations given in Brett (1995). In calculating ventilation volumes it was assumed that the fish respired air-saturated water and that oxygen extraction efficiency was 70%.

Temperature (°C)	[O_2] mg l^{-1}	Metabolic rate mg kg^{-1} h^{-1}		Ventilation volume l kg^{-1} h^{-1}		Maximum ration % body wt day^{-1}
		Unfed	Maximum ration	Unfed	Maximum ration	
5	12.72	45	170	5.1	19.1	5.35
10	11.25	70	245	8.9	31.1	6.50
15	10.10	92	345	13.0	48.8	7.32
20	9.08	127	425	20.0	66.9	7.78
23	8.57	185	408	30.8	68.0	5.85

both feeding and metabolic rates. Irrespective of the approach adopted, it is almost invariably found that increases in temperature initially lead to increased rates of ingestion, feeding rates peak at some intermediate temperature and then decline precipitously as the temperature continues to rise (Fig. 1).

The metabolic rates of feeding fish may be several-fold higher than those of fish that have been deprived of food (Brett, 1979, 1995; Jobling, 1981a, 1994), and the increase in oxygen consumption imposes increased demands upon the respiratory and circulatory systems. Thus, both increasing temperature and feeding lead to an increase in metabolic rate, and as temperature increases the increased oxygen demand must be met under conditions of decreased oxygen solubility. Brett (1995) provides a series of equations that describe the influence of feeding upon the metabolic rates of juvenile sockeye salmon (*Oncorhynchus nerka*) reared at various temperatures. These equations have been used to estimate the ventilatory requirements of the fish under different temperature and feeding conditions (Table 1); in making the calculations it was assumed that the fish were ventilating the gills with air-saturated water and that oxygen extraction efficiency was 70%. The results show that both increased temperature and feeding lead to marked increases in ventilation volume, i.e. the volume of water passed

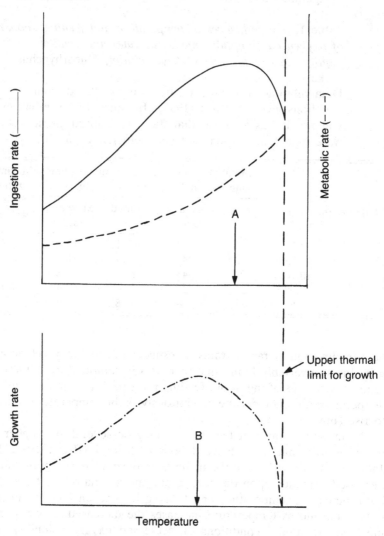

Fig. 1. Rate–temperature curves illustrating the effects of temperature on rates of ingestion, metabolism and growth. Note that the temperature at which ingestion rate reaches its maximum (A) is a few degrees higher than the optimum temperature for growth (B).

over the gills per unit time. At 23 °C the oxygen demand of a fish feeding maximally is lower than that of a fish consuming maximum rations at 20 °C because appetite is reduced at the higher temperature. Nevertheless ventilation volume continues to rise because of the decline in oxygen solubility that accompanies the temperature increase. Thus, it is possible that the suppression of appetite observed at high temperatures reflects limitations in the capacity of the respiratory and circulatory systems to deliver oxygen to the respiring tissues under conditions of very high oxygen demand.

The difference between the rate–temperature curves for ingestion rate and metabolic rate indicates the resources available for growth under different temperature conditions. The plotting of growth rate–temperature relationships indicates that growth rate reaches a peak at an intermediate temperature (Fig. 1). The optimum temperature for growth is, however, slightly lower than that at which ingestion rate reaches its maximum. If both ingestion rates and growth rates are known it is possible to examine the influence of temperature upon the efficiency with which the ingested food is used for growth. The temperature at which conversion efficiency is maximized is slightly lower than the optimum temperature for growth (Fig. 2). The generality of these relationships is demonstrated by the results obtained in a number of growth studies conducted on fish species (e.g. Brett, 1971, 1979; Cox & Coutant, 1981; Elliott, 1982, 1994; Wurtsbaugh & Cech, 1983; Keast, 1985; Xiao-Jun & Ruyung, 1992). For example, Woiwode and Adelman (1991) reported that the food consumption of hybrid bass (female *Morone saxatilis* × male *M. chrysops*) increased with temperature to 29.2 °C, whereas the optimum temperature for growth was 26.8 °C, and conversion efficiency peaked at 21.2 °C.

Behavioural thermoregulation studies suggest there is a general agreement between preferred temperatures and those that permit good growth. Optimum temperatures for growth often coincide with the final preferendum zone for temperature, i.e. the temperature zone in which individuals of a given species will ultimately congregate (Jobling, 1981b; Kellogg & Gift, 1983), and there may be a closer correspondence between thermal preferenda and optimum temperatures for growth, than between preferred temperatures and the temperatures at which either food consumption or conversion are maximized.

In behavioural thermoregulation studies it has often been assumed that the final preferendum is a fixed species-specific characteristic, and warmwater, coolwater and coldwater species have been distinguished on the basis of temperature selection and growth performance criteria (Magnuson *et al.*, 1979; Jobling, 1981b; Coutant, 1987). Some modifi-

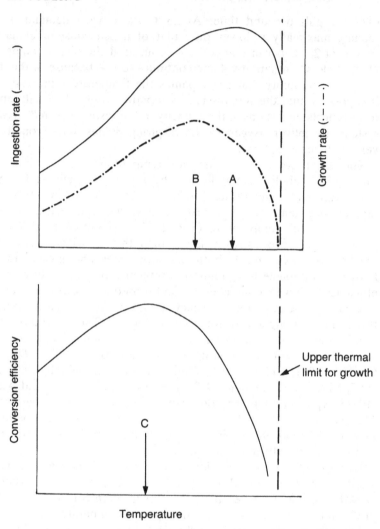

Fig. 2. Rate–temperature curves illustrating the effects of temperature on rates of ingestion and growth, and the influence of temperature upon conversion efficiency. Conversion is defined as growth per unit food ingested. Note that most efficient conversion (C) is achieved at a lower temperature than those at which ingestion (A) and growth (B) rates are at their respective maxima.

cation of the definition of the final preferendum may, however, be required because there is convincing evidence that there are ontogenetic changes in the thermal preferenda of some species of fish. Thus, there may be differences in selected temperatures at different stages of the life history, with the juveniles of several temperate zone freshwater species tending to select higher temperatures than adults (McCauley & Huggins, 1979; Coutant, 1987; Jobling, 1994). If it is taken as axiomatic that there is a close correspondence between preferred temperatures and optimum temperatures for growth, then effects of temperature on growth performance will also be expected to differ among the different ontogenetic stages within any given species. Only a few studies have been conducted in which the influence of fish size or age on the optimum temperature for growth has been examined, but there is evidence that the thermal optimum may change with increasing age and size of the fish (Coutant & Cox, 1976; Hogendoorn *et al.*, 1983; Keast, 1985).

Constant versus fluctuating temperatures

There are several reports that growth rates of fish and other aquatic organisms may be increased by exposure to fluctuating temperature regimes (Biette & Geen, 1980; Spigarelli, Thommes & Prepejchal, 1982; Diana, 1984; Konstantinov & Zdanovich, 1986; Vondracek, Wurtsbaugh & Cech, 1988; Berg *et al.*, 1990; Miao, 1992), the 'acceleration effect' of fluctuating temperatures being known as the *Kaufmann effect* (Cossins & Bowler, 1987). The growth acceleration resulting from exposure to fluctuating temperatures is usually assessed by comparing the rate of growth observed under conditions that cycle regularly between two temperatures with that obtained at the 'mean' temperature (Table 2).

The apparent growth-promoting effect of thermocycling is not universal, and there are also reports that growth rate is either depressed or unaffected by fluctuating temperature regimes (Hokanson, Kleiner & Thorslund, 1977; Cox & Coutant, 1981; Vondracek, Cech & Buddington, 1989; Woiwode & Adelman, 1991). The extent to which growth rate will be either depressed or accelerated, or be unaffected by a fluctuating temperature regime, will depend both upon the amplitude of the thermocycle and the 'mean' temperature about which the temperature fluctuates (for discussion see Cossins & Bowler, 1987). The nature of rate–temperature curves leads to the expectation that growth promotion will be greatest when cycles fluctuate around a low 'mean' temperature (Fig. 3). This is, in fact, the case. Growth is

Table 2. *Influence of thermocycling upon the growth of various fish species. The growth ratio refers to the rate of growth observed under thermocycling conditions relative to that attained under constant temperature at the cycle 'mean'*

Species	Temperature (°C)		Growth ratio cyclic:constant	Source
	Constant	Cyclic		
Salmo salar Atlantic salmon	13	10 ⇌ 16	1.06	Berg *et al.*, 1990
Gambusia affinis mosquitofish	25	20 ⇌ 30	1.07	Vondracek *et al.*, 1988
Salmo trutta brown trout	13	8 ⇌ 18	1.39	Spigarelli *et al.*, 1982
Morone saxatilis striped bass	18	14 ⇌ 22	1.26	Cox &
	24	20 ⇌ 28	0.83	Coutant,
	30	26 ⇌ 34	0.84	1981
Oncorhynchus mykiss rainbow trout	12	8 ⇌ 16	1.16	Hokanson
	17	13 ⇌ 21	0.95	*et al.*, 1977
	19	15 ⇌ 23	0.99	
	22	18 ⇌ 26	Cyclic groups lost weight	
Morone saxatilis ×	20	16 ⇌ 24	1.07	Woiwode &
M. chrysops	24	20 ⇌ 28	0.99	Adelman,
hybrid bass	28	24 ⇌ 32	0.87	1991

generally accelerated by temperatures that fluctuate around a low 'mean', thermocycling has less effect on growth when the 'mean' is close to the optimum temperature, and growth is markedly depressed when temperatures fluctuate around temperatures above the optimum (Fig. 3; Table 2).

Studies carried out under fluctuating temperature regimes indicate that, in general, the recorded growth response is equivalent to that expected at a constant temperature that lies between the 'mean' and the maximum to which the fish is exposed. The exception may be when temperatures fluctuate around the optimum. Furthermore, the results of thermal tolerance studies show that fish exposed to fluctuating temperature regimes are more tolerant of high temperatures than fish exposed to constant temperatures at the 'mean' of the thermocycle

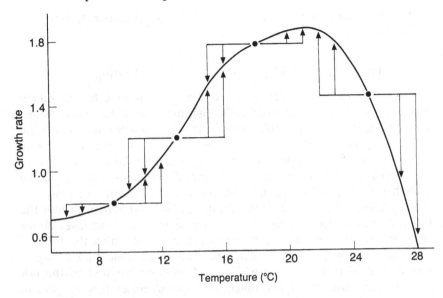

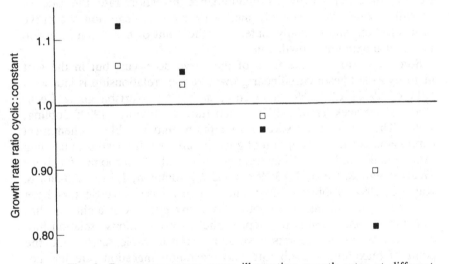

Fig. 3. Rate–temperature curve illustrating growth rates at different constant temperatures (●) and the possible influences of thermocycling (↑ ↓) upon rates of growth. The growth rate ratio refers to the rate of growth observed under thermocycling conditions relative to that attained under constant temperature at the cycle 'mean'. Comparisons were made using growth rate data for different constant temperatures and thermocycles of 'mean' ± 2 °C (□) and 'mean' ± 3 °C (■).

(Feldmeth, Stone & Brown, 1974; Otto, 1974; Threader & Houston, 1983; Woiwode & Adelman, 1992).

Temperature and growth – restricted feeding

Up to this point it has been assumed that the fish have had unlimited access to food. In other words, the fish have been able to consume the maximum ration possible under the given temperature condition (Fig. 1). It should, however, be obvious that growth rates will also be influenced by food supply, and there is a close relationship between ingestion rate and growth rate – the *growth–rations* relationship. The growth–rations relationship may not be a simple linear one (Fig. 4a) (Brett, 1979; Elliott, 1982, 1994; Jobling, 1994). At zero ingestion the fish starve and lose weight. As the rate of ingestion increases above zero, the curve rises steeply to reach the point at which the fish are ingesting food at the maintenance rate. The maintenance rate of ingestion is defined as the rate at which food must be ingested by the fish in order to maintain bodily functions without either loss or gain in weight. In other words, at maintenance ingestion rate the rate of growth is zero. With steadily increasing rates of ingestion the curve begins to flex, and it finally plateaus at the point of maximum ingestion rate and maximum growth rate.

Some variations in the form of the curve do occur, but in the vast majority of studies a curvilinear growth–rations relationship is indicated (Elliott, 1982, 1994; Hogendoorn et al., 1983; Wurtsbaugh & Cech, 1983; Vondracek et al., 1989; Xiao-Jun & Ruyung, 1992; Jobling, 1994). There have been several attempts to find suitable mathematical expressions for the description of growth–rations relationships, including exponential and quadratic equations, and hyberbolic functions (Wurtsbaugh & Cech, 1983; Woiwode & Adelman, 1991; Xiao-Jun & Ruyung, 1992; Jobling, 1994), but in many cases investigators have resorted to the fitting of smooth curves 'by eye'. A straight line has sometimes been used to depict the growth–rations relationship, especially when data points have shown considerable scatter and the point of maximum growth rate and maximum ingestion rate has not been well defined.

Some of the variations in growth–rations relationships that have been noted are clearly related to differences in experimental protocol, and the ways in which data have been analysed. Factors that will influence the shape of the growth–rations relationship include the way in which growth rates are expressed, and the durations over which growth trials are conducted. The shape of the growth–rations curve is also dependent upon whether growth rates are expressed in terms of wet weight, dry weight or the energetic content of the fish. This is because body composition changes with ingestion rate, altering the protein, lipid and water ratios and hence affecting relative dry weight and energy content. For example, lipid deposition tends to increase with increasing food supply, such that the bodies of fish fed maximally will generally contain proportionately more lipid than those fed on restricted rations. Thus, growth–rations relationships plotted on the basis of growth in terms of wet weight may be more curvilinear than those based upon growth determined as rates of energy gain (Elliott, 1982, 1994; Jobling, 1994). A curvilinear growth–rations relationship would also be expected if absorption efficiency became reduced at very high feeding rates. An increasing proportion of the ingested nutrients may be lost in the faecal wastes as ingestion rates increase and, depending upon food type and rearing conditions, absorption efficiency can decline by 10–15% as ingestion increases from low rates to approach the maximum (Elliott, 1982, 1994; Jobling, 1994).

Since temperature influences metabolic rate, and governs the processes involved in ingestion and growth, the growth–rations relationship will vary according to temperature. A theoretical model describing the interactions of ration, growth and temperature is outlined in Fig. 4a, illustrating the growth–rations relationships for four different temperatures. Fig. 4b shows the growth rate–temperature curve for fish on unlimited rations. The growth–rations relationships shift to the right with increasing temperature ($T_1 \rightarrow T_4$) as maintenance requirements increase (Fig. 4a). Thus, for a fixed restricted ration, growth will decrease with increasing temperature. The curves also become elevated with increasing temperature ($T_1 \rightarrow T_3$) (Fig. 4a) as the optimum temperature for growth is approached (Fig. 4b), and there is a decline thereafter (T_4) due to the inhibition of ingestion at high temperatures (Fig. 1).

This model predicts that when food supply is limited, i.e. when ingestion rates are low, the best growth and most efficient food conversion will be achieved at low temperature, but as food supply is increased growth and conversion efficiency are best at higher temperatures. Thus, in the example shown in Fig. 4, conditions that restrict ingestion to low rates result in the best growth and conversion being shown by fish held at T_1. At intermediate rates of ingestion (denoted by A↔B) fish held at T_2 would grow most rapidly, but if food supplies permitted ingestion rates to be high the most rapid growth and most efficient food conversion would be observed in fish held at T_3. These predictions are borne out by the results of a number of studies conducted on a

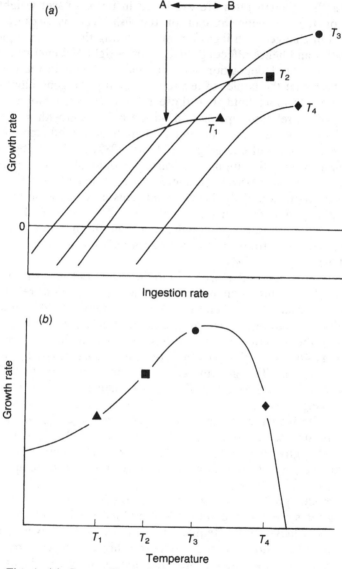

Fig. 4. (a) Curves illustrating the relationships between growth rate and ingestion rate (growth–rations curves) at increasing temperatures ($T_1 \rightarrow T_4$). The symbols indicate the points of maximum ingestion and maximum growth at each of the four temperatures. (b) The rate–temperature curve illustrating the growth rates attained when ingestion rates are at their respective maxima for each of the four temperatures. See text for further details.

variety of fish species; a reduction in the food supply leads to a decrease in the temperature at which best growth performance is observed (Brett, 1971, 1979; Elliott, 1982, 1994; Woiwode & Adelman, 1991; Jobling, 1994).

If fish elect to spend most time in waters at temperatures at which their growth rates are highest, it would be predicted that lower temperatures would be selected under conditions of severe food limitation than under those that permitted ingestion rates to be high. This prediction is supported by behavioural thermoregulation studies: well-fed fish select higher temperatures than do those that are severely food-restricted (Javaid & Anderson, 1967; Stuntz & Magnuson, 1976; Mac, 1985). Selection of higher temperatures by well-fed fish would lead to the promotion of rapid digestion and food processing, thereby permitting high rates of ingestion and growth to be maintained (Neverman & Wurtsbaugh, 1994). On the other hand, the selection of low temperatures under conditions of food restriction would lead to energy conservation via reductions in maintenance requirements (Fig. 4a). In some cases, however, the magnitude of the reduction in preferred temperature in response to food limitation appears to be less than would be required for the fish to achieve the highest possible growth under the given food restriction (Mac, 1985).

Feeding, growth and behavioural thermoregulation

A number of workers have discussed the possible energetic advantages of behavioural thermoregulation in water bodies with discontiguous thermal environments and localized food supplies, but there have been few experimental tests of the ideas (Brett, 1971; Biette & Geen, 1980; Diana, 1984; Clark & Levy, 1988; Wurtsbaugh & Neverman, 1988; Bevelhimer & Adams, 1993; Neverman & Wurtsbaugh, 1994). For example, Brett (1971) provided a review of the influences of temperature on feeding, metabolism and growth of juvenile sockeye salmon (*Oncorhynchus nerka*) and discussed the findings in relation to patterns of vertical migration in thermally stratified lakes. Brett (1971) hypothesized that energetic advantages might accrue if fish migrated between the warm epilimnion, where they fed on zooplankton, and the cooler waters of the hypolimnion where metabolic costs would be reduced. The energetic advantages of making vertical migrations between warm and cool waters were expected to be greatest under conditions of limited food supply (Brett, 1971). Biette and Geen (1980) attempted to test the hypothesis by conducting experiments under conditions that simulated those found in thermally stratified lakes.

They were able to demonstrate that juvenile sockeye salmon usually grew more rapidly under conditions of thermal cycling than when they were held in either warm or cool water throughout the experiment. There were two exceptions to the general trend of better growth under thermal cycling: severe food restriction resulted in growth being best under conditions of constant low temperature, while high food supply led to the greatest rates of growth under constant warmwater conditions. Thus, it was concluded that under feeding conditions prevailing in thermally stratified lakes there would be a growth advantage to be gained by making vertical migrations between the warm waters of the epilimnion and the cooler hypolimnetic waters (Biette & Geen, 1980).

The above discussion assumes that the growth advantage that accrues from migrating between waters of different temperature relates to the fact that metabolic rates will be low in cool waters, thereby leading to 'energy conservation' and improved growth. More recently, Wurtsbaugh and Neverman (1988) proposed an alternative viewpoint: it may be advantageous for fish to move into warmer waters after feeding if this stimulates digestion and thereby promotes increased rates of ingestion during subsequent feeding periods. Their hypothesis arose following observations that juvenile sculpin (*Cottus extensus*) fed on benthic organisms during the day and then migrated vertically, to spend the night in water that was 10 °C warmer than that occurring close to the lake bottom. The results of laboratory experiments provided evidence that there could be a substantial feeding and growth advantage to be gained by feeding in cool waters and then moving into warmer waters, rather than residing permanently in water at low temperature (Wurtsbaugh & Neverman, 1988; Neverman & Wurtsbaugh, 1994).

The two views of the mechanisms underlying the growth advantages to be gained by fish that display behavioural thermoregulation – 'minimization of metabolic costs' and 'maximization of feeding' – are not in conflict because both are essentially part of an hypothesis based upon 'growth maximization'. This can be demonstrated by reference to the growth-rations relationships shown in Fig. 4a. If food resources were localized in areas of high temperature, fish could gain a growth advantage if they moved into areas of lower temperature following the completion of a feeding bout. The improved growth at the lower temperatures would arise because the fish would incur very high metabolic costs if they stayed permanently in areas with high water temperatures.

In cases where food supplies occur in waters of intermediate temperature (e.g. T_3 in Fig. 4) the fish should remain within the area in which they feed provided that high rates of ingestion can be maintained. On the other hand, should food supplies be limited, the fish would grow better by moving to waters of lower temperature following feeding. For example, if food availability resulted in an ingestion rate lower than that indicated by point B in Fig. 4a, the fish would gain a growth advantage by moving from waters at T_3 to those at T_2 following completion of a feeding bout.

When food occurs within areas of low water temperature the fish should both feed and reside within such areas if food availability imposes a severe restriction on rates of ingestion. In other words, at ingestion rates below point A in Fig. 4a fish would grow most rapidly by remaining in cool water at a temperature of T_1. Under better feeding conditions the fish would gain a growth advantage by moving into warmer water to digest its food during inter-meal intervals, i.e. at ingestion rates above those indicated by A, the fish would increase it growth rate by migrating from water at T_1, to water at a temperature of T_2 during non-feeding periods. For these predictions to hold true there is the proviso that the energetic costs involved in moving between waters at different temperatures are not sufficient to negate the growth advantage associated with making the migration.

Thus, the general rules for behavioural thermoregulation, as determined from considerations of growth maximization, are as follows:

1 When food supply is limited, fish should feed in the area where there is most food then move into cooler water where energetic costs are reduced.

2 Movement into cooler water following the completion of a feeding bout should only occur provided that the digestion and processing of food continues sufficiently rapidly to ensure that consumption during the next feeding period is not limited by the amount of undigested food remaining in the stomach.

3 When rates of ingestion are limited by low temperature *per se*, rather than food availability, the fish should migrate into warmer waters during non-feeding periods, thereby ensuring increased rates of food processing and improved growth.

Growth at different latitudes: thermal compensation of growth?

A number of fish species are distributed over a wide range of latitudes, and different populations inhabit environments that differ in both mean annual temperature and length of the growing season. For many of these species, populations at low latitudes grow faster than those in high latitudes. In the northern hemisphere, for example, there are numerous temperate zone freshwater fish species in which the annual growth increment is lower for fish from northern populations than for those living further south. The conventional wisdom appears to be that high latitude populations tend to grow more slowly than those at lower latitudes as a result of the low temperature conditions experienced at high latitudes.

However, when reared under the same conditions, fish from high latitude environments sometimes grow faster than do low latitude conspecifics (Isely *et al.*, 1987; Conover & Present, 1990; Williamson & Carmichael, 1990; Philipp & Whitt, 1991; Nicieza, Reyes-Gavilán & Braña, 1994) suggesting some form of growth rate compensation in high latitude populations. There are currently two models to explain how latitudinal compensation may evolve (Conover, 1990, 1992; Conover & Schultz, 1995).

One way is through local *thermal adaptation*, such that rates of physiological functions are maximized or optimized within the temperature range most commonly experienced within the environment. According to this model there would be shifts in the rate-temperature curves for growth leading to changes in the optimum temperature for growth amongst populations inhabiting different thermal environments (Fig. 5a). Since there appears to be a close link between thermal preferenda and optimum temperatures for growth (Jobling, 1981b; Kellogg & Gift, 1983), this model predicts that the compensation should be accompanied by a change in preferred temperatures. However, intraspecific differences in temperature preferences and tolerances are usually found to be small among populations from different thermal environments, and thermal preferenda may not differ significantly among high and low latitude fish populations (McCauley, 1958; Kaya, 1978; Winkler, 1979; Hirshfield, Feldmeth & Soltz, 1980; McCormick & Wegner, 1981; Matthews, 1986; Fields *et al.*, 1987; Koppelman, Whitt & Philipp, 1988; McGeer, Baranyi & Iwama, 1991; Kaya *et al.*, 1992). Koppelman *et al.* (1988) cautioned, however, that the failure to find significant differences among populations in thermal preferenda might be related to the combined effects of the general insensitivity of behavioural thermoregulation tests and a failure to control adequately for the various biotic and abiotic factors that are known to influence the temperatures selected by fish. Nevertheless, the impression is gained that there may be only small intraspecific differences in temperature tolerances and preferenda among fish populations, and this would in

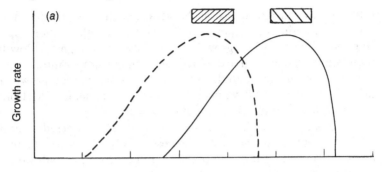

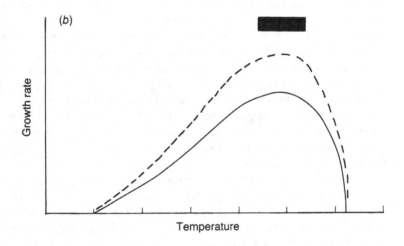

Fig. 5. Rate–temperature curves illustrating two ways in which compensatory adjustments could be made to counteract the effects of environmental differences upon growth. (a) The rate–temperature curves are displaced in relation to each other; fish from low-temperature environments have a lower optimum temperature for growth, and grow faster at low temperatures, than do fish from warmwater populations (*thermal adaptation*). (b) Compensation is envisaged to occur by means of an elevation of the rate–temperature curve; fish with the highest growth capacity are found in environments that have the greatest inhibitory effects upon growth (*countergradient variation*). Bars indicate the zones of thermal preferenda predicted for fish from different environments.

turn tend to call into question the idea that there is marked downward shift in the optimum temperature for growth in high-latitude populations (Fig. 5a). Indeed, when growth rates of fish from high and low latitude populations have been examined at various temperatures, and rate–temperature curves plotted, optimum temperatures for growth for the different populations have been found to be very similar (McCormick & Wegner, 1981; Conover & Present, 1990).

These observations led Conover (1990, 1992) to conclude that individuals of a given species grow at essentially the same temperatures independent of latitude, and that any differences in growth were related to latitudinal differences in the length of the growing season. Thus, the second model for latitudinal compensation focuses upon latitudinal differences in seasonality rather than temperature *per se* (Conover, 1990, 1992; Conover & Present, 1990; Conover & Schultz, 1995). In this model, high-latitude fish have a higher capacity for growth to compensate for short growing seasons, rather than low average temperatures, and the latitudinal compensation is observed as an elevation in the growth rate–temperature curve (Fig. 5b). This is an example of *countergradient variation*, in that genetic and environmental influences are hypothesized to oppose one another along a gradient of decreasing length of the growing season, i.e. the fastest growing genotypes are found in environments that have the most depressive effect on growth (Conover & Schultz, 1995).

Latitudinal compensation for growth via countergradient variation might evolve to offset the potential disadvantages of being small at the end of the growing season (Conover, 1990; Conover & Present, 1990). For fish, especially juveniles, this may be of particular importance at high latitudes where mortality rates are high during the winter, and mortality is often size-dependent (Conover & Present, 1990). Thus, whilst the ability to grow rapidly within a short growing season is clearly advantageous for high latitude fish, it is by no means obvious how differences in growth performance between high and low latitude populations could be maintained. Genotypes that maximize growth should be favoured in all environments, so fast-growing genotypes should be favoured in both high and low latitude populations, and there would need to be other selective forces that offset the contribution of rapid growth if latitudinal variation in growth rate capacity were to be maintained (Conover & Schultz, 1995).

Latitudinal compensations in metabolic rate have been observed in many species. In other words when compared at the same temperature, individuals from high latitude populations usually have higher metabolic rates than do conspecifics from low latitudes (Cossins & Bowler, 1987). Whilst this metabolic compensation has often been interpreted as being a response to temperature *per se*, the basis of the compensatory response remains a matter of debate (Cossins & Bowler, 1987; Hochachka, 1988; Clarke, 1991, 1993). The links between metabolic rates, feeding and growth have been pointed out by several workers (for discussion see Clarke, 1991, 1993; Jobling, 1994) and the increase in growth capacity of high latitude fish may be accompanied by increased rates of ingestion (Present & Conover, 1992). It has also been reported (Nicieza, Reiriz & Braña, 1994) that the digestive performance of Atlantic salmon (*Salmo salar*) from a northern population is potentially greater than that of salmon from a low latitude population. Absorption efficiency increased with increasing temperature, but fish from both populations were equally efficient at digesting and absorbing nutrients at the three temperatures tested (5, 12 and 20 °C). By contrast, the times taken for food to be passed through the gut were longer for the salmon from the southern population. When absorption efficiencies and passage rates were integrated and expressed as digestion rates (food digested and absorbed per unit time), it was found that salmon from the high latitude population had digestion rates potentially as much as 1.6 times greater than those of the low latitude fish (Nicieza *et al.*, 1994). Thus, it is possible that the high rates of ingestion, food processing and growth in high latitude populations may be coupled to high metabolic rates. The maintenance of high metabolic rates may be a prerequisite if the fish are to exploit the available food resources over the course of a short growth season. Thus, the possibility arises that differential metabolic costs might represent an important trade-off leading to the maintenance of latitudinal variation in growth rate capacity.

A latitudinal compensation of growth, resulting from an elevation of rate–temperature curves in fish from high latitude populations, has been demonstrated for the Atlantic silverside (*Menidia menidia*), but there is evidence that this type of growth rate compensation may also occur in other species (Conover, 1990, 1992; Nicieza *et al.*, 1994; Conover & Schultz, 1995). Clearly, further research is required in order to fully elucidate the mechanisms of latitudinal compensations in fish species. Laboratory studies of growth rate–temperature curves, coupled to the study of feeding, digestive performance and temperature preference of fish populations from different latitudes, may prove to be a fruitful approach for the examination of compensatory mechanisms at the whole animal level.

Thermal limits for somatic and reproductive growth

Temperature influences growth processes in a number of ways, but one consistent feature that emerges is that somatic and reproductive growth are possible only within temperature ranges that are much narrower than that which permits short-term survival. Furthermore, somatic growth is usually possible over a wider temperature range than that under which reproduction can be successfully completed (Cossins & Bowler, 1987; Blaxter, 1992; Kamler, 1992; Elliott, 1994; Jobling, 1994). This illustrates one of the problems in defining the thermal requirements of a species. It would be meaningless to define the thermal requirements of a species simply in terms of the critical limits for short-term survival of adults. There are narrower limits for feeding, even narrower limits for growth, and the thermal requirements for gametogenesis and spawning (Van Der Kraak & Pankhurst, this volume) may differ from those for successful egg and larval development (Rombough, this volume). Thus, temperature may have a profound influence upon various aspects of reproduction, from the hormonal regulation of gonadal growth to the timing of spawning, and the relative timing of organogenesis in developing embryos. Unfavourable temperatures experienced by the adult fish may lead to disorders in gonad development, oocyte atresia and degeneration, inhibition of ovulation or a delay in the timing of spawning (Gillet, 1991; Gillet & Breton, 1992; Luksiene & Sandström, 1994; Jobling et al., 1995; Van Der Kraak & Pankhurst, this volume), whereas incubation temperatures experienced by developing eggs and embryos may have profound effects upon the differentiation of different organ systems and the timing of organogenesis (Johnston, 1993; Brooks & Johnston, 1994). In turn, any negative influences of unfavourable temperatures upon reproductive and developmental events would be expected to lead to reduced reproductive performance, and in extreme cases could lead to recruitment failure to the population.

An overview of the effects of temperature on reproduction is given by Van Der Kraak and Pankhurst in this volume. Consequently, only a brief summary of the influences of temperature on reproductive and somatic growth, and the thermal limits of these processes, will be given here. Studies carried out on salmonids of the genus *Salvelinus*, and on the desert pupfish (*Cyprinodon nevadensis*), are used as examples. The Arctic charr (*Salvelinus alpinus*) and the brook trout (*S. fontinalis*) are coldwater salmonid species, whereas the desert pupfish is a cyprinodont species that is adapted to survive the high temperatures that occur in water bodies of Death Valley, California.

The Arctic charr is a cold-tolerant stenothermal species; rates of growth of juvenile fish are depressed at temperatures above 15 °C (Jobling et al., 1993) and reproductive development is disrupted at even lower temperatures (Jobling et al., 1995). Ovulation may be inhibited at temperatures over 10 °C, and there may be rapid 'over-ripening' of eggs produced by females held at high temperature (Gillet, 1991; Gillet & Breton, 1992). Fertilized eggs and developing embryos do not tolerate exposure to high temperatures, and mortality is high in eggs incubated at 10 °C and over (Jungwirth & Winkler, 1984; Steiner, 1984; Gillet, 1991). Thus, temperatures of 10 °C and above can inhibit reproduction in the Arctic charr, whereas somatic growth is possible in waters that are 5 °C or so warmer.

The brook trout is a little more tolerant of high temperatures than the Arctic charr, and juvenile brook trout can maintain quite good rates of growth at temperatures within the range 9–18 °C (McCormick, Hokanson & Jones, 1972). Reproductive growth and development are, however, restricted to lower temperatures. Ovulation is restricted to temperatures of 16 °C and below, but at 16 °C not all the eggs produced are viable. The upper thermal limit for egg development is approximately 12 °C, but the optimum temperature for development is 6 °C (Hokanson *et al.*, 1973). Thus, the numbers of embryos surviving to hatch depend upon temperatures experienced by maturing broodstock, and spawning and incubation temperatures.

Thermal limits for successful reproduction are influenced by the effects of temperature on both egg production and egg viability. For example, the desert pupfish produces many eggs at 32 °C but viability of these eggs is low; egg production is greatest at 30 °C and egg viability peaks at about 28 °C. Consequently, reproductive performance is optimized at 28–29 °C. When the thermal limits for successful reproduction were arbitrarily defined as those delimited by 50% of peak performance, it was found that the upper and lower thermal limits for reproduction were 26 and 30 °C, respectively (Gerking, 1979). On the other hand, using the same performance criteria, the upper and lower thermal limits for somatic growth were calculated to be 17 and 31 °C (Gerking & Lee, 1983). These thermal limits for reproduction and growth are both considerably narrower than the temperature tolerance limits for short-term survival of adults of the species (critical thermal limits: 2–44 °C) (Gerking, 1979).

Reproductive performance – oogenesis, ovulation, egg and embryonic development – is highly temperature-sensitive (Van Der Kraak & Pankhurst, this volume), much more so than is somatic growth. Both reproductive and somatic growth have thermal limits that are considerably narrower than the temperature tolerance range, and as a generalization it can be said that successful reproduction can be expected only over about 10–20% of the range of temperatures tolerated by adult fish (Gerking, 1979). Thus, there will probably be much more profound effects of environmental temperature change upon reproductive success and recruitment to populations than upon survival and growth of individuals within the population.

Concluding comments

Temperature has a profound influence upon growth, and since growth is an important parameter with obvious implications at the population level it is often considered to be a useful indicator in the study of the effects of thermal stress on fish and other aquatic organisms. The value of using growth performance as an indicator in studies of thermal stress is that the temperature range over which growth is possible is narrower than that permitting short-term survival. Nevertheless, the thermal limits for somatic growth are often broad, and fish also display considerable growth plasticity, so an examination of reproductive performance will generally provide a more sensitive indication of thermal stress in fish than will the study of somatic growth.

One major conclusion to be drawn from this chapter is that temperature effects upon growth should never be viewed in isolation, i.e. any discussion of the influences of temperature on growth should also include considerations of food availability. This is important because feeding conditions will obviously have profound influences upon growth. Unfortunately this has frequently been ignored in studies of the growth characteristics of wild populations: attempts have often been made to correlate rates of growth with environmental temperature without any reference being made to potential differences in the food base available to fish from different geographic locations. Thus, increased attention should be directed towards the examination of feed availability–temperature interactions amongst populations of wild fish, especially since there is evidence that behavioural thermoregulation may represent a means by which fish can counteract the suppressive effects of high or low temperatures on growth performance under different conditions of food availability.

Many species display latitudinal clines in growth across their geographic range, but these could result from genetic variation and/or phenotypic differences related to environmental effects. Careful experimentation, involving the rearing of fish in common controlled environments, is required to distinguish between phenotypic differences arising from genotypic and environmental variations (Conover & Present, 1990; Belk, 1995; Conover & Schultz, 1995), but few such controlled studies have been performed on fish from different latitudes or thermal environments. Nevertheless, it is becoming increasingly apparent that there may be genetic differences between populations that result in growth compensation across the geographic range of the species. There is, however, debate as to what this compensation represents. Geographic, latitudinal variations in growth characteristics could reflect thermal adaptation; adaptations for good growth performance in one thermal environment compromise growth in another, and the geographic variability in growth rates reflects adaptation to local thermal conditions (Fig. 5a). If populations show adaptations of this type it would be predicted that long-term temperature increases, such as might be envisaged under a general global warming, would lead to genotypic changes with warm-type genotypes being favourably selected.

The alternative model for latitudinal compensation, involving countergradient variation, focuses on latitudinal differences in seasonality: high latitude fish have a high capacity for growth to compensate for short growing seasons (Fig. 5b). In other words, this model is not based simply upon a genotype–environmental temperature interaction. If this second model is applicable it is much more difficult to predict the outcome of a long-term increase in environmental temperature. For example, although genotypes displaying maximum allowable rates of growth should be favoured irrespective of environmental temperature, a temperature increase that led to changes in seasonal availability of resources could favour the development of populations dominated by slow-growing genotypes. Thus, data indicating the existence of countergradient variation offer a new perspective on how selection may influence phenotypic variations in growth, and they present a challenge to traditional concepts of selective adaptations to changes in the thermal environment.

References

Beitinger, T. L. & Fitzpatrick, L. C. (1979). Physiological and ecological correlates of preferred temperature in fish. *American Zoologist*, **19**, 319–29.

Belk, M. C. (1995). Variation in growth and age at maturity in bluegill sunfish: genetic or environmental effects? *Journal of Fish Biology*, **47**, 237–47.

Berg, O. K., Finstad, B., Grande, G. & Wathne, E. (1990). Growth of Atlantic salmon (*Salmo salar* L.) in a variable diel temperature regime. *Aquaculture*, **90**, 261–6.

Bevelhimer, M. S. & Adams, S. M. (1993). A bioenergetics analysis of diel vertical migration by kokanee salmon, *Oncorhynchus nerka*. *Canadian Journal of Fisheries and Aquatic Sciences*, **50**, 2336–49.

Biette, R. M. & Geen, G. H. (1980). Growth of underyearling sockeye salmon (*Oncorhynchus nerka*) under constant and cyclic temperatures in relation to live zooplankton ration size. *Canadian Journal of Fisheries and Aquatic Sciences*, **37**, 203–10.

Binkowski, F. P. & Rudstam, L. G. (1994). Maximum daily ration of Great Lakes bloater. *Transactions of the American Fisheries Society*, **123**, 335–43.

Blaxter, J. H. S. (1992). The effect of temperature on larval fishes. *Netherlands Journal of Zoology*, **42**, 336–57.

Brett, J. R. (1971). Energetic responses of salmon to temperature. A study of some thermal relations in the physiology and freshwater ecology of sockeye salmon (*Oncorhynchus nerka*). *American Zoologist*, **11**, 99–113.

Brett, J. R. (1979). Environmental factors and growth. In *Fish Physiology*, Vol. VIII, ed. W. S. Hoar, D. J. Randall & J. R. Brett, pp. 599–675. London: Academic Press.

Brett, J. R. (1995). Energetics. In *Physiological Ecology of Pacific Salmon*, ed. C. Groot, L. Margolis & W. C. Clarke, pp. 1–68. Vancouver: UBC Press.

Brill, R. W., Dewar, H. & Graham, J. B. (1994). Basic concepts relevant to heat transfer in fishes, and their use in measuring the physiological thermoregulatory abilities of tunas. *Environmental Biology of Fishes*, **40**, 109–24.

Brooks, S. & Johnston, I. A. (1994). Temperature and somitogenesis in embryos of the plaice (*Pleuronectes platessa*). *Journal of Fish Biology*, **45**, 699–702.

Clark, C. W. & Levy, D. A. (1988). Diel vertical migrations by juvenile sockeye salmon and the antipredation window. *American Naturalist*, **131**, 271–90.

Clarke, A. (1991). What is cold adaptation and how should we measure it? *American Zoologist*, **31**, 81–92.

Clarke, A. (1993). Seasonal acclimatization and latitudinal compensation in metabolism: do they exist? *Functional Ecology*, **7**, 139–49.

Conover, D. O. (1990). The relation between capacity for growth and length of growing season: evidence for and implications of countergradient variation. *Transactions of the American Fisheries Society*, **119**, 416–30.

Conover, D. O. (1992). Seasonality and the scheduling of life history at different latitudes. *Journal of Fish Biology*, **41** (Supplement B), 161–78.

Conover, D. O. & Present, T. M. C. (1990). Countergradient variation in growth rate: compensation for length of the growing season among Atlantic silversides from different latitudes. *Oecologia*, **83**, 316–24.

Conover, D. O. & Schultz, E. T. (1995). Phenotypic similarity and the evolutionary significance of countergradient variation. *Trends in Ecology and Evolution*, **10**, 248–52.

Cossins, A. R. & Bowler, K. (1987). *Temperature Biology of Animals*. London: Chapman & Hall.

Coutant, C. C. (1987). Thermal preference: when does an asset become a liability? *Environmental Biology of Fishes*, **18**, 161–72.

Coutant, C. C. & Cox, D. K. (1976). Growth rates of subadult largemouth bass at 24 to 35.5 °C. In *Thermal Ecology II*, ed. G. W. Esch & R. W. McFarlane, pp. 118–20. Springfield, Ill: National Technical Information Service.

Cox, D. K. & Coutant, C. C. (1981). Growth dynamics of juvenile striped bass as functions of temperature and ration. *Transactions of the American Fisheries Society*, **110**, 226–38.

Diana, J. S. (1984). The growth of largemouth bass, *Micropterus salmoides* (Lacepede), under constant and fluctuating temperatures. *Journal of Fish Biology*, **24**, 165–72.

Elliott, J. M. (1982). The effects of temperature and ration size on the growth and energetics of salmonids in captivity. *Comparative Biochemistry and Physiology*, **73B**, 81–91.

Elliott, J. M. (1994). *Quantitative Ecology and the Brown Trout*. Oxford: Oxford University Press.

Feldmeth, C. R., Stone, E. A. & Brown, J. H. (1974). An increased scope for thermal tolerance upon acclimating pupfish (*Cyprinodon*) to cycling temperatures. *Journal of Comparative Physiology*, **89**, 39–44.

Fields, R., Lowe, S. S., Kaminski, C., Whitt, G. S. & Philipp, D. P. (1987). Critical and chronic thermal maxima of northern and Florida largemouth bass and their reciprocal F_1 and F_2 hybrids. *Transactions of the American Fisheries Society*, **116**, 856–63.

Gerking, S. D. (1979). Fish reproduction and stress. In *Environmental Physiology of Fishes*, ed. M. A. Ali, pp. 569–87. New York: Plenum Press.

Gerking, S. D. & Lee, R. M. (1983). Thermal limits for growth and reproduction in the desert pupfish *Cyprinodon n. nevadensis*. *Physiological Zoology*, **56**, 1–9.

Gillet, C. (1991). Egg production in an Arctic charr (*Salvelinus alpinus* L.) brood stock: effects of temperature on the timing of spawning and the quality of eggs. *Aquatic Living Resources*, **4**, 109–16.

Gillet, C. & Breton, B. (1992). Research work on Arctic charr (*Salvelinus alpinus*) in France – broodstock management. *Icelandic Agricultural Sciences*, **6**, 25–45.

Hirshfield, M. F., Feldmeth, C. R. & Soltz, D. L. (1980). Genetic differences in physiological tolerances of Amargosa pupfish (*Cyprinodon nevadensis*) populations. *Science*, **207**, 999–1001.

Hochachka, P. W. (1988). Channels and pumps – determinants of metabolic cold-adaptation strategies. *Comparative Biochemistry and Physiology*, **90B**, 515–19.

Hogendoorn, H., Jansen, J. A. J., Koops, W. J., Machiels, M. A. M., van Ewijk, P. H. & van Hees, J. P. (1983). Growth and production of the African catfish, *Clarias lazera* (C. & V.) II. Effects of body weight, temperature and feeding level in intensive tank culture. *Aquaculture*, **34**, 265–85.

Hokanson, K. E. F., Kleiner, C. F. & Thorslund, T. W. (1977). Effects of constant temperatures and diel temperature fluctuations on specific growth and mortality rates and yield of juvenile rainbow trout, *Salmo gairdneri*. *Journal of the Fisheries Research Board of Canada*, **34**, 639–48.

Hokanson, K. E. F., McCormick, J. H., Jones, B. R. & Tucker, J. H. (1973). Thermal requirements for maturation, spawning, and embryo survival of the brook trout, *Salvelinus fontinalis*. *Journal of the Fisheries Research Board of Canada*, **30**, 975–84.

Isely, J. J., Noble, J. B., Koppelman, J. B. & Philipp, D. P. (1987). Spawning period and first-year growth of northern, Florida and intergrade stocks of largemouth bass. *Transactions of the American Fisheries Society*, **116**, 757–62.

Javaid, M. Y. & Anderson, J. M. (1967). Influence of starvation on selected temperatures of some salmonids. *Journal of the Fisheries Research Board of Canada*, **24**, 1515–19.

Jobling, M. (1981a). The influences of feeding on the metabolic rate of fishes: a short review. *Journal of Fish Biology*, **18**, 385–400.

Jobling, M. (1981b). Temperature tolerance and the final preferendum – rapid methods for the assessment of optimum growth temperatures. *Journal of Fish Biology*, **19**, 439–55.

Jobling, M. (1994). *Fish Bioenergetics*. London: Chapman & Hall.

Jobling, M., Jørgensen, E. H., Arnesen, A. M. & Ringø, E. (1993). Feeding, growth and environmental requirements of Arctic charr: A review of aquaculture potential. *Aquaculture International*, **1**, 20–46.

Jobling, M., Johnsen, H. K., Pettersen, G. W. & Henderson, R. J. (1995). Effect of temperature on reproductive development in Arctic charr, *Salvelinus alpinus* (L.). *Journal of Thermal Biology*, **20**, 157–65.

Johnston, I. A. (1993). Temperature influences muscle differentiation and the relative timing of organogenesis in herring (*Clupea harengus*) larvae. *Marine Biology*, **116**, 363–79.

Jungwirth, M. & Winkler, H. (1984). The temperature dependence of embryonic development of grayling (*Thymallus thymallus*), Danube salmon (*Hucho hucho*), Arctic char (*Salvelinus alpinus*) and brown trout (*Salmo trutta fario*). *Aquaculture*, **38**, 315–27.

Kamler, E. (1992). *Early Life History of Fish: An Energetics Approach*. London: Chapman & Hall.

Kaya, C. M. (1978). Thermal resistance of rainbow trout from a permanently heated stream, and of two hatchery strains. *Progressive Fish-Culturist*, **40**, 138–42.

Kaya, C. M., Brussard, P. F., Cameron, D. G. & Vyse, E. R. (1992). Biochemical genetics and thermal tolerances of Kendall Warm Springs dace (*Rhinichthys osculus thermalis*) and Green River speckled dace (*R. o. yarrowi*) *Copeia*, 1992(2), 528–35.

Keast, A. (1985). Growth responses of the brown bullhead (*Ictalurus nebulosus*) to temperature. *Canadian Journal of Zoology*, **63**, 1510–15.

Kellogg, R. L. & Gift, J. J. (1983). Relationship between optimum temperatures for growth and preferred temperatures for the young of four fish species. *Transactions of the American Fisheries Society*, **112**, 424–30.

Konstantinov, A. S. & Zdanovich, V. V. (1986). Peculiarities of fish growth in relation to temperature fluctuation. *Journal of Ichthyology*, **26**(4), 65–74.

Koppelman, J. B., Whitt, G. S. & Philipp, D. P. (1988). Thermal preferenda of northern, Florida, and reciprocal F_1 hybrid largemouth bass. *Transactions of the American Fisheries Society*, **117**, 238–44.

Luksiene, D. & Sandström, O. (1994). Reproductive disturbance in a roach (*Rutilus rutilus*) population affected by cooling water discharge. *Journal of Fish Biology*, **45**, 613–25.

Mac, M. J. (1985). Effects of ration size on preferred temperature of lake charr *Salvelinus namaycush*. *Environmental Biology of Fishes*, **14**, 227–31.

McCauley, R. W. (1958). Thermal relations of geographic races of *Salvelinus*. *Canadian Journal of Zoology*, **36**, 655–62.

McCauley, R. W. & Huggins, N. W. (1979). Ontogenetic and non-thermal seasonal effects on thermal preferenda of fish. *American Zoologist*, **19**, 267–71.

McCormick, J. H., Hokanson, K. E. F. & Jones, B. R. (1972). Effects of temperature on growth and survival of young brook trout, *Salvelinus fontinalis*. *Journal of the Fisheries Research Board of Canada*, **29**, 1107–12.

McCormick, J. H. & Wegner, J. A. (1981). Responses of largemouth bass from different latitudes to elevated water temperatures. *Transactions of the American Fisheries Society*, **110**, 417–29.

McGeer, J. C., Baranyi, L. & Iwama, G. K. (1991). Physiological responses to challenge tests in six stocks of coho salmon (*Oncorhynchus kisutch*). *Canadian Journal of Fisheries and Aquatic Sciences*, **48**, 1761–71.

Magnuson, J. J., Crowder, L. B. & Medvick, P. A. (1979). Temperature as an ecological resource. *American Zoologist*, **19**, 331–43.

Matthews, W. J. (1986). Geographic variation in thermal tolerance of a widespread minnow *Notropis lutrensis* of the North American mid-west. *Journal of Fish Biology*, **28**, 407–17.

Miao, S. (1992). Thermocycle effect on the growth rate of redtail shrimp *Penaeus penicillatus* (Alock). *Aquaculture*, **102**, 347–55.

Neverman, D. & Wurtsbaugh, W. A. (1994). The thermoregulatory function of diel vertical migration for a juvenile fish, *Cottus extensus*. *Oecologia*, **98**, 247–56.

Nicieza, A. G., Reiriz, L. & Braña, F. (1994). Variation in digestive performance between geographically disjunct populations of Atlantic salmon: countergradient in passage time and digestion rate. *Oecologia*, **99**, 243–51.

Nicieza, A. G., Reyes-Gavilán, F. G. & Braña, F. (1994). Differentiation in juvenile growth and bimodality patterns between northern and southern populations of Atlantic salmon (*Salmo salar L.*) *Canadian Journal of Zoology*, **72**, 1603–10.

Otto, R. G. (1974). The effects of acclimation to cyclic thermal regimes on heat tolerance of the western mosquitofish. *Transactions of the American Fisheries Society*, **103**, 331–5.

Philipp, D. P. & Whitt, G. S. (1991). Survival and growth of northern, Florida, and reciprocal F_1 hybrid largemouth bass in central Illinois. *Transactions of the American Fisheries Society*, **120**, 58–64.

Present, T. M. C. & Conover, D. O. (1992). Physiological basis of latitudinal growth differences in *Menidia menidia*: variation in consumption or efficiency? *Functional Ecology*, **6**, 23–31.

Spigarelli, S. A., Thommes, M. M. & Prepejchal, W. (1982). Feeding, growth and fat deposition by brown trout in constant and fluctuating temperatures. *Transactions of the American Fisheries Society*, **111**, 199–209.

Steiner, V. (1984) Experiments towards improving the culture of Arctic charr (*Salvelinus alpinus*). In *Biology of the Arctic Charr*, ed. L. Johnson & B. Burns, pp. 509–21. Winnipeg: University of Manitoba Press.

Stuntz, W. E. & Magnuson, J. J. (1976). Daily ration, temperature selection, and activity of bluegill. In *Thermal Ecology II*, ed. G. W. Esch & R. W. McFarlane, pp. 180–5. Springfield, Ill: National Technical Information Service.

Threader, R. W. & Houston, A. H. (1983). Heat tolerance and resistance in juvenile rainbow trout acclimated to diurnally cycling temperatures. *Comparative Biochemistry and Physiology*, **75A**, 153–5.

Vondracek, B., Cech, J. J. Jr & Buddington, R. K. (1989). Growth, growth efficiency, and assimilation efficiency of the Tahoe sucker in cyclic and constant temperature. *Environmental Biology of Fishes*, **24**, 151–6.

Vondracek, B., Wurtsbaugh, W. A. & Cech, J. J. Jr (1988). Growth and reproduction of the mosquitofish, *Gambusia affinis*, in relation to temperature and ration level: consequences for life history. *Environmental Biology of Fishes*, **21**, 45–57.

Williamson, J. H. & Carmichael, G. J. (1990). An aquacultural evaluation of Florida, northern, and hybrid largemouth bass, *Micropterus salmoides*. *Aquaculture*, **85**, 247–58.

Winkler, P. (1979). Thermal preference of *Gambusia affinis affinis* as determined under field and laboratory conditions. *Copeia*, *1979(1)*, 60–4.

Woiwode, J. G. & Adelman, I. R. (1991). Effects of temperature, photoperiod, and ration size on growth of hybrid striped bass × white bass. *Transactions of the American Fisheries Society*, **120**, 217–29.

Woiwode, J. G. & Adelman, I. R. (1992). Effects of starvation, oscillating temperatures, and photoperiod on the critical thermal maximum of hybrid striped × white bass. *Journal of Thermal Biology*, **17**, 271–5.

Wurtsbaugh, W. A. & Cech, J. J. Jr (1983). Growth and activity of juvenile mosquitofish: temperature and ration effects. *Transactions of the American Fisheries Society*, **112**, 653–60.

Wurtsbaugh, W. A. & Neverman, D. (1988). Post-feeding thermotaxis and daily vertical migration in a larval fish. *Nature*, **333**, 846–8.

Xiao-Jun, X. & Ruyung, S. (1992) The bioenergetics of the southern catfish (*Silurus meridionalis* Chen): growth rate as a function of ration level, body weight, and temperature. *Journal of Fish Biology*, **40**, 719–30.

KEITH BRANDER

Effects of climate change on cod (*Gadus morhua*) stocks

Introduction

Cod have provided one of the major North Atlantic fisheries for more than 500 years (Cushing, 1982, 1986; de La Villemarque, 1994) and have been more intensively studied than any other marine fish species. The existence of catch records going back several hundred years in some fisheries makes cod an excellent species on which to investigate the effects of environmental change. This was recognised in 1992, when the International Council for the Exploration of the Sea (ICES) established a Working Group on Cod and Climate Change to coordinate research and to encourage participation by a wider range of scientific disciplines than is normally involved in marine fisheries assessment. To date, insufficient attention has been paid to the application of results from biological oceanography and from experimental studies in assessing fish stocks and it is worth exploring some of the possibilities. ICES held a symposium on Cod and Climate Change in 1993, the proceedings of which (Jakobsson *et al.*, 1994) provide an excellent coverage of recent research.

The total catch of cod rose from the early years of the 20th century to a peak of 3.9 million tonnes in 1968 and has declined steadily since then (Fig. 1). A number of stocks, which have separate spawning sites and different migration patterns, contribute to the total catch (Fig. 2). International fisheries assessments use data from commercial catches and research surveys to estimate stock abundance. Variability in abundance is principally ascribed to the effects of fishing and one of the main aims is to construct quantitative models of the population dynamics of individual stocks, which are useful for sustainable management of fisheries. The fact that most North Atlantic cod stocks are currently considered to be overfished is due to the difficulty of devising and implementing effective fisheries management systems, but in some cases

Society for Experimental Biology Seminar Series 61: *Global Warming: Implications for freshwater and marine fish*, ed. C. M. Wood & D. G. McDonald. © Cambridge University Press 1996, pp. 255–278.

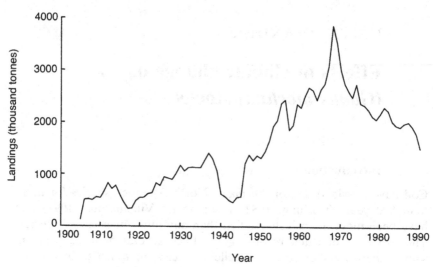

Fig. 1. Total North Atlantic cod catch. (Data from Garrod & Schumacher, 1994.)

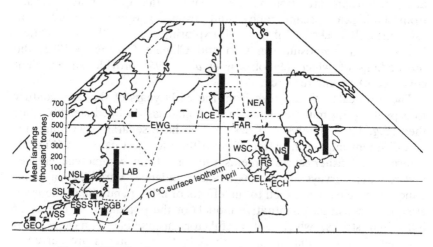

Fig. 2. Distribution and average landings of cod stocks throughout the North Atlantic. Broken lines indicate stock boundaries. Full names of stocks corresponding to abbreviated codes are given in Brander (1995).

it is becoming clear that environmental factors have also contributed to the decline in abundance. Thus, for successful fishery regulation it is important to be able to distinguish between changes in fish abundance due to fishing and those due to environmental factors.

The questions which this chapter will address are:

1 What is the evidence that cod stocks fluctuate as a result of environmental change?
2 What processes are involved in causing stocks to fluctuate due to environmental factors?
3 What role can experimental studies play in improving our ability to explain and predict such stock fluctuations?

What do we mean by 'environmental factors' and 'climate change'?

Environmental factors which affect cod at some stage in their life history include temperature, salinity, winds and currents. Global warming will affect sea temperatures, but is also expected to result in changes in winds, fresh water input and mixing of the surface layer of the ocean, with direct and indirect consequences for cod. Warming may also lead to changes in large-scale advection, with consequences for transport and retention of eggs and larvae.

The climate record shows that there have been periods of warming and cooling of the North Atlantic, but the changes have not been uniform over all areas and neither have the local consequences for cod stocks (Dickson & Brander, 1993). Over time periods of 60–200 years the temperature of the whole North Atlantic behaves in a similar way, but over shorter periods some areas may be out of phase, or completely uncorrelated, with others. A good example of this is the winter air temperature at Greenland and over northern Europe, which seesaws in opposite directions at short time scales (van Loon & Rogers, 1978).

Environmental factors, such as temperature, show variability at all scales. Diurnal, seasonal, annual and long-term fluctuations in temperature are the norm and it is not easy to detect the signal of global warming within this 'normal' variability. Tree ring widths respond to differences in temperature and can be used as a proxy for temperature in order to reconstruct long-term temperature series. For example, a 1400 year reconstruction of annual northern Fennoscandian summer temperature anomalies, based on the analysis of tree rings, shows periods of warming similar to recent changes during many previous centuries (Fig. 3). This record has been used to determine how soon we will know if the recent warming differs significantly from past warm

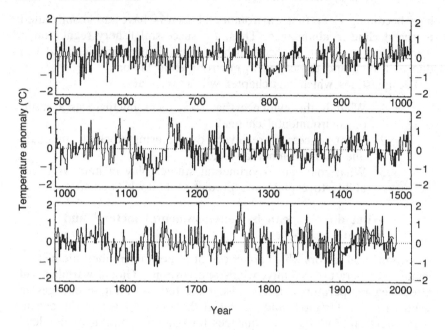

Fig. 3. Northern Fennoscandia temperature anomalies in summer reconstructed from tree-ring densities from the northern treeline of Scandinavia (from Briffa *et al.*, 1990). The instrumental record, smoothed with a low-pass filter, is superimposed after 1876 (smooth curve).

periods (Briffa *et al.*, 1990) and hence to identify the signal of global warming. Previous effects of warming on cod stocks can also be studied by identifying historic warm periods (e.g. in the mid-19th century) and exploring archive records of fisheries and their changes at that time (Anon., 1995).

Looking for field evidence of the effects of environmental change on cod stocks is more difficult than carrying out controlled experiments. For example, in a laboratory setting it is axiomatic that temperature will be measured in an appropriate and accurate way and that other factors will be controlled. In field studies the temperature regime actually experienced by the fish (i.e. the ambient temperature) is rarely known and a proxy or average has to be used, particularly when analysing long-term records of stocks that inhabit large areas of ocean. Even our present coverage is very patchy and one often has to start by using correlation methods with proxy data in order to explore possible cause and effect relationships affecting the stocks. Much of the published work on environmental effects on fish stocks has been based on poorly explained correlations, but as the examples later in this chapter show, we are gradually moving on to a more detailed, process-orientated approach, in which experimental science can help.

Distribution, life history and abundance of the major cod stocks

Cod are distributed over most of the shelf areas of the North Atlantic, with their most southerly spawning sites occurring just north of 40° N, on Georges Bank in the western North Atlantic, and their most northerly spawning grounds at about 70° N off Norway (Fig. 4). Juvenile and adult cod occur north of 80° N in the Arctic, and south of 40° N off the US coast. The southerly limit for cod corresponds roughly to the 10 °C surface isotherm in April (Fig. 2). That particular isotherm is used to represent the geographic temperature pattern, not the thermal tolerance limit for cod. Thermal tolerance varies seasonally and with age, so that it is difficult to give a general figure for cod, but the species occurs in temperatures below zero and up to 20 °C. Juvenile fish have wider tolerance than adults and their geographic range is slightly greater.

Spawning occurs at depths from near surface to several hundred metres, mainly in coastal locations, but also on offshore banks, such as Georges Bank, the banks off Nova Scotia and Labrador, and Faeroe Bank and Plateau. For almost all stocks spawning is concentrated in space and time, with the same locations and seasonal timing recurring every year (Brander, 1994b). The eggs are buoyant and remain within the surface mixed layer. The larvae also remain mainly within 30 m of the surface and are dispersed as they develop. This may be an adaptation to reduce competition for food among the developing larvae (Economou, 1991). The larvae metamorphose while still pelagic and then settle to the bottom after about 100–200 days, during which time they may be transported a long or short distance from their origin, depending on the current system in which they are spawned (Fig. 4). On Georges Bank and Faeroe Plateau the eggs occur within a 'retentive' circulation pattern: eggs and developing larvae are transported tens to hundreds of kilometres, but they remain mainly within a gyre, which keeps them close to their point of origin. Eggs and larvae of the northeast Arctic, and Labrador cod stocks are transported up to 1600 km away from their spawning sites, but remain mainly within shelf circulation systems. At Iceland eggs and larvae generally remain in the coastal circulation around the island, but in some years they are transported across the Denmark Strait to East Greenland. There is also evidence that eggs and larvae may very occasionally be carried across the Davis Strait from West Greenland to Labrador (Dickson & Brander, 1993).

The fact that cod and many other species migrate over long distances, in order to concentrate their spawning in space and time, suggests that this strategy must be particularly suitable for the survival of eggs and larvae. Since plankton production is seasonal and geographically patchy in the sea it seems likely that concentrated spawning is a means of placing larvae in a favourable feeding environment, where survival

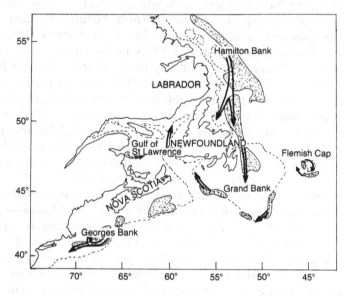

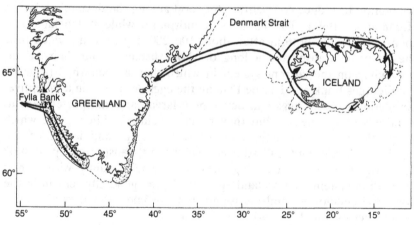

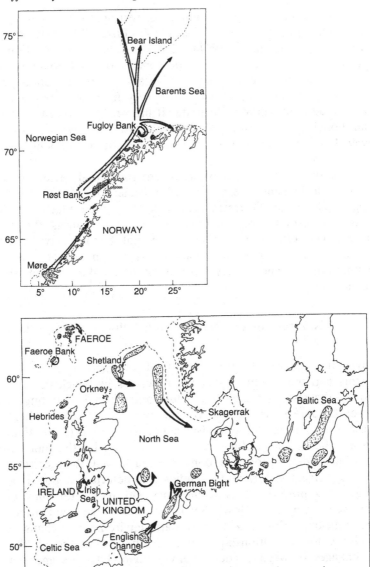

Fig. 4. Spawning areas of cod (stippled) and general pattern of egg
and larval transport (arrows) for all major stocks. (Data from Bagge &
Thurow, 1994; Brander, 1992, 1994a, b; Brander & Hurley, 1992;
Buch *et al.*, 1994; Frank, Drinkwater & Page, 1994; Heath, Rankine &
Cargill, 1994, Taggart *et al.*, 1994.)

rates will be high and large numbers of recruits to the adult population will result. There is evidence (Brander & Hurley, 1992; Myers, Mertz & Bishop, 1993; Brander, 1994a) that the timing and location of cod spawning is coupled to spring phytoplankton and zooplankton production, but because the timing of plankton production depends on several environmental factors, which vary from year to year, survival and recruitment will also fluctuate. It has proved difficult to determine and model the processes in this 'match–mismatch' hypothesis (Cushing, 1969, 1982) sufficiently well to test whether the recruitment is governed in this way.

Some of the processes whereby environmental factors may affect population dynamics are presented in Table 1 for different life history stages. The table illustrates the widely held view that environmental factors governing stock abundance act mainly during the early life stages. The effects of the environmental factors may be classified as direct (e.g. temperature-dependent development) or indirect (e.g. temperature-dependent development of zooplankton, upon which cod larvae feed).

Examples of environmentally induced changes in cod stocks

Given the scarcity of appropriate field information on environmental factors and the complexity of the processes by which they affect cod stocks, the task of establishing cause and effect is daunting. We may never be able to model the relationships in detail, but there are at least three reasons for optimism about the prospects of gaining some understanding and useful predictive capability. Firstly, there are cases in which simple processes dominate, such that their effects can be predicted from experimental studies and detected at population level (e.g. the combined effects of buoyancy and oxygen tolerance on Baltic cod; see example 1 below). Secondly, some large population changes are sufficiently closely related to major environmental change that the evidence for a relationship is overwhelming even though the details of linkages between particular environmental factors and population processes are not known with certainty (e.g. the presence or absence of cod at Greenland; see example 2 below). Thirdly, for stocks which have been studied intensively over many years, the complex inter-active effects of several environmental (and biological) factors have gradually been revealed (see example 3 below, on the Northeast Arctic cod).

Table 1. *Effects of environmental factors on processes in the life history of cod. Indirect effects shown in italics*

Life history stage	Process	Environmental factor	Comment	References
Eggs	Buoyancy	Salinity	Baltic, Labrador	Nissling, 1994; Anderson & de Young, 1994
	Respiration	Oxygen		Nissling, 1994
	Development rate	Temperature		Thompson & Riley, 1981
	Transport	Wind, freshwater	Vestfjord, Georges Bank	Lough et al., 1994; Adlandsvik & Sundby, 1994
Larvae	Growth	Temperature		van der Meeren et al., 1994
	Feeding			van der Meeren et al., 1994
	Encounter rate	Turbulence		Sundby, Ellertsen & Fossum, 1994; *Brander, 1992, 1994a*
	Food production	*Wind, light, tidal mixing*	*Effects on phytoplankton and copepod production*	
	Vertical migration	Mixing, light		Skiftesvik, 1994
	Transport	Wind, freshwater	Retentive (Georges Bank) non-retentive (Norwegian shelf) systems	Lough et al., 1994; Adlandsvik & Sundby, 1994
Juveniles	Settlement	Water depth	Georges Bank	Lough & Potter, 1993
Adults	Growth	Temperature	Inter-stock comparison	Brander, 1995
	Migration	Temperature		Rose et al., 1994

Example 1   Effects of declining salinity and low oxygen levels on Baltic cod

The halocline in the deep basins of the Baltic, where cod spawn, tends to become progressively deeper following occasional periods of inflow, when the deep bottom water is flushed out by more saline, oxygen-rich water from the Skagerrak. Eggs of the Baltic cod are buoyant at salinities greater than 11 ppt and their survival decreases at low oxygen levels. Experimental studies have shown that higher salinity benefits the survival of cod eggs in two ways: first, by ensuring that they are sufficiently buoyant to remain in the upper mixed layer where oxygen levels are higher and second, by improving the survival of eggs kept at low oxygen levels (Nissling & Westin, 1991; Nissling, 1994).

Major inflows into the Baltic in the winters of 1974–6 and 1976–7 (Fonselius, 1988) increased the proportion of the Baltic with favourable spawning conditions and led to an increase in the stock and catches. Since then, continuing heavy fishing and almost two decades without a major inflow have resulted in a decline in the stock and greatly reduced catches. In this case the direct relevance of experimental studies of egg buoyancy and survival to population modelling is clear.

Similar experimental work on egg buoyancy of Newfoundland cod showed that, unlike Norwegian coastal cod and Baltic cod, the density of the eggs decreased during development (Anderson & de Young, 1994). Reduced buoyancy of Newfoundland cod seemed to be related to egg condition and may result in higher mortality due to exposure to low oxygen levels.

Example 2   The 'cod period' at Greenland from 1920 to 1970

The cod stock at Greenland is the most northerly of the northwest Atlantic. Spawning takes place inshore in many of the West Greenland fiords and also offshore along the southwest and east coasts of Greenland. In addition to these locally spawned components of the stock a large number of developing larvae drift across from Iceland in years when transport conditions are favourable.

Intermittent records of cod catches at Greenland go back to the mid-17th century. Cod were always present in inshore areas, but not always offshore. Sufficient numbers were discovered offshore to support a brief fishery in the 1820s and late 1840s, ending in 1850. Very few cod occurred offshore at West Greenland during the first decade of the 20th century, but from 1917 they spread gradually northward and

the Greenland Administration established a succession of fishing sta-
tions to prepare fish for export. The northward spread of cod and the
increase in abundance were part of a poleward extension of the range
of many boreal marine species during the 1920s and 1930s (Jensen,
1939). This coincided with a prolonged warm period, which lasted until
the end of the 1960s (Fig. 5). Cod catches rose to 460 000 tonnes in
1962 as a result of regular good recruitment of the offshore spawning
component and occasional large numbers of larvae arriving from Iceland. The variability in recruitment at West Greenland is greater than
observed for any other major cod stock, with the number of 3 year
old fish ranging from fewer than 10 million to more than 500 million
(Brander, 1994b).

Since the late 1960s the stock has again retreated south and declined
to levels probably last seen before 1920. The West Greenland offshore

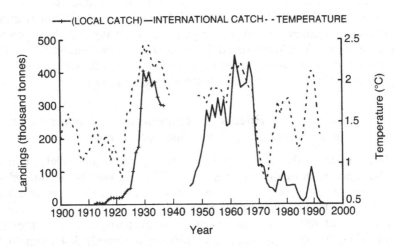

Fig. 5. Cod catches and water temperature at West Greenland. Temperature data are running 5 year means of surface values up to 1972
(from Smed, 1980) and running 5 year means for Fylla Bank 0–40 m
(from Buch *et al.*, 1994) since then. The two series were combined
to the Fylla Bank temperature scale using the regression for the 22
year period (1950–72) during which they overlapped ($r^2 = 0.92$). They
are plotted as the mean of the previous 5 years. Local catches are
shown for the period from 1912–37 (from Jensen 1939), because
they reacted to local changes in abundance more quickly than the
international fleet. The scaling of the local catch is $\frac{1}{20}$ of the international catch. The International catch is for ICNAF (International
Commission for Northwest Atlantic Fisheries) Div. 1.

component has almost disappeared and the remaining stock is sustained by recruitment from Iceland.

The 45 year period when cod were abundant at Greenland during this century was due to large-scale warming throughout much of the northern hemisphere, but local conditions do not always coincide with large-scale changes and the cooling periods at Greenland since 1968 seem anomalous, when viewed against the larger pattern of global warming. Although the collapse of the Greenland fishery was accelerated by heavy fishing pressure, some decline in the size of the stock was probably inevitable (Buch, Horsted & Hovgard, 1994).

The processes which resulted in the reduction of the range of the West Greenland stock and the failure of offshore recruitment are not known, but field and experimental studies on distribution, migration, recruitment and growth of cod stocks in other areas may provide some insights. For example, the effect of low temperature on distribution has been studied for the cod stocks off Labrador and Newfoundland (Rose *et al.*, 1994) and recruitment of Northeast Arctic cod is greatly affected by temperature in the range 1–3.5 °C during and after spawning (see example 3 below). In addition to reduced recruitment, the West Greenland stock suffered from a reduced growth rate, which was probably due to the declining temperature (Fig. 6a).

Example 3   Effects of temperature and other factors on recruitment of Northeast Arctic cod

Recruitment to this stock has been studied more intensively than any other and much of the observed variability can now be accounted for, in spite of the fact that the processes which govern recruitment are complex. The main spawning ground of the Northeast Arctic cod, in Vestfjord at 68° N, is one of the most northerly and it is therefore not surprising that the number of recruits is roughly 3–4 times higher in warm years than in cold. For the period 1946–88 about 28% of the observed variability in recruitment was probably due to temperature and about the same proportion was due to the abundance of spawning fish (Nilssen *et al.*, 1994; Ottersen, Loeng & Raknes, 1994).

As usual with population studies, it is not easy to devise indices to represent interannual variability in the stock and in the environment. Year class abundance for Northeast Arctic cod is mainly determined prior to settlement and is estimated by midwater fishing surveys in August and September of the first year, before the juvenile fish settle to the bottom. A second estimate of the abundance of a year class as 3 year old recruits is obtained from virtual population analysis (VPA) of commercial catch at age data. During the first 5–6 months of life, before settlement, the eggs and larvae are transported out of the semi-enclosed Vestfjord and are dispersed over a wide area of the Norwegian shelf, with the result that they are subject to different environmental regimes. The numbers caught in the midwater surveys in August–September do not correlate with temperature at a depth of 10 m, but do correlate with the 0–200 m depth-integrated time series. Temperatures in the upper layer depend mainly on local summer heating, whereas the depth-integrated temperature reflects large-scale variability in Atlantic water flowing into the Barents Sea. This suggests that the depth-integrated temperature is the best measure of what cod actually experience, but there are a number of other characteristics of Atlantic water (e.g. zooplankton abundance) that may also affect survival of the larvae.

The development rate of cod eggs and larvae is controlled by temperature, and this may affect survival either directly or by reducing exposure to predation. Experimental studies, field studies, exploratory statistics (using correlation and regression methods) and foraging theory also point to a number of other factors which affect recruitment of Northeast Arctic cod, including plankton production and timing, wind-induced mixing and advection (Sundby, Ellertsen & Fossum, 1994; Adlandsvik & Sundby, 1994), cannibalism by older ages (Nilssen *et al.*, 1994), spawning stock biomass, fecundity and egg quality. There is also evidence that the stock may be sensitive to environmental effects when it is small and consists mainly of young fish, perhaps because younger fish have a short spawning period and produce eggs with a narrow range of vertical (and hence horizontal) distribution (Kjesbu *et al.*, 1992).

Example 4   Effects of temperature on migration and distribution of cod in Newfoundland waters

The three examples of environmentally induced change so far described have concentrated on the effects on recruitment, in other words directly on population number. In the West Greenland example the distribution of the stock also changed and recent events off Labrador and Newfoundland also show how migration and distribution are affected by temperature.

Rose *et al.* (1994) found a significant positive correlation between the mean latitude of cod distribution during fall research surveys off Newfoundland and the annual sea temperature during the period 1981–92. The mean latitude shifted south by about 200 km for each degree

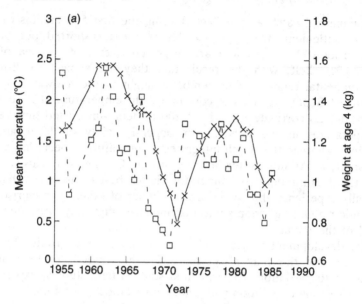

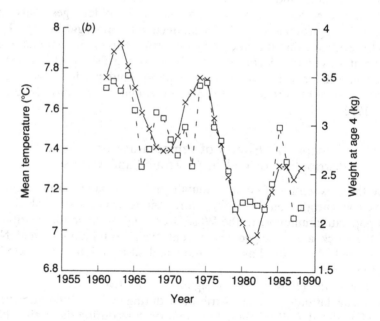

Celsius decrease in temperature. The authors ascribed this to migration by the fish, rather than to the effects of mortality or fishing, but the response may be an indirect one if changing temperature also causes a change in distribution of prey species, such as capelin (*Mallotus villosus*). The southerly shift in distribution of the adult cod may also have an effect on recruitment, if maturing adults are no longer able to reach the northern spawning sites and utilize the full potential range for producing recruits (de Young & Rose, 1993).

Young cod (<3 years old) inhabit colder water than do older fish and there is clear experimental and field evidence that they have greater tolerance of cold temperatures and react to decreasing temperatures by producing plasma antifreeze (Goddard & Fletcher, 1994). Laboratory results help to interpret the field data on distribution and can be used to deduce the recent thermal history of wild fish.

Effects of temperature on growth rate

Growth rates of cod vary considerably between different stocks, such that a 4 year old fish at Labrador weighs about 0.6 kg, whereas in the Celtic Sea it weighs about 7.2 kg. A comparative study of mean weights at age from commercial catches and mean annual temperatures for each area suggests that most of the variability in growth rate between stocks can be accounted for by differences in temperature (Brander, 1995). A logistic model, in which mean weight is a function of the product of age and temperature, accounts for 92% of the variance for age 2–9 for 12 of the North Atlantic cod stocks (Fig. 7). An exponential model fitted to the 2–4 year olds, which weighed 0.5–7.3 kg, gave a Q_{10} for growth of 2 (Brander, 1995 – Model 3, temperature range 2–11 °C). Year to year differences in temperature cause changes in the growth rate within a stock (Fig. 6a, b) and may therefore be responsible for some of the observed fluctuations in catches and in stock biomass.

Fig. 6. (*a*) Mean weight at age 4 (□) and mean lifetime temperature (×) for cod at West Greenland in the period 1955–90 (from Brander, 1995, Fig. 10). The temperature data are for the upper 40 m on Fylla Bank. The temperature is several degrees higher in deeper water and the values shown here represent the interannual variability, but not necessarily the temperature actually experienced by the cod.

(*b*) Mean weight at age 4 (□) and mean lifetime temperature (×) for cod at Faeroe in the period 1955–90 (from Brander, 1995, Fig. 7).

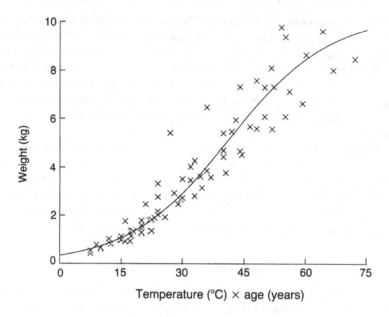

Fig. 7. Mean weight-at-age for 12 cod stocks as a function of temperature × age (from Brander, 1995, Fig. 5). The fitted function is: Weight-at-age = $10.28/(1 + \exp(0.082(41.273 - \text{temperature} \times \text{age})))$.

A comparison between field estimates of the effects of temperature on growth and laboratory studies is useful to test whether differences in growth, such as those observed between different stocks, can be reproduced under controlled temperature conditions, and whether other factors, such as food limitation, may be involved. Jobling (1988) summarized published work on the effect of temperature on growth rate of laboratory-reared cod, which were fed to satiation. The cod weighed 100–700 g and over the temperature range 2–10 °C the Q_{10} for growth was 7.3 (Fig. 8). This value is remarkably high, but is comparable to other Q_{10} values for laboratory-reared fish (e.g. sockeye salmon, *Oncorhynchus nerka*; Brett & Higgs, 1970). Above 11 °C the growth rate for laboratory-reared cod slows and above about 14 °C it declines (Fig. 8).

Determining whether wild fish show comparable growth rates and temperature effects is problematic, because growth rates and Q_{10} values decline with age or size and the data for wild fish are mainly for older, larger fish. The ambient temperature is not well estimated for wild fish

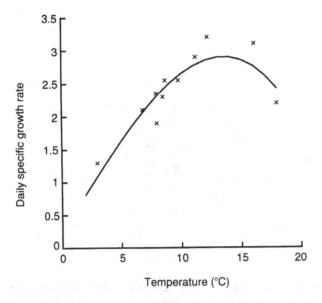

Fig. 8. Effects of temperature on growth rates of cod held under controlled conditions (from Jobling, 1988, Fig. 4). The fitted function is:
$$\ln(G) = (0.216 + 0.297 \times \text{temp} - 0.000538 \times \text{temp}^3) - 0.441 \times \ln(\text{weight})$$
where $G = 100 \times \ln$ (weight on day $(x+1)$/weight on day (x)) (i.e. G is the daily specific growth rate $\times 100$) and weight is set at 1 g (i.e. last term of equation $= 0$).

and if annual averages are used then, as Fig. 9 shows, the range in temperature experienced within a stock is narrow. In Fig. 10 the growth rate for laboratory-reared cod (from Fig. 8) is extrapolated to compare it with the wild growth rate (from Fig. 7) and also the growth rate from the Faeroese commercial fishery. One might infer that cod are food-limited in the wild from the fact that the laboratory rates are higher, but these rates may be unnaturally high because of the artificial diet (see Jobling, 1988, on the hepato-somatic index) and because energy consumption for feeding and migration is abnormally low. At this stage the principal conclusions are that better data are needed on ambient temperature in the wild and that future comparisons should seek to minimize differences in conditions and size of fish between experimental systems and the wild.

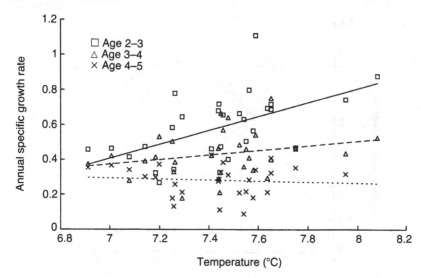

Fig. 9. Annual specific growth rates for Faeroe cod aged 2–5 as a function of temperature. Correlations for 2–3 and 3–4 year olds are significant ($p < 0.01$ and $p < 0.05$, respectively), but not significant for 4–5 year olds (or older ages, which are not shown) (from Brander, 1995, Fig. 8).

Conclusions and suggestions for future research

Most North Atlantic fish stocks, including cod, have been subject to excessive levels of fishing for several decades, with the result that catches and biomass have declined. Given the severe and widespread effects of fishing it is not easy to distinguish the effects of other factors, such as environmental change, particularly when our data on environmental factors are not very good and we are uncertain of the processes by which they act on fish stocks. In a few cases (e.g. the increase and decline in cod at Greenland) the pattern and scale of the changes makes it obvious that they are due to the concurrent large-scale climatic warming and cooling.

In other cases a combination of experimental studies, field observations and long-term data on population change allows the factors and processes to be identified (e.g. Northeast Arctic cod). Experimental studies on their own can indicate which processes are likely to have a critical effect on stocks, but it is risky to extrapolate from an artificial system, in which only a few factors are allowed to vary. Field studies of course suffer from the opposite problem – that many factors

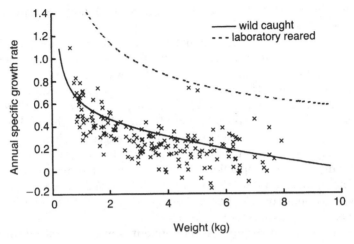

Fig. 10. The data points are annual specific growth rates for Faeroe cod, which have been adjusted to 7.4 °C (the average water temperature for Faeroe) using the equation $W_{t+1}/Wt = \exp(0.06821 \times$ temperature) (from Brander, 1995, Model 3). The exponential model is used (with a temperature of 7.4 °C) to represent the 'laboratory reared' growth rate and the equation is given in the caption to Fig. 8. Note that the observations from which Jobling estimated his parameters were mainly for fish in the size range 100–700 g. The logistic model is used to represent the 'wild caught' growth rate. It is fitted to the data shown in Fig. 7, with an extra term to bring it closer to the origin:

$$\text{Weight-at-age} = (10.54 + 0.596)/$$
$$(1 + \exp(0.071(40.39 - 7.4 \times \text{age}))) - 0.596$$

vary simultaneously and that it is not easy to measure the environmental variables in an appropriate way. A judicious combination of the two approaches is required.

Cod stocks have fluctuated in the past due to changes in temperature and other environmental factors and some of the recent changes may be due to global warming, although it is too early to identify that particular signal in a noisy data set. Future changes in temperature, wind, rainfall, mixing and advection can be expected to result in further shifts in population distribution and abundance. At Greenland and perhaps other areas where cod are close to their lower thermal limit, the stocks may increase due to range expansion, higher recruitment and more rapid growth. Evidence of the upper thermal limit for cod from population studies is less clear, and it would be interesting to look for field confirmation of the experimental findings that growth rate declines above 14 °C.

The rapid growth of cod produced under farmed, food-satiated conditions and the high Q_{10} for growth also raise interesting questions about food limitation in the wild and the way in which temperature affects growth rate. If growth in the wild is currently limited by food then rising temperature can only increase growth rate if it allows cod to increase their food consumption. It may therefore be necessary to investigate the food supply for cod at all stages of their life history and perhaps also how temperature affects overall production in the ecosystem. Since phytoplankton production is not generally thought to be temperature limited, the scope for a change in ecosystem productivity is limited, so the alternative is that consumption by cod increases at the expense of other species.

The effect of temperature on recruitment is well-established for some stocks (e.g. Northeast Arctic) but the processes involved are complex. Cold temperatures slow the rates of egg and larval development, but also slow the production of *Calanus finmarchicus*, the principal prey species for the larvae. Once again it would be useful to compare growth rates of larvae under different temperature and feeding conditions in experiments and in the wild, to try to establish whether growth is being limited by food intake and how this affects survival.

The effects of low temperature on migration and distribution are becoming clearer due to a combination of field studies and experimental work on low temperature tolerance (Rose *et al.*, 1994; Goddard & Fletcher, 1994). Off Labrador and Newfoundland the mean distribution of cod shifted south by about 250 km between 1987 and 1992, associated with a drop of about 1 °C in the annual mean temperature. At West Greenland, during the warming period in the 1920s, the distribution of cod extended northward at the rate of about 80 km per year over a 15 year period, associated with an increase in annual mean temperature of more than 2 °C. The rapid increase in the population size at West Greenland seems to have been due to the offshore spawning component. It would be useful to know whether the increase in the offshore stock was self-generated or the result of an influx from the fiord stocks and fish transported from Iceland as larvae. In either case, the evidence suggests that changes in temperature can lead to rapid changes in population distribution and abundance, whether positive or negative.

Finally, in any evaluation of the likely effects of climate change on cod stocks one has to acknowledge that the continued existence of adequate stock sizes is a necessary precondition and this depends on reducing the prevailing excessive levels of fishing on many stocks. The experimental and field work on growth, recruitment and migration which allows us to evaluate the effects of climate change also helps to improve the short-term modelling of cod stocks and thus provide more reliable advice on which to base fisheries management.

References

Adlandsvik, B. & Sundby, S. (1994). Modelling the transport of cod larvae from the Lofoten area. *ICES Marine Science Symposia*, **198**, 379–92.

Anderson, J. T. & de Young, B. (1994). Stage-dependent density of cod eggs and larvae (*Gadus morhua* L.) in Newfoundland waters. *ICES Marine Science Symposia*, **198**, 654–65.

Anon. (1995). Report of the Cod and Climate Backward-Facing Workshop. *ICES CM* 1995/A:7 23pp. (mimeo).

Bagge, O. & Thurow, F. (1994). The Baltic cod stock: fluctuations and possible causes. *ICES Marine Science Symposia*, **198**, 254-68.

Brander, K. M. (1992). A re-examination of the relationship between cod recruitment and *Calanus finmarchicus* in the North Sea. *ICES Marine Science Symposia*, **195**, 393–401.

Brander, K. M. (1994a). The location and timing of cod spawning around the British Isles. *ICES Journal of Marine Science*, **51**, 71–89.

Brander, K. M. (ed.) (1994b). Spawning and life history information for North Atlantic cod stocks. *ICES Cooperative Research Report*, **205**, 150pp.

Brander, K. M. (1995). The effect of temperature on growth of Atlantic cod (*Gadus morhua* L.). *ICES Journal of Marine Science*, **52**, 1–10.

Brander, K. M. & Hurley, P. C. F. (1992). Distribution of early stage Atlantic cod (*Gadus morhua*), haddock (*Melanogrammus aeglefinus*), and witch flounder (*Glyptocephalus cynoglossus*) eggs on the Scotian Shelf: a reappraisal of evidence on the coupling of cod spawning and plankton production. *Canadian Journal of Fisheries and Aquatic Sciences*, **49**, 238–51.

Brett, J. R. & Higgs, D. A. (1970). Effect of temperature on the rate of gastric digestion in fingerling sockeye salmon, *Oncorhynchus nerka*. *Journal of the Fisheries Research Board of Canada*, **27**, 1767–79.

Briffa, K. R., Bartholin, T. S., Eckstein, D. *et al.* (1990). A 1,400-year tree-ring record of summer temperatures in Fennoscandia. *Nature (London)*, **346**, 434–9.

Buch, E., Horsted, S. A. & Hovgard, H. (1994). Fluctuations in the occurrence of cod in Greenland waters and their possible causes. *ICES Marine Science Symposia*, **198**, 158–74.

Cushing, D. H. (1969). The regularity of the spawning season of some fishes. *Journal du Conseil International pour l'Exploration de la Mer*, **33**, 81–97.

Cushing, D. H. (1982). *Climate and Fisheries*. London: Academic Press.

Cushing, D. H. (1986). The Northwest Atlantic Cod Fishery. *International Symposium on Long-Term Changes in Marine Fish Populations*, Vigo, 83–94.

de La Villemarque, J. H. (1994). French cod fisheries from the sixteenth to the middle of the twentieth century. *ICES Marine Science Symposia*, **198**, 56–8.

de Young, B. & Rose, G. A. (1993). On recruitment and distribution of Atlantic cod (*Gadus morhua*) of Newfoundland. *Canadian Journal of Fisheries and Aquatic Sciences*, **50**, 2729–2741.

Dickson, R. R. & Brander, K. M. (1993). Effects of a changing windfield on cod stocks of the North Atlantic. *Fisheries Oceanography*, 2(3/4), 124–53.

Economou, A. N. (1991). Is the dispersal of fish eggs, embryos and larvae an insurance against density dependence? *Environmental Biology of Fishes*, **31**, 313–21.

Fonselius, S. (1988). Long-term trends of dissolved oxygen, pH and alkalinity in the Baltic deep basins. *ICES Doc. CM* C:23, 8pp.

Frank, K. T., Drinkwater, K. F. & Page, F. H. (1994). Possible causes of recent trends and fluctuations in Scotian Shelf/Gulf of Maine cod stocks. *ICES Marine Science Symposia*, **198**, 110–20.

Garrod, D. J. & Schumacher, A. (1994). North Atlantic cod: the broad canvas. *ICES Marine Science Symposia*, **198**, 59–76.

Goddard, S. V. & Fletcher, G. L. (1994). Antifreeze proteins: their role in cod survival and distribution from egg to adult. *ICES Marine Science Symposia*, **198**, 676–83.

Heath, M., Rankine, P. & Cargill, L. (1994). Distribution of cod and haddock eggs in the North Sea in 1992 in relation to oceanographic features and compared with distributions in 1952–1957. *ICES Marine Science Symposia*, **198**, 438–9.

Jakobsson, J. *et al.* (ed.) (1994). Cod and Climate Change. *ICES Marine Science Symposia*, **198**, 693pp.

Jensen, A. S. (1939). Concerning a change of climate during recent decades in the Arctic and SubArctic regions, from Greenland in the west to Eurasia in the east, and contemporary biological and geophysical changes. *Det Kgl. Danske Videnskabernes Selskab. Biologiske Medd.*, **XIV**(8), 75pp.

Jobling, M. (1988). A review of the physiological and nutritional energetics of cod, *Gadus morhua* L., with particular reference to growth under farmed conditions. *Aquaculture*, **70**, 1–19.

Kjesbu, O. S, Kryvi, H., Sundby, S. *et al.* (1992). Buoyancy variation in eggs of Atlantic cod (*Gadus morhua* L.) in relation to chorion thickness and egg size: theory and observations. *Journal of Fish Biology*, **41**, 581–99.

Lough, R. G. & Potter, D. C. (1993). Vertical distribution patterns and diel migrations of larval and juvenile haddock, *Melanogrammus aeglefinus*, and Atlantic cod, *Gadus morhua*, on Georges Bank. *Fisheries Bulletin U.S.*, **91**, 281–303.

Lough, R. G, Smith, W. G. Werner, F. E. *et al.* (1994). Influence of wind-driven advection on interannual variability in cod egg and larval distributions on Georges Bank: 1982 vs 1985. *ICES Marine Science Symposia*, **198**, 357–78.

Myers, R. A, Mertz, G. & Bishop, C. A. (1993). Cod spawning in relation to physical and biological cycles of the northern North-west Atlantic. *Fisheries Oceanography*, **2**(3/4), 154–65.

Nilssen, E. M. Pedersen, T., Hopkins, C.C.E. *et al.* (1994). Recruitment variability and growth of Northeast Arctic cod: influence of physical environment, demography, and predator–prey energetics. *ICES Marine Science Symposia*, **198**, 449–70.

Nissling, A. (1994). Survival of eggs and yolk-sac larvae of Baltic cod (*Gadus morhua* L.) at low oxygen levels in different salinities. *ICES Marine Science Symposia*, **198**, 626–31.

Nissling, A. & Westin, L. (1991). Egg buoyancy of Baltic cod (*Gadus morhua*) and its implications for cod stock fluctuations in the Baltic. *Marine Biology*, **111**, 33–5.

Ottersen, G., Loeng, H. & Raknes, A. (1994). Influence of temperature variability on recruitment of cod in the Barents Sea. *ICES Marine Science Symposia*, **198**, 471–81.

Rose, G. A., Atkinson, B. A., Baird, J. *et al.* (1994). Changes in distribution of Atlantic cod and thermal variations in Newfoundland waters, 1980–1992. *ICES Marine Science Symposia*, **198**, 542–52.

Skiftesvik, A. B. (1994). Impact of physical environment on the behaviour of cod larvae. *ICES Marine Science Symposia*, **198**, 646–53.

Smed, J. (1980). Temperature of the waters off southwest and south Greenland during the ICES/ICNAF Salmon Tagging Experiment in 1972. *Rapports et Proces Verbaux des Réunions du Conseil International pour l'Exploration de la Mer*, **176**, 18–21.

Sundby, S., Ellertsen, B. & Fossum, P. (1994). Encounter rates between first-feeding cod larvae and their prey during moderate to strong turbulent mixing. *ICES Marine Science Symposia*, **198**, 393–405.

Taggart, C. T., Anderson, J., Bishop, C. *et al* (1994). Overview of cod stocks. biology and environment in the Northwest Atlantic region of Newfoundland, with emphasis on northern cod. *ICES Marine Science Symposia*, **198**, 140–57.

Thompson, B. M. & Riley, J. D. (1981). Egg and larval development studies in the North Sea cod (*Gadus morhua* L.). *Rapports et*

Proces-Verbaux des Réunions du Conseil International pour l'Exploration de la Mer, **178**, 553–9.

van der Meeren, T., Jorstad, K. E. Solemdal, P. *et al.* (1994). Growth and survival of cod larvae (*Gadus morhua* L.): comparative enclosure studies of Northeast Arctic cod and coastal cod from western Norway. *ICES Marine Science Symposia*, **198**, 633–45.

van Loon, H. & Rogers, J. C. (1978). The seesaw in winter temperatures between Greenland and northern Europe. Part 1. General description. *Monthly Weather Review*, **106**, 296–310.

S.D. McCORMICK, J.M. SHRIMPTON and
J.D. ZYDLEWSKI

Temperature effects on osmoregulatory physiology of juvenile anadromous fish

Introduction

The anadromous life history entails early development in fresh water followed by movement to the ocean and subsequent return to fresh water for spawning. In contrast to other euryhaline fish, in which movement from fresh water to sea water is often frequent, anadromy usually results in seasonal movement of juveniles from fresh water to sea water. For many (but not all) anadromous species seaward migration occurs only once in an individual's life. Some anadromous salmonids have evolved a preparatory adaptation (known as the parr–smolt transformation or smolting) in which salinity tolerance and other adaptations for ocean life develop at the time of seaward migration (McCormick & Saunders, 1987; Hoar, 1988). The relative 'strength' and developmental stage of the parr–smolt transformation varies widely among salmonids. Although smolting is by definition strictly a salmonid phenomenon, similar preparatory adaptations may exist in other anadromous species (Youson, 1980) but to date their presence has not been widely examined.

For most anadromous species seaward migration is highly seasonal, often spring or fall, at a time of rapid temperature change. Under these conditions temperature may be an important factor for determining the timing of development and migration, and may impose further physiological challenge to animals whose osmotic tolerances are being altered. Here we will review the known effects of temperature on the osmoregulatory physiology of anadromous fish with emphasis and speculation on aspects that may affect the survival and distribution of these important fish.

Anadromous salmonids

Smolt development

Most anadromous salmonids undergo a distinct transformation prior to migration from fresh water to the ocean. Behavioural and physical

Society for Experimental Biology Seminar Series 00: *Global Warming: Implications for freshwater and marine fish*, ed. C. M. Wood & D. G. McDonald. © Cambridge University Press 1996, pp. 279–301.

changes are accompanied by a series of physiological and biochemical alterations that are crucial to the successful rapid entry of smolts to the marine environment. The parr–smolt transformation is mediated by hormonal changes in response to environmental stimuli. Among the hormones that have been found to participate in smolting are cortisol, thyroid hormones, growth hormone and prolactin (see review by Hoar, 1988). In response to spring hormonal changes there is a proliferation of chloride cells on the gill lamellae (Langdon & Thorpe, 1985; Richman *et al.*, 1987), which are involved in salt extrusion in sea water teleosts (Foskett & Scheffey, 1982). Na$^+$,K$^+$-ATPase, the primary enzyme for excretion of Na$^+$ and Cl$^-$, has been shown to be located in chloride cells (Karnaky *et al.*, 1976; McCormick, 1990). The specific activity of Na$^+$,K$^+$-ATPase increases during the parr–smolt transformation coincident with the onset of migration and the development of saltwater tolerance in several salmonid species (McCormick & Saunders, 1987).

Timing of the parr–smolt transformation is regulated by seasonal changes in environmental variables, especially cyclical variations in photoperiod and temperature. Early experiments demonstrated that photoperiod was the primary environmental factor controlling timing of the parr–smolt transformation (see reviews by Wedemeyer, Saunders & Clarke, 1980; Hoar, 1988). Seasonal changes in coloration and saltwater tolerance can be accelerated or delayed by manipulating photoperiod. Although photoperiod plays a pivotal role in the timing of smolting, there is also evidence that water temperature influences the timing and onset of smolting. For many salmonids the parr–smolt transformation is size dependent. Since temperature is an important factor in determining growth rates in salmonids it will directly affect the age at which smolting occurs. Temperature may also act directly as a cue for smolting and the onset of migration. We will examine evidence from the literature for these roles of temperature in affecting the initiation and timing of smolting in salmonids.

Temperature effects on growth

Achieving a minimum size appears to be necessary before smolting can occur (Clarke, Shelbourn & Brett, 1978; Thorpe *et al.*, 1980; Hoar, 1988). The minimum size for smolting is species-specific, and Clarke (1982) reported ranges from 10 g in coho salmon to greater than 50 g in steelhead trout. Growth is therefore important for smolt development, as higher water temperatures generally result in greater growth rates until an optimal temperature is exceeded (Brett, Shelbourn & Shoop, 1969).

The benefit of rearing in warmer water on smolt production has been clearly demonstrated for Atlantic salmon where considerable differences in growth rates result in a bimodal size distribution within populations (Thorpe, 1977; Thorpe *et al.*, 1980). The upper mode (UM) of the population exhibit a growth spurt in early autumn, while the lower mode (LM) of the population has a reduced appetite leading to lower growth rate and requires an additional year in fresh water before undergoing the parr–smolt transformation (Metcalfe & Thorpe, 1992). In general, the better the opportunity for growth in the spring and summer, the higher the percentage of fish that enter the UM and smolt the following spring (Thorpe, 1994).

An overall increase in water temperature of rivers due to global warming may benefit many stocks of fish due to greater growth opportunity. In northern latitudes increased growth will shorten the duration of river life and reduce mortality (Power & Power, 1994). Acceleration of growth and smolt production have already been documented in streams where water temperatures have increased due to removal of overhead cover (Holtby, 1988) or discharge of thermal effluent (Thorpe *et al.*, 1989). In southern latitudes, increases above optimal temperature for growth during the summer may lead to a decline in growth and smolt production. The consequence of decreased growth is longer river life, greater parr density and increased exposure to density-dependent mortality (Power & Power, 1994). Consequently, global warming will be beneficial if it brings the fish closer to its optimal temperature and deleterious if it raises temperature beyond the optimum.

Temperature effects on smolting

Warmer rearing temperatures during the late winter and spring have been demonstrated to advance the timing of the parr–smolt transformation in coho salmon (Zaugg & McLain, 1976) and Atlantic salmon (Solbakken, Hansen & Stefansson, 1994). We have used data from a number of studies on Atlantic salmon to model the date at which peak gill Na^+,K^+-ATPase activity is observed as a function of mean water temperature and plotted this relationship in Fig. 1a. Data were included from experiments that maintained fixed or ambient (seasonally changing) temperature regimes and natural or simulated natural photoperiod. Warmer rearing temperature significantly advanced the date of peak gill Na^+,K^+-ATPase activity, with an estimated 5 week difference in peak levels when mean daily temperature is increased from 2 to 10 °C. Another approach to examining the role that temperature plays on increasing development rate of smolting is to plot maximal gill Na^+, K^+-ATPase activity as a function of degree days, the accumulated

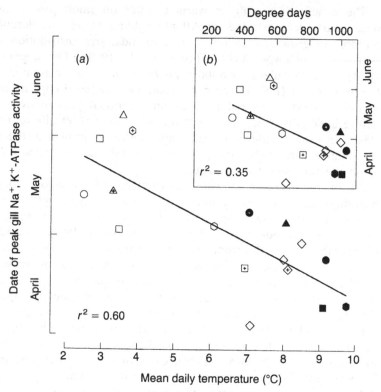

Fig. 1. (a) Effect of mean water temperature on the date of maximal Na⁺,K⁺-ATPase activity in Atlantic salmon smolts. Mean water temperature was calculated from 1 January to the peak of Na⁺,K⁺-ATPase activity. Data for this analysis included studies conducted at ambient (seasonally changing) and constant temperatures. (*b*) The cumulative degree days experienced until maximal Na⁺,K⁺-ATPase activity for the same data as shown in (a). Degree days are calculated as the additive daily temperatures experienced since 1 January to the peak in gill Na⁺,K⁺-ATPase activity measured in each study. First order regression lines are shown on each plot. Symbols represent the following studies and conditions: Atlantic salmon reared at ambient (increasing) temperature (⬧, Boeuf & Prunet, 1985); ambient temperature (◇, Boeuf *et al.*, 1985); ambient temperature (△, Duston *et al.*, 1991); 6–8 °C (⊙, McCormick *et al.*, 1987); 10 °C from 9 January (●) or ambient temperature (○, S. McCormick *et al.*, unpublished data); 10 °C from early January (■), ambient temperature (□) or ambient temperature advanced 6 weeks (⊡, S. McCormick, unpublished data); ambient temperature (⊕, Olsen *et al.*, 1993); and ambient temperature (⬡) or 12 °C from 6 February (⬣, Solbakken *et al.*, 1994).

daily temperature in degrees Celsius (Fig. 1*b*). If smolting was strictly temperature dependent, degree days to the peak in gill Na^+,K^+-ATPase activity would be constant and independent of date. From this relationship, however, it is clear that at warmer temperatures more degree days are required to achieve an advance in timing of peak gill Na^+,K^+-ATPase activity. This limited capacity of temperature to accelerate smolting supports previous work indicating that photoperiod is the primary environmental factor driving the parr–smolt transformation.

Although increases in rearing temperature will accelerate smolt development, an upper temperature exists for successful smolting. This range, however, differs from that for growth. Rearing temperatures of 17.5 °C in sockeye salmon impaired smolt development, whereas a lower temperature (10 °C) resulted in higher seawater tolerance (Clarke *et al.*, 1978). Steelhead trout reared at 6.5 or 10 °C showed a 2-fold increase in gill Na^+,K^+-ATPase activity in March, whereas, fish reared at 15 or 20 °C showed no increase in gill Na^+,K^+-ATPase activity (Adams, Zaugg & McLain, 1973). However, in Atlantic and coho salmon, higher temperatures (16–20 °C) resulted in normal increases in gill Na^+,K^+-ATPase activity, but these decreased rapidly after the peak (Johnston & Saunders, 1981; Zaugg & McLain, 1976) (see section on Loss of smolt characteristics). The temperatures found to suppress smolting in these species are near or less than the optimum temperature for growth (Fig. 2). Different temperature optima exist for development of seawater tolerance and for growth. This does not necessarily indicate a 'conflict' between the two as smolting is a winter/spring phenomenon and temperatures for optimal growth will occur later in the year.

Seasonal changes in temperature as a stimulus for smolting
Increasing day-length, in combination with increasing temperature, is a stronger stimulus to smolting than each alone, as shown by increases in gill Na^+,K^+-ATPase activity (Muir *et al.*, 1994) and migration rate (Wagner, 1974; Muir *et al.*, 1994). Photoperiod manipulation at constant temperature, however, still produces smolts (see Hoar, 1988), indicating that seasonal changes in temperature are not crucial. In the absence of a photoperiod stimulus, the evidence for temperature directly stimulating smolting is equivocal. There was no significant effect on smolting in sockeye reared under increasing (7–13 °C), constant (10 °C) or decreasing (13–7 °C) temperature regimes (Clarke *et al.*, 1978). Temperature raised in February from 5 to 12 °C resulted in earlier development of saltwater tolerance in Atlantic salmon compared with controls held at ambient water temperature (5–6 °C) (Solbakken *et al.*, 1994).

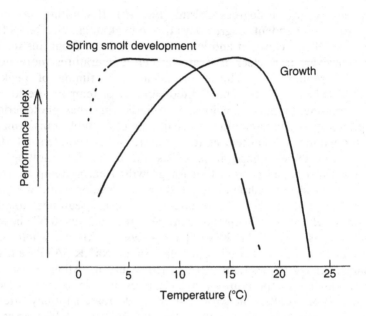

Fig. 2. A theoretical relationship between water temperature, smolt development and growth. The growth curve was modified from that of Brett *et al.* (1969) and is typical for salmonids. An optimal temperature for growth exists, but is greater than the optimal temperature for smolting. At temperatures that favour smolt development, an increase in temperature will accelerate the parr–smolt transformation as shown in Fig. 1. At temperatures where smolt development is impaired, peak gill Na^+,K^+-ATPase activity levels and saltwater tolerance are reduced or occur only transiently.

McCormick *et al.* (1987, 1989) found that exposure to continuous light inhibited the increase in salinity tolerance and gill Na^+,K^+-ATPase activity that normally occur in the spring. In these studies, water temperature during the spring remained constant (6–8 °C). In contrast, Atlantic salmon reared in constant light with seasonally changing water temperature that increased rapidly in late May (2–12 °C) showed typical smolt characteristics, with marked silvering, sea water tolerance and high Na^+,K^+-ATPase activity (Staurnes, Sigholt & Gulseth, 1994). These limited studies indicate that water temperature may be a cue for development of saltwater tolerance in juvenile Atlantic salmon in the absence of a photoperiod stimulus.

Temperature may also play an important role in stimulating behavioural changes associated with smolting. In Atlantic salmon, the start of migration has been found to be controlled by a combination of rate of increase in temperature and absolute temperature in the river during the spring (Jonsson & Ruud-Hansen, 1985). Timing of chinook salmon migration is related to water temperature, with 7 °C appearing to be a thermal threshold for migration (Raymond, 1979). Temperature may not be the only factor stimulating downstream movement of smolts, as river discharge is linked to migration of Atlantic salmon (Solomon, 1978) and steelhead trout (Raymond, 1979).

Although winter temperatures are unlikely to exceed an upper thermal limit for successful smolting, temperatures high enough to impair smolting may be reached in the spring. Water temperatures above 12 °C in steelhead (Adams *et al.*, 1973) and 17 °C in sockeye (Clarke *et al.*, 1978) are detrimental for smolting due to suppressed increases in gill Na^+,K^+-ATPase activity and impaired development of salinity tolerance. If global warming causes an overall 2 °C increase in temperature or shifts the spring increase in temperature to an earlier date, the onset of smolting may be advanced. Conducting experiments where temperature increases earlier or is moderately elevated above ambient river temperature (e.g. 2 °C) throughout the winter and spring is essential to understand fully impacts of global warming on physiological changes related to the parr–smolt transformation in juvenile salmonids.

Loss of smolt characteristics

Increases in salinity tolerance that occur during smolting are known to be reversible; fish that are maintained in fresh water beyond the period of normal spring migration lose their elevated capacity for hypo-osmoregulation and growth in sea water (Folmar *et al.*, 1982; McCormick & Saunders, 1987; Hoar, 1988). Although exhaustive studies on the environmental factors that might affect loss of smolt characteristics have not been conducted, studies to date clearly indicate that higher temperatures will increase the rate at which these characters are lost. Zaugg and McLain (1976) found that coho salmon previously reared at 6 or 10 °C experienced increasingly rapid losses of gill Na^+,K^+-ATPase activity upon exposure to higher temperatures (15 or 20 °C). Juvenile steelhead exposed to 13 °C for 20 days at the peak of smolting had lower gill Na^+,K^+-ATPase and reduced migratory behaviour compared with fish maintained at 6 °C (Zaugg & Wagner, 1973). Photoperiod can also influence this process: exposure of masu salmon to short photoperiod increased the number of fish showing morphological indications of loss of smolt characteristics (Kurokawa, 1990).

Recent studies have found that as in Pacific salmon, Atlantic salmon respond to elevated temperatures with reductions in salinity tolerance and gill Na^+,K^+-ATPase activity (Duston, Saunders & Knox, 1991; S. McCormick, M. O'Dea & J. Carey, unpublished data). Figure 3 presents a summary of these data, demonstrating a relationship between degree days experienced by smolts and loss of gill Na^+,K^+-ATPase activity. Treatment groups in these studies included constant temperature (10, 13 and 16 °C) and ambient conditions (seasonal change in water temperature) under which temperature increased rapidly in spring. This analysis indicates that after the peak in gill Na^+,K^+-ATPase activity there is a period of stability (around 100 degree days) followed by a period of rapid decline. There is no indication from this analysis that changes experienced under ambient temperatures were qualitatively different than constant temperature, nor does it appear that higher temperatures have a greater effect than that predicted by degree days.

Research on Atlantic salmon reared in the wild indicates that loss of smolt characteristics also occurs under some conditions in naturally migrating smolts. As part of a restoration programme in northeastern USA, Atlantic salmon have been stocked into tributaries as fry where they remain for 2–3 years before migrating as smolts. In the Connecticut River, near the historical southern distribution of Atlantic salmon, smolts generally migrate in the mainstem of the river from early May to early June. Migrating smolts captured early in May at a dam bypass facility 198 km from the ocean had high salinity tolerance and gill Na^+,K^+-ATPase activity (S. McCormick *et al.*, unpublished data). Fish sampled in late May had significant reductions in both salinity tolerance and gill Na^+,K^+-ATPase activity. Earlier reductions in gill Na^+,K^+-ATPase occurred in 1993 when river temperatures were warmer relative to the migratory period in 1994. In addition, fish captured in early May and maintained in river water under laboratory conditions exhibited a period of stable salinity tolerance and gill Na^+,K^+-ATPase followed by a decline. As in studies on laboratory and hatchery fish, this decline is correlated with the degree days experienced by these fish (Fig. 3). Because these declines in salinity tolerance and gill Na^+,K^+-ATPase activity were detected 198 km from the mouth of the Connecticut River, it seems likely that further losses would occur in the time required to complete migration to the ocean (probably a minimum of 7 days). Reductions in physiological smolt characters may result in decreased survival, growth and swimming performance in sea

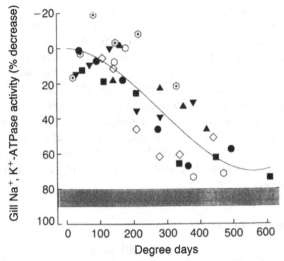

Fig. 3. Effect of degree days experienced on loss of gill Na$^+$,K$^+$-ATPase activity in Atlantic salmon smolts. Percentage loss of gill Na$^+$,K$^+$-ATPase activity is calculated as percentage change from peak levels between 20 April and 3 May. Degree days are calculated as the additive daily temperature experienced since peak gill Na$^+$,K$^+$-ATPase levels. Symbols represent the following conditions: laboratory-reared Atlantic salmon (Duston *et al.*, 1991) at 16°C (■), 13 °C (●), 10 °C (▼) and ambient (increasing) temperature (⊕); laboratory-reared Atlantic salmon (S. McCormick, unpublished data) at 10 °C (▲) and ambient temperature (○) and Atlantic salmon reared in the wild, captured early in migration and maintained at ambient temperature (◇). Shaded bar represents levels of gill Na$^+$,K$^+$-ATPase typical of parr with low salinity tolerance. Line is a third-order regression forced to go through the origin, $r^2 = 0.70$.

water, increased predator susceptibility and reduced migratory behaviour (McCormick & Saunders, 1987; Hoar, 1988).

If scenarios of global warming which predict increases in river temperatures of 2 °C are realized, temperature-related loss of smolt characteristics will have a negative impact on Atlantic salmon populations. Figure 3 indicates that Atlantic salmon have a period of 100–200 degree days in which to migrate without loss of gill Na$^+$,K$^+$-ATPase activity; a 2 °C increase in average river temperature would decrease this period by 10–20%. Temperature increases greater than 2 °C may occur through earlier and more rapid increases in spring temperature that may accompany global warming (Hengeveld, 1990), resulting in even shorter periods for downstream migration. Although loss of smolt characters may be partly compensated by earlier migration from tributaries (since temperature appears to act as a migratory cue; Jonsson & Ruud-Hansen, 1985), the mainstems of rivers are likely to be more significantly affected by increases in air temperature (Meisner, 1990). Fish in long river systems may be more susceptible to loss of smolt characteristics due to inherently longer periods of migration. Other factors such as changes in food availability, growth rate, loss of rearing habitat and overall smolt production (Meisner, 1990; Mangel, 1994) must enter into overall considerations of the effects of global warming on anadromous salmonid populations.

Anadromous clupeids

Though details differ between species and geographic locations, the anadromous clupeids all possess a common life history in which spawning occurs in spring or early summer, larvae and juveniles grow for 1–6 months in rivers, lakes or estuaries and then migrate to the ocean in summer or autumn. In contrast to salmonids, little is known of the ontogeny of salinity tolerance in these species or even basic aspects of osmoregulatory physiology.

We have undertaken a preliminary investigation of the alewife (*Alosa pseudoharengus*) to examine survival and the response of gill Na^+,K^+-ATPase activity to changes in salinity and temperature. Juvenile alewives were collected from Lake Utopia, New Brunswick, Canada in September and maintained in 1 m diameter tanks with flow-through sea water (24–33 ppt) with seasonally varying temperature (13 °C in September, 5 °C in January). Between January and April fish were acclimated to 5 or 20 °C in either fresh water or sea water (28–30 ppt) over a period of one week. After 4–8 weeks, gill samples were taken and gill Na^+,K^+-ATPase activity was measured as outlined by McCormick *et al.* (1987). There were no mortalities except for fish acclimated to fresh water at 5 °C in which two of 10 fish died in the first week of exposure to 'full' fresh water. For alewives at 20 °C, gill Na^+,K^+-ATPase activity was approximately 4-fold higher for fish in sea water than those in fresh water (Table 1). Results were similar whether the activity was assayed at 5 or 20 °C. In contrast, fish acclimated to 5 °C had gill Na^+,K^+-ATPase activity in fresh water that was almost as high as fish in sea water. Gill Na^+,K^+-ATPase was 3-fold higher for alewives at 5 °C in fresh water than for those at 20 °C in fresh water. Higher levels of gill Na^+,K^+-ATPase in cold-acclimated freshwater fish have been described previously for pupfish (*Cyprinodon salinus*), roach

Table 1. *Gill Na+,K+-ATPase activity in freshwater- and seawater-adapted alewife acclimated to 5 or 20 °C.*
Crude homogenates were assayed at both 5 and 20 °C. Values are mean ± SE ($n = 6$–8 in each group). For each assay temperature, two-way ANOVA indicated a significant effect of salinity ($P < 0.001$), no effect of temperature ($P > 0.1$), but a significant interaction ($P < 0.01$). *, significant difference from the freshwater, 20 °C group ($P < 0.05$, Student–Newman–Keuls test).

| | | Na+,K+-ATPase activity | |
	Acclimation temperature	Fresh water	Sea water
Assay temperature 20 °C	20 °C	7.3 ±1.6	27.2*±1.5
	5 °C	18.9*±0.8	22.5*±3.5
Assay temperature 5 °C	20 °C	1.0 ±0.3	5.2*±0.4
	5 °C	3.0*±0.1	4.4*±0.7

(*Rutilus rutilus*) and Arctic char (*Salvelinus alpinus*) (Stuenkel & Hillyard, 1980; Schwarzbaum, Wieser & Niederstatter, 1991), but this response does not occur in all species (Paxton & Umminger, 1983). Higher gill Na+,K+-ATPase following cold acclimation may compensate for the lower transport capacity of enzymes at low temperatures (Q_{10} effect). Absence of increased gill Na+,K+-ATPase following cold acclimation of some species may reflect their ability to affect ion transport by other mechanisms such as membrane fluidity and ion permeabilities (Schwarzbaum, Wieser & Cossins, 1992). This may also explain why gill Na+,K+-ATPase does not increase in alewives following cold acclimation in sea water.

The mortality of alewives in fresh water at cold temperatures is consistent with the findings of Stanley and Colby (1971) in which fresh water alewives exposed to 3 °C experienced greater loss of plasma and muscle sodium than fish at higher temperatures. Alewives at 3 °C in sea water also experienced ionic perturbations (high levels of plasma and muscle sodium). These authors noted that ion regulation was not significantly affected by temperatures up to 31 °C in either fresh water or sea water, indicating that in their normal geographic range osmotic tolerances of alewives are not likely to be negatively affected by global warming.

We have also examined the development of hypo-osmoregulatory ability in another anadromous clupeid, the American shad (*Alosa sapidissima*), an abundant species indigenous to the eastern coast of the United States and Canada. The autumn emigration of juvenile shad peaks when declining river temperatures reach 14–9 °C (Leggett & Whitney, 1972; O'Leary & Kynard, 1986). These temperature-related behaviours may reflect a preference pattern due in part to physiological constraints as mortality of shad occurs at temperatures below 4 °C (J. Zydlewski & S. McCormick, unpublished data).

Information on the ontogeny of salinity tolerance of young American shad has been sparse and conflicting. Leim (1924) reported that shad eggs are able to hatch and survive through the yolk-sac larva stage in brackish water but are unable to develop in salinities significantly greater than isosmotic. Tagatz (1961) reported high mortality of juvenile shad (5.6–7.9 cm fork length) in isothermal direct transfers to 33 ppt sea water at 21 °C, and complete mortality when accompanied by a decrease in temperature to 13 and 7 °C. Due to this apparent lack of hypo-osmoregulatory ability, analogies were drawn between development of salinity tolerance in shad and the salmonid parr–smolt transformation. Although Chittenden (1973) found 100% survival in isothermal (17 °C) salinity transfers of juvenile shad (4.4–6.1 cm), these fish were transported in 5 ppt sea water, arguably allowing acclimation. Recent work in our laboratory has shown that tolerance of young shad to sea water develops long before the migratory period in the Connecticut River (Zydlewski & McCormick, 1996). In this study young shad were reared from eggs and subjected to direct transfers from 0 ppt fresh water to 35 ppt sea water at approximately 10 day intervals. Development of tolerance to full strength sea water was found to be coincident with the completion of the larval–juvenile metamorphosis (3–5 weeks after hatching). This period of development is characterized by the formation of gill filaments. Salinity tolerance developed 3 months prior to the peak of migration. While gill Na^+,K^+-ATPase activity is initially low in fish with good hypo-osmoregulatory ability, exposure of juvenile shad to sea water elicits an increase in gill Na^+,K^+-ATPase activity (Zydlewski & McCormick, 1996).

The ability to hypo-osmoregulate early in their development is consistent with their variable life history, as the adults may spawn within tidal influence hence exposing the young to elevated salinities well before their seaward migration. Early development of hypo-osmoregulatory ability possibly augments fresh water nursery areas during years of high juvenile populations (Crecco, Savoy & Gunn, 1983). Early migration in some rivers may reflect a 'spread the risk' migration strategy. The fact remains, however, that most young shad utilize fresh water habitat as a nursery ground and are required to

integrate developmental and/or environmental information for a successful autumn seaward migration.

One significant physiological change observed in migrating juvenile
shad is a loss of hyperosmoregulatory ability. During the course of
migration plasma chloride levels decline more than 20%, from 113 mM
to less than 90 mM over a 5 week period, coincident with the decline
in river temperatures (J. Zydlewski & S. McCormick, unpublished
data). Juvenile shad held in fresh water and subjected to a decline in
temperature corresponding to that of the river also exhibited decreases
in plasma chloride, falling from 117 mM to 38 mM over 2 months.
Juvenile shad cease feeding between 13 and 12 °C and high mortality
was observed as temperature declined further. Plasma chloride levels
and survival of shad maintained in fresh water at 23–24 °C also decrease,
but the rate of decline is lower than in shad in cooler, ambient fresh
water. High mortality of shad maintained in fresh water (16–20 °C)
past the migratory period has also been observed by Howey (1985).
It seems likely that high mortalities in these studies were caused by
loss of hyper-osmoregulatory ability. Shad acclimated to 32 ppt sea
water do not exhibit perturbations in plasma chloride at either constant
(23–24 °C) or decreasing temperatures, and no mortality was observed
in either seawater group until temperatures fell below 4 °C in the
declining temperature group.

Pre-migratory young shad in the Connecticut River maintain relatively low levels of gill Na^+,K^+-ATPase, and there is up to a 3-fold
increase in enzyme activity during migration (J. Zydlewski & S. McCormick, unpublished data). This increase is probably not related to salinity
tolerance (as measured by survival after a 35 ppt sea water challenge)
which developed earlier, though we cannot rule out the possibility that
there are further, undetected increases in salinity tolerance during
migration. Increased gill Na^+,K^+-ATPase activity during migration may
be related to decreasing hyper-osmoregulatory ability or cold acclimation. The rate of increase in gill Na^+,K^+-ATPase, like the decline in
plasma chloride, is also hastened by declining water temperature in
fresh water.

Both alewives and shad appear to have increased mortality at temperatures below 4 °C in both fresh water and sea water. Shad lose
their capacity to regulate ions in fresh water at the end of the migratory
period even at high temperatures, and cannot be reintroduced into
fresh water from sea water after the migratory period. This is not the
case for alewives, which can be reintroduced into fresh water after the
migratory period. This difference in loss of hypo-osmoregulatory ability
may explain why only one population of non-anadromous, freshwater

shad has been reported (in California, outside its native range; Lambert et al., 1980), whereas non-anadromous populations of alewives are relatively common. The loss of hypo-osmoregulatory ability in shad has a partial parallel in smolting salmonids in which greater rates of sodium exchange have been reported for smolts in fresh water (Primmett et al., 1988). Although most studies indicate that smolts do not lose their capacity to regulate ions in fresh water during migration (McCormick & Saunders, 1987), they may expend more energy counteracting increased ion effluxes in fresh water and maintaining mechanisms for ion regulation in both fresh water and sea water. Whether physiologically constraining as occurs in shad, or energetically demanding as potentially occurs in salmon, this osmoregulatory conflict of anadromy may be more widespread than currently appreciated.

A perturbation in the annual temperature regime may cause conflicting messages for the physiological processes of migration in juvenile clupeids. If the physiological changes discussed above are linked to migratory behaviour, as is likely, an increase in global temperature would extend the length of time during which juvenile shad may reside in fresh water by delaying migration. If the summer growing season currently limits the northern distribution of shad, temperature perturbation could result in northern range expansion. In currently populated rivers, a delay of migration could result in a larger average size of migrants, which might be beneficial to survival. Although the impact of increased temperature on food resources and predators is difficult to predict, from a strictly physiological viewpoint global warming may have positive effects on anadromous clupeids by allowing northern range expansion and increased growth opportunities in fresh water.

Other anadromous fishes

Lamprey

Many species of lamprey, including the sea lamprey (Petromyzon marinus), European river lamprey (Lampetra fluviatilis) and Australian lamprey (Mordacia mordax), are anadromous. Marine adults enter fresh water to spawn, and the eggs hatch into benthic, filter-feeding ammocoete larvae. These larvae remain in fresh water for 3–7 years until metamorphosing into juveniles. Metamorphosis is highly synchronized within populations and is marked by morphological and physiological changes (Purvis, 1980; Potter, 1980) followed by migration to the ocean. The main cue for seaward migration is water flow; consequently the migration is not precisely timed and may be protracted through the autumn into the next spring (Potter, 1970).

Lamprey metamorphosis is size dependent. Warmer rearing temperatures, except where optimal temperatures are exceeded, accelerate growth of ammocoetes and lower the mean age of metamorphosis (Potter, 1980). Low temperatures inhibit metamorphosis in ammocoetes of similar size and age. The proportion of metamorphosing sea lamprey is 11-fold higher in ammocoetes reared at 20–21 °C versus 7–11 °C from July to September (Purvis, 1980). Increased spring temperature appears to be an important cue for metamorphosis. Elevated winter temperatures (22–25 vs 15 °C) advance the transformation 4–5 weeks in Australian lamprey (Potter, 1970). The rate of temperature change and increasing day length may also be important cues for metamorphosis (Potter & Beamish, 1977; Eddy, 1969).

Sea water tolerance develops upon completion of metamorphosis (Morris, 1980; Beamish, 1980). This leads to the supposition that the factors that modify the timing and incidence of metamorphosis should also affect osmoregulatory performance; however, the relevant studies have not been done. Similar to the situation in teleosts, the gills, gut and kidney of lamprey are important organs of osmoregulation (Beamish, 1980; Youson, 1980). Mitochondrion-rich chloride cells are present in the gills of metamorphosed sea lamprey, and these cells enlarge and gill Na^+,K^+-ATPase activity increases after sea water acclimation. Studies on the effects of temperature on these and other osmoregulatory parameters are required to characterize adequately the effect of temperature on osmoregulation.

Elevated temperatures, as might occur from global warming, are likely to influence metamorphosis in three major ways. Increased growth will lower the mean age of ammocoetes at metamorphosis except where thermal growth optima are exceeded. Elevated temperatures would also increase the proportion of metamorphosing individuals in higher latitudes. Both of these effects would shorten river residence time of ammocoetes, and shift the species ranges to higher latitudes. Additionally, earlier rising spring temperatures will advance the timing of metamorphosis. Assuming the development of hypo-osmoregulatory ability is linked to metamorphosis, this would allow for earlier seaward emigration.

Sturgeon

The shortnose (*Acipenser brevirostrum*), Atlantic (*A. oxyrhynchus*) and white sturgeon (*A. transmontanus*) are anadromous, using seawater habitat to different degrees. In general, juvenile sturgeon migrate seaward over a period of years (Vladykov & Greeley, 1963). Atlantic sturgeon may remain 1–6 years in fresh and brackish water before migrating to the ocean for an extended period of time (Smith, 1985). The riverine movements of young sturgeon are linked to temperature and include seasonal use of estuarine habitat, but the patterns are quite complex and poorly understood. A general effect of global warming might be a change in the use of the estuarine environment by juveniles.

Studies on the ontogeny of seawater tolerance in sturgeon are limited. Survival of white sturgeon (0.4–56 g) in 15 ppt sea water increases with size, but even the largest juveniles tested show ion perturbations and no survival in 25 ppt sea water (McEnroe & Cech, 1985). It is not clear whether sturgeon develop hypo-osmoregulatory ability during a distinct phase of development or if osmotic tolerance develops evenly with increasing size. Studies which delineate the ontogeny of salinity and thermal tolerances (and preferences) of juvenile sturgeon are required before the impact of global warming on sturgeon juveniles can be predicted.

Striped bass

Striped bass (*Morone saxatilis*) is a commercially and recreationally important anadromous fish indigenous to the Atlantic coast of North America. Adults spawn in fresh and brackish water along the coast at 10–25 °C in the spring. Larval development is temperature sensitive, with an optimum for hatching and survival occurring at 18 °C and malformations occurring above 24 °C (Doroshev, 1970). Optimal salinity for larval survival is 2–3 ppt (Kane, Bennett & May, 1990). Salinity tolerance develops early (5 day old striped bass survive in 20 ppt sea water) and is relatively insensitive to temperature (Otwell & Merriner, 1975). However, significant mortalities occur if transfer to sea water is accompanied by a large change in temperature, either an increase or decrease. Juveniles (50–100 g) acclimated to sea water maintain plasma ions close to that of freshwater controls, and gill Na^+,K^+-ATPase activity is unaffected by seawater acclimation (Madsen *et al.*, 1994).

Extreme temperature changes are more detrimental to young striped bass than are salinity changes. The ability to hypo-osmoregulate appears to be affected only when increases in salinity are accompanied by considerable temperature changes and mortalities in these situations probably reflect thermal tolerance limits. While global warming may affect striped bass, the effect is not likely to be through osmoregulatory abilities.

Summary and projections

There are clear differences in the effect of temperature change on osmoregulatory physiology in different families of anadromous fish, and it can be presumed that responses are likely to be species specific. The possibility that adaptations related to thermal physiology and behaviour may even be population specific is particularly important to consider for anadromous fishes in which reproductive isolation may occur. For these reasons it is difficult to make generalized statements about responses of anadromous fish to global warming. It will be important to examine and consider each species to determine precisely the impact of temperature change in the aquatic environment.

The effects of increased temperature and altered seasonality on anadromous salmonids will be complex. The number of smolts may be increased through greater growth opportunities in fresh water, but there will be a complex relationship between higher growth rates (more smolts) and lethal summer temperatures (fewer smolts). Although photoperiod is the primary factor regulating the timing of smolting, higher temperatures will result in earlier development of salinity tolerance. In addition, higher temperatures also result in more rapid losses of salinity tolerance and gill Na^+,K^+-ATPase activity, and these losses are directly related to the degree days experienced by downstream migrants. Temperature-related loss of smolt characteristics will have a negative impact on Atlantic salmon populations by shortening the period for successful downstream migration. Overall, the negative impacts of temperature on anadromous salmonids will be greatest in the southern range of their distribution; positive effects may occur in their northern range.

In spite of recent research, less is known of the osmoregulatory physiology of anadromous clupeids than that of salmonids. In American shad, high salinity tolerance develops at the time of larval–juvenile metamorphosis (July), several months before the peak of downstream migration (October). At the end of the migratory period ion losses occur in laboratory-reared and wild fish, coincident with increased gill Na^+,K^+-ATPase activity. Ion losses are delayed in fish maintained at elevated (summer) temperature, indicating that higher autumn temperatures will permit a longer period of fresh water residence for shad. Currently there is too little known about other anadromous species to predict the effects of global warming on osmoregulatory physiology. More research is needed on all aspects of the development of salinity tolerance and changes that occur during migration.

It would appear that global warming may have its greatest and most complex effect on anadromous salmonids. This may be the result of their longer period of residence in fresh water and complex developmental changes, but may also reflect our greater knowledge of this group of fish. In spite of this knowledge, there are several areas that require more research for salmonids and all anadromous species. Information on the physiological changes that occur in anadromous fish under different thermal regimes in nature are necessary to confirm and extend laboratory studies. Connecting physiological responses with behavioural changes (such as migratory speed) and overall survival and return rates will increase our understanding of the environmental factors that limit migratory fish populations.

Unfortunately for anadromous fish populations, global warming is not the only anthropogenic source of environmental perturbation. Obstacles to migration such as dams and water diversions not only present a physical hindrance to migration but can alter water temperatures (and other aspects of water quality). The effects of global warming on water temperatures may be exacerbated in rivers with dams, and delays in migration imposed by dams may increase the detrimental effects of high temperature in both juvenile and adult fish. Pollution effects may also be greater at higher temperatures. Predicting the effects of global warming will necessitate incorporation of many environmental changes, both natural and manmade.

Acknowledgements

We thank Dr Richard L. Saunders for encouragement and facilities for studies of alewives. Drs Henry Booke, Anthony Farrell and Gordon McDonald made many helpful comments in review.

References

Adams, B. L., Zaugg, W. S. & McLain, L. R. (1973). Temperature effect on parr–smolt transformation in steelhead trout (*Salmo gairdneri*) as measured by gill sodium–potassium stimulated ATPase. *Comparative Biochemistry and Physiology*, **44**A, 1333–9.

Beamish, F. W. H. (1980). Osmoregulation in juvenile and adult lampreys. *Canadian Journal of Fisheries and Aquatic Sciences*, **37**, 1739–50.

Boeuf, G. & Prunet, P. (1985). Measurements of gill (Na^+–K^+)-ATPase activity and plasma thyroid hormones during smolt-ification in Atlantic salmon (*Salmo salar* L.). *Aquaculture*, **45**, 111–19.

Boeuf, G., Le Roux, A., Gaignon, J. L. & Harache, Y. (1985). Gill (Na^+–K^+)-ATPase acitivity and smolting in Atlantic salmon (*Salmo salar* L.) in France. *Aquaculture*, **45**, 73–81.

Brett, J. R., Shelbourn, J. E. & Shoop, C. T. (1969). Growth rate and body composition of fingerling Sockeye salmon, *Oncorhynchus nerka*, in relation to temperature and ration size. *Journal of the Fisheries Research Board of Canada*, **26**, 2363–94.

Chittenden, M. E. Jr (1973). Salinity tolerance of young American shad, *Alosa sapidissima. Chesapeake Science*, **14**, 207–10.

Clarke, W. C. (1982). Evaluation of the sea water challenge test as an index of marine survival. *Aquaculture*, **28**, 177–83.

Clarke, W. C., Shelbourn, J. E. & Brett, J. R. (1978). Growth and adaptation to sea water in underyearling sockeye (*Oncorhynchus nerka*) and coho (*Oncorhynchus kisutch*) salmon subjected to regimes of constant or changing temperature and daylength. *Canadian Journal of Zoology*, **56**, 2413–21.

Crecco, V., Savoy, T. & Gunn, L. (1983). Daily mortality rates of larval and juvenile American shad (*Alosa sapidissima*) in the Connecticut River with changes in year-class strength. *Canadian Journal of Fisheries and Aquatic Sciences*, **40**, 1719–28.

Doroshev, S. I. (1970). Biological features of the eggs, larvae, and young of the striped bass (*Roccus saxatilis* (Walbaum)) in connection with the problem of its acclimatization in the USSR. *Journal of Ichthyology*, **10**, 235–48.

Duston, J., Saunders, R. L. & Knox, D. E. (1991). Effects of increases in fresh water temperature on loss of smolt characteristics in Atlantic salmon (*Salmo salar*). *Canadian Journal of Fisheries and Aquatic Sciences*, **48**, 164–9.

Eddy, J. (1969). Metamorphosis and the pineal complex in the brook lamprey, *Lampetra planeri. Journal of Endocrinology*, **44**, 451–2.

Folmar, L. C., Dickhoff, W. W., Mahnken, C. V. W. & Waknitz, F. W. (1982). Stunting and parr-reversion during smoltification of coho salmon (*Oncorhynchus kisutch*). *Aquaculture*, **28**, 91–104.

Foskett, J. K. & Scheffey, C. (1982). The chloride cell: definitive identification as the salt-secretory cell in teleosts. *Science*, **215**, 164–6.

Hengeveld, H. G. (1990). Global climate change: implications for air temperature and water supply in Canada. *Transactions of the American Fisheries Society*, **119**, 176–82.

Hoar, W. S. (1988). The physiology of smolting salmonids. In *Fish Physiology*, Vol. XIB, ed. W. S. Hoar & D. Randall, pp. 275–343. New York: Academic Press.

Holtby, L. B. (1988). Effects of logging on stream temperatures in Carnation Creek, British Columbia, and associated impacts on the coho salmon (*Oncorhynchus kisutch*). *Canadian Journal of Fisheries and Aquatic Sciences*, **45**, 502–15.

Howey, R. G. (1985). Intensive culture of juvenile American shad. *Progressive Fish-Culturist*, **47**, 203–12.

Johnston, C. E. & Saunders, R. L. (1981). Parr–smolt transformation of yearling Atlantic salmon (*Salmo salar*) at several rearing temperatures. *Canadian Journal of Fisheries and Acquatic Sciences*, **38**, 1189–98.

Jonsson, B. & Ruud-Hansen, J. (1985). Water temperature as the primary influence on timing of seaward migrations of Atlantic salmon (*Salmo salar*). *Canadian Journal of Fisheries and Aquatic Sciences*, **42**, 593–5.

Kane, A. S., Bennett, R. O. & May, E. B. (1990). Effect of hardness and salinity on survival of striped bass larvae. *North American Journal of Fisheries Management*, **10**, 67–71.

Karnaky, K. J., Kinter, L. B., Kinter, W. B. & Stirling, C. E. (1976). Teleost chloride cell II. Autoradiographic localization of gill Na,K-ATPase in killifish *Fundulus heteroclitus* adapted to low and high salinity environments. *Journal of Cell Biology*, **70**, 157–77.

Kurokawa, T. (1990). Influence of the date and body size at smolt-ification and subsequent growth rate and photoperiod on desmolt-ification in underyearling masu salmon (*Oncorhynchus masou*). *Aquaculture*, **86**, 209–18.

Lambert, T. R., Toole, C. L., Handley, J. M., Mitchell, D. F., Wang, J. C. S. & Koeneke, M. A. (1980). Environmental conditions associated with spawning of a landlocked American shad (*Alosa sapidissima*) population. *American Zoologist*, **20**, 813 (Abstract).

Langdon, J. S. & Thorpe, J. E. (1985). The ontogeny of smoltification: developmental patterns of gill Na$^+$/K$^+$-ATPase, SDH, and chloride cells in juvenile Atlantic salmon, *Salmo salar* L. *Aquaculture*, **45**, 83–96.

Leggett, W. C. & Whitney, R. R. (1972). Water temperature and the migrations of the American shad. *Fishery Bulletin*, **70**(3), 659–70.

Leim, A. H. (1924). The life history of the shad (*Alosa sapidissima*) with special reference to factors limiting its abundance. *Contributions to Canadian Biology*, **2**(11), 163–284.

McCormick, S. D. (1990). Fluorescent labelling of Na$^+$,K$^+$-ATPase in intact cells by use of a fluorescent derivative of ouabain: salinity and teleost chloride cells. *Cell and Tissue Research*, **260**, 529–33.

McCormick, S. D. & Saunders, R. L. (1987). Preparatory physiological adaptations for marine life in salmonids: osmoregulation, growth and metabolism. *American Fisheries Society Symposium*, **1**, 211–29.

McCormick, S. D., Saunders, R. L., Henderson, E. B. & Harmon, P. R. (1987). Photoperiod control of parr–smolt transformation in Atlantic salmon (*Salmo salar*): changes in salinity tolerance, gill Na$^+$,K$^+$-ATPase activity and plasma thyroid hormones. *Canadian Journal of Fisheries and Aquatic Sciences*, **45**, 1462–8.

McCormick, S. D., Saunders, R. L. & MacIntyre, A. D. (1989). Mitochondrial enzyme activity, and ion regulation during parr–smolt transformation of Atlantic salmon (*Salmo salar*). *Fish Physiology and Biochemistry*, **6**, 231–41.

McEnroe, M. & Cech, J. J. Jr (1985). Osmoregulation in juvenile and adult white sturgeon, *Acipenser transmontanus*. *Environmental Biology of Fishes*, **14**, 23–30.

Madsen, S. S., McCormick, S. D., Young, G. & Endersen, J. S. (1994). Physiology of sea water acclimation in the striped bass, *Morone saxatilis* (Walbaum). *Fish Physiology and Biochemistry*, **13**, 1–11.

Mangel, M. (1994). Climate change and salmonid life history variation. *Deep Sea Research*, **41**, 75–106.

Meisner, J. D. (1990). Potential loss of thermal habitat of brook trout, due to climatic warming, in two southern Ontario streams. *Transactions of the American Fisheries Society*, **119**, 282–91.

Metcalfe, N. B. & Thorpe, J. E. (1992). Anorexia and defended energy levels in over-wintering juvenile salmon. *Journal of Animal Ecology*, **61**, 175–81.

Morris, R. (1980). Blood composition and osmoregulation in ammocoete larvae. *Canadian Journal of Fisheries and Aquatic Sciences*, **37**, 1665–79.

Muir, W. D., Zaugg, W. S., Giorgi, A. E. & McCutcheon, S. (1994). Accelerating smolt development and downstream movement in yearling chinook salmon with advanced photoperiod and increased temperature. *Aquaculture*, **123**, 387–99.

O'Leary, J. A. & Kynard, B. (1986). Behavior, length, and sex ratio of seaward-migrating juvenile American shad and blueback herring in the Connecticut River. *Transactions of the American Fisheries Society*, **115**, 529–36.

Olsen, Y. A., Reitan, L. J. & Roed, K. H. (1993). Gill Na^+, K^+-ATPase activity, plasma cortisol level, and non-specific immune response in Atlantic salmon (*Salmo salar*) during parr–smolt transformation. *Journal of Fish Biology*, **43**, 559–73.

Otwell, W. S. & Merriner, J. V. (1975). Survival and growth of juvenile striped bass, *Morone saxatilis*, in a factorial experiment with temperature, salinity and age. *Transactions of the American Fisheries Society*, **104**, 560–6.

Paxton, R. & Umminger, B. L. (1983). Altered activities of branchial and renal Na/K- and Mg-ATPases in cold-acclimated goldfish (*Carassius auratus*). *Comparative Biochemistry and Physiology*, **74B**, 503–6.

Potter, I. C. (1970). The life cycles and ecology of Australian lampreys of the genus *Mordacia*. *Journal of Zoology (London)*, **161**, 487–511.

Potter, I. C. (1980). Ecology of larval and metamorphosing lampreys. *Canadian Journal of Fisheries and Aquatic Sciences*, **37**, 1641–57.

Potter, I. C. & Beamish, F. W. H. (1977). The fresh water biology of adult anadromous sea lampreys (*Petromyzon marinus*). *Journal of Zoology (London)*, **181**, 113–30.

Power, M. & Power, G. (1994). Modelling the dynamics of smolt production in Atlantic salmon. *Transactions of the American Fisheries Society*, **123**, 53–48.

Primmett, D. R. N., Eddy, F. B., Miles, M. S., Talbot, C. & Thorpe, J. E. (1988). Transepithelial ion exchange in smolting atlantic salmon (*Salmo salar* L.). *Fish Physiology and Biochemistry*, **5**, 181–6.

Purvis, H. A. (1980). Effects of temperature on metamorphosis and the age and length at metamorphosis in the sea lamprey (*Petromyzon marinus*) in the Great Lakes. *Canadian Journal of Fisheries and Aquatic Sciences*, **37**, 1827–34.

Raymond, H. L. (1979). Effects of dams and impounds on migration of juvenile chinook salmon and steelhead from the Snake River, 1966 to 1975. *Transactions of the American Fisheries Society*, **108**, 50–29.

Richman, N. H. I., Tai de Diaz, S., Nishioka, R. S., Prunet, P. I. & Bern, H. A. (1987). Osmoregulatory and endocrine relationships with chloride cell morphology and density during smoltification in coho salmon, (*Oncorhynchus kisutch*). *Aquaculture* **60**, 265–285.

Schwarzbaum, P. J., Wieser, W. & Cossins, A. R. (1992). Species-specific responses of membranes and the Na^+-K^+ pump to temperature change in the kidney of 2 species of fresh water fish, roach (*Rutilus rutilus*) and arctic char (*Salvelinus alpinus*). *Physiological Zoology* **65**, 17–34.

Schwarzbaum, P. J., Wieser, W. & Niederstatter, H. (1991). Contrasting effects of temperature acclimation on mechanisms of ionic regulation in a eurythermic and a stenothermic species of fresh water fish (*Rutilus rutilus* and *Salvelinus alpinus*). *Comparative Biochemistry and Physiology*, **98A**, 483–9.

Smith, T. I. J. (1985). The fishery, biology, and management of Atlantic sturgeon, *Acipenser oxyrhynchus*, in North America. *Environmental Biology of Fishes*, **14** 61–72.

Solbakken, V. A., Hansen, T. & Stefansson, S. O. (1994). Effects of photoperiod and temperature on growth and parr-smolt transformation in Atlantic salmon (*Salmo salar* L.) and subsequent performance in sea water. *Aquaculture*, **121**, 13–27.

Solomon, D. J. (1978). Migration of smolts of Atlantic salmon (*Salmo salar* L.) and sea trout (*Salmo trutta* L.) in a chalkstream. *Environmental Biology of Fishes*, **3**, 223–9.

Stanley, J. G. & Colby, P. J. (1971). Effects of temperature on electrolyte balance and osmoregulation in the alewife (*Alosa pseudoharengus*) in fresh and sea water. *Transactions of the American Fisheries Society*, **100**, 624–38.

Staurnes, M., Sigholt, T. & Gulseth, O. A. (1994). Effects of seasonal changes in water temperature on the parr-smolt transformation of Atlantic salmon and anadromous Arctic char. *Transactions of the American Fisheries Society*, **123**, 408–15.

Stuenkel, E. L. & Hillyard, S. D. (1980). Effects of temperature and salinity on gill Na^+-K^+-ATPase activity in the pupfish, *Cyprinodon salinus*. *Comparative Biochemistry and Physiology*, **67A**, 179–82.

Tagatz, M. E. (1961). Tolerance of striped bass and American shad to changes in temperature and salinity. *U.S. Fish and Wildlife Service Special Scientific Report Fisheries*, **388**, 8 pp.

Thorpe, J. E. (1977). Bimodal distribution of length of juvenile Atlantic salmon (*Salmo salar* L.) under artificial rearing conditions. *Journal of Fish Biology*, **11**, 175–84.

Thorpe, J. E. (1994). Salmonid flexibility: responses to environmental extremes. *Transactions of the American Fisheries Society*, **123**, 606–612.

Thorpe, J. E., Adams, C. E., Miles, M. S. & Keay, D. S. (1989). Some photoperiod and temperature influences on growth opportunity in juvenile Atlantic salmon, *Salmo salar* L. *Aquaculture*, **82**, 119–26.

Thorpe, J. E., Morgan, R. I. G., Ottaway, E. M. & Miles, M. S. (1980). Time of divergence of growth groups between potential 1+ and 2+ smolts among sibling Atlantic salmon. *Journal of Fish Biology*, **17**, 13–21.

Vladykov, V. D. & Greeley, J. R. (1963). *Order Acipenseroidei. Fishes of the Western North Atlantic.* New Haven, Conn: Sears Foundation of Marine Research, Yale University.

Wagner, H. H. (1974). Photoperiod and temperature regulation of smolting in steelhead trout (*Salmo gairdneri*). *Canadian Journal of Zoology*, **52**, 219–34.

Wedemeyer, G. A., Saunders, R. L. & Clarke, W. C. (1980). Environmental factors affecting smoltification and early marine survival of anadromous salmonids. *Marine Fisheries Review*, **42**, 1–14.

Youson, J. H. (1980). Morphology and physiology of lamprey metamorphosis. *Canadian Journal of Fisheries and Aquatic Sciences*, **37**, 1687–710.

Zaugg, W. S. & McLain, L. R. (1976). Influence of water temperature on gill sodium, potassium-stimulated ATPase activity in juvenile coho salmon (*Oncorhynchus kisutch*). *Comparative Biochemistry and Physiology*, **54**, 419–21.

Zaugg, W. S. & Wagner, H. H. (1973). Gill ATPase activity related to parr–smolt transformation and migration in steelhead trout (*Salmo gairdneri*): influence of photoperiod and temperature. *Comparative Biochemistry and Physiology*, **45B**, 955–65.

Zydlewski, J. & McCormick, S. D. (1996). The ontogeny of osmotic tolerances in the American shad, *Alosa sapidissima. Canadian Journal of Fisheries and Aquatic Sciences* (in press).

CHRISTOPHER J. KENNEDY and
PATRICK J. WALSH

Effects of temperature on xenobiotic metabolism

Introduction

Through evolutionary history, aquatic organisms have been challenged by a vast array of natural *foreign chemicals*, or *xenobiotics*, of biogenic, pyrogenic and diagenic origin. More recently, aquatic systems have become the ultimate sinks of anthropogenic inputs of contaminants into the environment as well, increasing the threat to individual organisms and populations. Acute releases of toxic substances by either human activity (McEwen & Stephenson, 1979, pp. 312–15) or natural causes (Steidinger, Burklew & Ingle, 1972) have resulted in many incidences of high mortality in fish populations over short periods of time. More subtle chronic exposures to lower concentrations of xenobiotics, however, can prove to be equally devastating to fish populations. For example, an increased incidence of tumours in feral fish populations has been linked to contaminants such as the polycyclic aromatic hydrocarbons (PAHs) (Malins *et al.*, 1985), and reproductive impairment has been shown in fish which reside in waters receiving pulp and paper mill effluent (Munkittrick *et al.*, 1991).

The susceptibility of fish to xenobiotic action can be modulated by a variety of abiotic factors including water pH, dissolved oxygen content and temperature. In toxicology, as in other areas of biology such as ecology, physiology and biochemistry, the influence of temperature at all levels of biological organization is pervasive and often of dominant importance (Hochachka & Somero, 1984). Fish habitats vary in temperature from close to the freezing point of sea water ($-1.86\,°C$) in polar regions to temperatures in the range of 40–45 °C in desert springs and streams, while temperate and tropical seas possess surface temperatures in the range of 10–20 °C and *c.* 30 °C, respectively (Hazel, 1993). Because of the wide ranging and fluctuating thermal habitats of fishes, there is considerable information regarding piscine thermal biology;

Society for Experimental Biology Seminar Series 61: *Global Warming: Implications for freshwater and marine fish*, ed. C. M. Wood & D. G. McDonald. © Cambridge University Press 1996, pp. 303–324.

however, there has been a profound lack of toxicological research with respect to the thermal modulation of chemical action. This information is important in other areas of study: for example, in comparative pharmacology and toxicology, basic fish physiology and biochemistry, the use of fish as supplements or alternatives to mammalian models (Hoover, 1984; Powers, 1989), in biomonitoring programmes (Payne et al., 1987), in predictive modelling (Law, Abedini & Kennedy, 1991) and in ecological risk assessment (Bartell, Gardner & O'Neill, 1992). Lastly, the effects of future global climate change will almost certainly be against a backdrop of a more polluted aquatic environment. Therefore, interactive studies and models of the effects of temperature on toxic action will be especially powerful predictive tools. To these ends, this chapter will review the current knowledge regarding the modulating effects of temperature on xenobiotic toxicity with special reference to the role of temperature in xenobiotic metabolism in fishes.

Thermal modulation of xenobiotic toxicity

The toxicity of many xenobiotics to fish is altered by changes in water temperature; however, its influence on chemical toxicity is complex. This is because temperature alone may be a lethal factor with thermal limits that may be altered by a specific toxicant. This chapter, however, deals specifically with the combined effects of temperature and xenobiotics within the thermal tolerance zone of fish. With this in mind, temperature may act as a modulator of toxicity through effects on chemical availability (e.g. solubility in water), toxicokinetics (uptake, distribution, metabolism and excretion), or toxicant–receptor interactions. No consistent pattern for the effects of temperature on toxicity emerges from the literature; a temperature change may increase, decrease or result in no change in toxicity. For example, a general increase in the susceptibility of trout and bluegills to many pesticides including dieldrin, chlordane and malathion was noted as temperature increased (Macek, Hutchinson & Cope, 1969). Other studies with fish have implicated temperature as a modulating factor in carcinogenesis. In European eels (Anguilla anguilla), an increase of epidermal papilloma incidence and tumour size coincided with the onset of summer (Peters & Peters, 1977). In laboratory studies, temperature significantly altered the effects of the carcinogen diethylnitrosamine (DENA) in the medaka (Oryzias latipedes); treatment at 22–25 °C led to tumour development in 95% of fish, but most fish had no tumours when exposed to DENA at 5–8 °C (Kyono-Hamaguchi, 1984). Similar responses were noted in zebra fish (*Danio rerio*), guppies (*Poecilia reticulata*) (Khudoley, 1984) and rainbow trout (*Oncorhynchus mykiss*) (Hendricks *et al.*, 1984). Contrary to these results, several studies have also shown that decreasing temperature leads to an increase in the toxicity of organic compounds such as phenol (Brown, Jordan & Tiller, 1967) and DDT (Cairns, Heath & Parker, 1975), as well as several forms of metals (Lemly, 1993; Brown, 1968). These examples indicate that although temperature can significantly alter the toxicity of specific chemicals, it is difficult to predict the course of temperature's modulating effects.

Mechanisms of toxicity modulation

The mechanisms of thermal modulation of chemical action are not yet clear; potentially temperature may act at many levels including xenobiotic uptake, distribution, metabolism and elimination. Temperature has profound effects on, and leads to substantial adaptations of, the routine physiology and metabolism of ectothermic species, including fish (Hazel & Prosser, 1974; Schmidt-Nielsen, 1983; Hochachka & Somero, 1984).

It is generally accepted that increased temperature increases the rate of uptake of xenobiotics through changes in ventilation rate in response to an increased metabolic rate and decrease in oxygen solubility. Even following a suitable acclimation period to high temperatures, xenobiotic uptake rates can remain significantly elevated (Kennedy, Gill & Walsh, 1989a). At the cellular level, an acute temperature increase leads to an increase in the uptake of the PAH, benzo[a]pyrene (BaP) in isolated gill cells of the gulf toadfish (*Opsanus beta*). This is probably through an increase in membrane fluidity since PAHs are believed to diffuse directly across the lipid bilayer (Kennedy & Walsh, 1994).

Other studies have demonstrated that increased temperature may also increase the excretion rate of xenobiotics via the biliary route (Curtis, 1983; Jimenez, Cirmo & McCarthy, 1987; Kennedy, Gill & Walsh, 1989b); however, no firm conclusions can be made regarding the direct effects on elimination due to the complicating effects of temperature on chemical uptake in these experiments.

Rates of xenobiotic metabolism have direct impact on the toxicity of xenobiotics to fish and there is evidence to suggest that temperature has substantial effects on these processes as well. Furthermore, xenobiotic metabolism is an energy-consuming process (e.g. especially through NADPH consumption) and it is thought that any general temperature-mediated effects on an organism's energy state may affect xenobiotic metabolism (Andersson, 1984) and toxicity.

Significance of xenobiotic metabolism

Xenobiotic metabolism or biotransformation is defined as the enzymatic conversion of one chemical into another and is differentiated from physical–chemical processes (such as photolysis) that can also effect chemical conversions. In the past it was generally believed that fish did not possess the ability to metabolize xenobiotics and excreted lipophilic compounds by diffusional processes through the gills and skin (Brodie & Maickel, 1962). Research has now demonstrated an appreciable ability of fish to metabolize both natural and anthropogenic chemicals. The general trend of these metabolic transformation processes is the conversion of lipophilic compounds to more polar hydrophilic metabolites through Phase I (oxidation, reduction and hydrolysis) and Phase II (conjugation) reactions (Fig. 1). This conversion usually results in reduced toxicity and an enhanced potential for excretion, the hepato-biliary system being the major excretory route.

The availability of xenobiotics for these reactions and the rates of metabolism have important implications in the bioaccumulation, persistence, dynamics and toxicity of xenobiotics in fish. For example, it has been shown that the four orders of magnitude difference in the bioaccumulation by *Gambusia affinis* of two structurally related compounds, DDT (2-bis(*p*-chlorophenyl)-1,1,1-trichloroethane) and 2-bis(*p*-methylthiophenyl)-1,1,1-trichloroethane, with similar octanol/water partition coefficients, was due to the latter compound's ability to be metabolized by this species (Kapoor *et al.*, 1973). In another

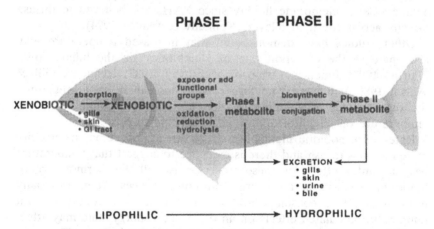

Fig. 1. Generalized scheme of the integration of Phase I and Phase II biotransformation reactions in fish.

study, the LC_{50} of the larvicide of sea lamprey, 3-trifluoromethyl-4-nitrophenol (TFM), to rainbow trout was reduced from 5.05 to 2.67 mg l^{-1} when trout were pretreated with an inhibitor of conjugative metabolism of the xenobiotic, increasing the half-life of TFM 2-fold (Lech, 1974).

Temperature effects on net *in vivo* metabolism of xenobiotics

The *in vivo* effects of temperature on xenobiotic metabolism in fish have received little attention. The data available are equivocal although they generally indicate that acute temperature change alters xenobiotic metabolism to a much greater extent than if fish are allowed to acclimate to the new temperature. For example, there were no significant differences in the amounts of ^{14}C-naphthalene metabolite-derived radioactivity in the bile of coho salmon at 4 and 10 °C when the fish were exposed to the parent compound at their respective acclimation temperatures (Varanasi, Uhler & Stranahan, 1978). Similarly, in steelhead trout, over a temperature acclimation range of 10–18 °C, a uniform *in vivo* biotransformation of phenolphthalein to its glucuronide conjugate was maintained. Curtis, Fredrickson and Carpenter (1990) reported similar results in rainbow trout exposed to ^{14}C-benzo[a]pyrene at 10 and 18 °C; however, acute temperature shifts to an intermediate temperature of 14 °C stimulated the production of polar metabolites in the 10 °C-acclimated fish and lowered metabolite production in the 18 °C-acclimated fish. These results indicate an ideal temperature compensation in the biotransformation of xenobiotics in these fish species through alterations in xenobiotic metabolizing enzyme activities. Other studies of this nature, however, show contrary results. Bluegill sunfish (*Lepomis macrochirus*) metabolized ^{14}C-benzo[a]pyrene to both polar and aqueous-soluble metabolites at a faster rate when fish were exposed at an acclimation temperature of 23 versus 13 °C (Jimenez *et al.*, 1987). This study also showed that temperature affects not only the rate of biotransformation but also the products formed. Tentative metabolite identification indicated that bluegills metabolized benzo[a]pyrene to more polar compounds at the lower temperature. Another study demonstrated that the metabolism and excretion of benzo[a]pyrene by the gulf toadfish is enhanced at higher temperatures, and that the effects of acute temperature changes are more pronounced than acclimatory temperature changes (Kennedy *et al.*, 1989b). However, in this study the similarity of Q_{10} values obtained for chemical uptake (Kennedy *et al.*, 1989a), to those obtained for accumulation in the bile, indicates that the effects of acute and acclimatory temperature change on rates of metabolism and excretion may derive largely from effects on the amount of chemical made available by uptake, and perhaps less so from direct effects of temperature on metabolism and excretion *per se*. In all of the experiments described it is difficult to isolate the effects of temperature on metabolism because of the method of chemical administration (intraperitoneal or water exposure). A comparative study on metabolism was performed which, for the first time, removed the complicating effects of temperature on uptake by administering the chemical intravenously (Kennedy & Walsh, 1991). In contrast to previous findings, the accumulation of biliary metabolites was not significantly different between fish exposed at different temperatures. Temperature changes did not lead to differences in the relative proportions of Phase I and Phase II metabolites of benzo[a]pyrene although the metabolic profile of individual Phase I metabolites was altered in fish *in vivo*: acclimation to high temperature produced more triols and tetrols (breakdown products of highly carcinogenic benzo[a]pyrene diol epoxides) than acclimation at low temperature regardless of exposure temperature (Fig. 2). Others have indicated that higher temperatures lead to an increased DNA-adduct formation by diol epoxides in the liver of fish exposed to benzo[a]pyrene (Shugart *et al.*, 1987). Thus, despite compensation in overall quantitative rates of metabolism, higher temperatures appear to lead to a shift in metabolism towards potentially more carcinogenic intermediates, an observation which may in part explain the enhanced tumorigenesis seen in fish at higher temperatures.

Temperature effects on the various steps in metabolism

Strictly speaking, biotransformation is the enzymatic conversion of a xenobiotic to another form. However, there are several steps towards chemical metabolism through which temperature may play the role of a modulator. The following describes the current knowledge regarding the effects on these steps including cellular uptake of xenobiotics, biotransformation enzyme activities and enzyme microenvironment.

Cellular uptake of xenobiotics

The liver of fish is the primary site of xenobiotic biotransformation, although realized and potential capacity has been shown to exist (mainly by the presence of enzyme activity) in other tissues such as the gill, kidney and gastrointestinal tract. Information regarding the effects of temperature on the cellular uptake of xenobiotics through diffusion across cell membranes is extremely limited.

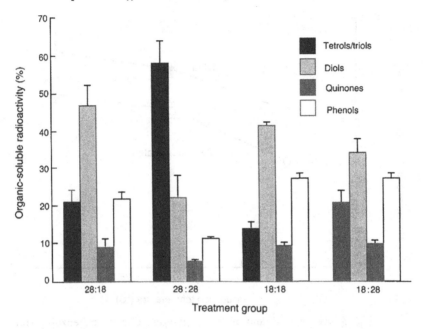

Fig. 2. The percentage of total organic-soluble radioactivity as the predominant metabolite groups in toadfish bile 72 h following an intravenous administration of ^{14}C-benzo[a]pyrene. The fish were acclimated to 18 or 28 °C (first number) and exposed to the chemical at 18 or 28 °C (second number). (From Kennedy & Walsh, 1991.)

Reduced temperatures may promote lipid cluster formation, phase separation of lipids into coexisting fluid and crystalline domains, and the ultimate crystallization of the entire lipid domain in cell membranes (Lee *et al.*, 1974; Wunderlick *et al.*, 1975). This diminished membrane fluidity has been shown to affect the rate of membrane-associated processes including non-electrolyte transport (Thilo, Trauble & Over-ath, 1977; Hazel, 1979). Studies with isolated gill cells from gulf toadfish acclimated to different temperatures show that at all incubation temperatures, benzo[a]pyrene uptake rates were higher in cells from 18 °C-acclimated fish when compared to cells from 28 °C-acclimated fish. However, when cells were exposed to BaP at the respective acclimation temperatures of the fish, uptake rates were similar (Fig. 3; Kennedy & Walsh, 1994). This observation is explained by measurements of plasma membrane order which indicated that increasing temperature increased the fluidity of the membranes in each group of fish.

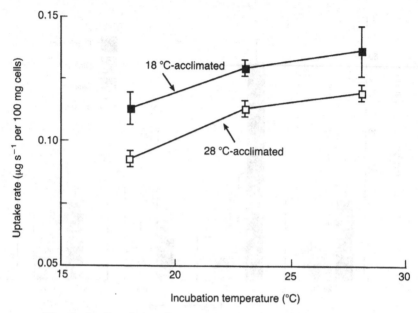

Fig. 3. Acclimation and acute temperature effects on benzo[a]pyrene uptake in isolated gill cells of gulf toadfish (*Opsanus beta*). Rates at 18 °C are significantly different from rates at 28 °C at all incubation temperatures at $p < 0.05$ level. Bars represent SE. (From Kennedy & Walsh, 1994.)

A partial compensation for the effect of temperature on membrane lipid order was seen so that when measurements were made at acclimation temperatures, membrane fluidities were similar (Kennedy & Walsh, 1994).

Alterations in the fatty acid components of phospholipids in plasma membranes, such as increases in the ratio of unsaturated to saturated fatty acids in cold acclimation, are believed to be the mechanism involved in the maintenance of a relatively constant fluidity of membranes at all temperatures (Hazel, 1979). It appears that the fluidity and possibly fatty acid composition of membranes are important determinants of passive diffusion of non-polar xenobiotics into cells. Other temperature effects on the accumulation of organic chemicals may include alterations of diffusion through the cytosol, solvation, and protein-binding constants (Barron, 1990). In this respect, the diffusion of small molecules such as D-lactate, 2-deoxyglucose and Ca^{2+} through fish muscle cytosol decreases with decreasing temperature, possibly due to increases in cytosolic viscosity, reduced kinetic energy of the system and decreases in bond strength of ionic and dipole interactions (Sidell & Hazel, 1987). The impacts of these processes on xenobiotic diffusion are virtually unknown.

Effects on enzymes

Together with alterations in membrane constituents, the molecular basis of temperature adaptation in the maintenance of biological activity may include quantitative or qualitative alterations in enzyme systems. Although investigations at the biochemical and molecular levels regarding temperature-induced alterations in xenobiotic metabolism are recent, there is considerable information regarding these events. The types of metabolic transformations (enzyme activities) are categorized by intracellular localization, i.e. microsomal (in the smooth endoplasmic reticulum) and non-microsomal (in mitochondria, cytosol, plasma) reactions, or by type of reaction and substrate. Phase I reactions unmask or introduce polar groups into a xenobiotic through oxidative, reductive and hydrolytic processes. Phase II reactions involve the conjugation of a xenobiotic or its phase I metabolite with polar endogenous molecules such as sulphate, glucuronic acid or glutathione.

Oxidative metabolism of xenobiotics and many endogenous substrates is usually catalysed by the cytochrome P450 monooxygenase system, or mixed-function oxidases (MFOs) located in the smooth endoplasmic reticulum. The MFO system is comprised of cytochrome P450, the phospholipid of the endoplasmic reticulum membrane, and the flavo-protein NADPH-cytochrome P450 reductase (NADPH-cytochrome *c* reductase). Many examples of xenobiotic monooxygenations by cytochrome P450 have been reported; this system catalyses a diversity of other types of chemical transformations including deamination, aliphatic hydroxylation, alkene and aromatic hydrocarbon epoxidation, *O*-dealkylation, *N*-dealkylation, N-oxide formation and dehalogenation reactions. Metabolic compensation to temperature has been shown to occur for MFO activities in several fish species. For example, in field studies rainbow trout showed ideal compensatory patterns in MFO activities (benzo[a]pyrene hydroxylase and 7-ethoxycoumarin-*O*-deethylase activities) in response to seasonal fluctuations in temperature, although other parameters such as reproductive state, availability of food and photoperiod may also be involved in the exhibited alterations (Koivusaari, Harri & Hanninen, 1981; Koivusaari, 1983; Koivusaari & Andersson 1984; Blank *et al.*, 1989). In laboratory studies utilizing microsomal (Dewaide, 1971; Ankley *et al.*, 1985; Stegeman, 1979) and isolated hepatocyte preparations (Andersson & Koivusaari, 1986), temperature compensatory mechanisms for P450-dependent activities have also been shown in many species of fish.

Both quantitative and qualitative changes occur in the MFO system of fish during adaptation to temperature. In a laboratory study with bluegills, the level of P450 varied inversely with acclimation temperature (Karr, Reinert & Wade, 1985). However, in a field study, the total amount of P450 remained constant in rainbow trout during summer and winter (Blank et al., 1989). In the first study, it is suggested that qualitative differences based on the nature of the enzyme present and its reactivity with various substrates are shown with temperature acclimation. For example, benzo[a]pyrene hydroxylase activity tested in vitro showed significant decreases in K_m (but not V_{max}) with decreasing incubation temperature, which facilitates an increase in reaction rate under non-saturating substrate conditions, suggesting that different isozymes are produced in response to temperature changes (Karr et al., 1985). Carpenter et al. (1990), using antibodies, determined that the content of the specific P450 LM_{4b} isozyme (equivalent to the mammalian form P450 1A1, the PAH-inducible form) was higher in trout acclimated to 10 °C vs 18 °C. In the second study, P450 levels remained constant with changing temperature although P450-mediated enzyme activity showed compensation. This is partly explained by temperature compensation in NADPH-cytochrome reductase activity which delivers reducing equivalents to cytochrome P450 (Blank et al., 1989). They also reported an increase in cytochrome P450 reduction at colder temperatures which was facilitated by an increase in the cytochrome P450 reductase: cytochrome P450 ratio.

Hydrolytic reactions occur in many species of fish and the enzymes involved are diverse, although in general, the types of xenobiotics which are involved in these reactions include esters, amides and epoxides. There is little direct evidence regarding the effects of temperature on hydrolysis reactions in fish, although one study reported a significantly higher hepatic epoxide hydrolase (EH) activity in trout when maintained at low temperatures (Egaas & Varanasi, 1982). Indirect evidence with environmental significance indicates that EH activity is modified by temperature. An increased accumulation of tetrols and triols in the bile of gulf toadfish exposed to benzo[a]pyrene at higher temperatures indicates that thermal modulation of EH activity may be important in chemical carcinogenesis (Kennedy & Walsh, 1991). The metabolism of the PAH benzo[a]pyrene is an example of 'bioactivation' in which a metabolite is more toxic than the parent compound. Via a sequence of epoxidation and hydrolysis reactions, benzo[a]pyrene is converted into an ultimate carcinogen. Detoxification of benzo[a]pyrene oxides by the formation of dihydrodiols by EH may occur, resulting in the excretion of these compounds (Kennedy *et al.*, 1989b) or their participation in subsequent conjugation reactions (Goksoyr, 1985). However, some dihydrodiols such as BaP *trans*-7,8-dihydrodiol may be recycled and further oxidized by the cytochrome P450 system forming the ultimate carcinogens, *anti*-, *syn*- BaP 7,8-diol 9,10-oxides (Gelboin, 1980). Correlations between the relative amount of binding of different PAH metabolites to DNA, RNA and proteins and their carcinogenicity (Brookes & Lawley, 1964) and the relative mutagenicity of BaP diol epoxides (Gelboin, 1980) points to the potential significance of PAH metabolism in carcinogenicity in fish. Indeed, chemical carcinogens such as PAHs have been implicated in epizootics of neoplasia in feral fish of both marine and fresh water species (Baumann, 1989).

Fewer data exist on the effects of temperature on Phase II enzyme systems. It has been reported that *in vitro* and *in vivo* UDP-glucuronosyl transferase activity is affected by temperature. Glucuronyl transfer or glucuronidation involves the conjugation of a metabolically activated UDP-glycoside (UDPGA) such as UDP-glucuronic acid or UDP-glucose, to a xenobiotic forming a D-glucuronide. UDPGT activity showed 'inverse compensation', i.e. a decrease in enzyme activity with a decrease in temperature, in toadfish and trout *in vitro* (Kennedy, Gill & Walsh, 1991; Andersson & Koivusaari, 1985) and *in vivo* (Koivusaari *et al.*, 1981; Koivusaari, 1983). It has been suggested that the regulation of hepatic glucuronidation by ambient temperature is important in the control of plasma steroid levels during the reproductive season (Andersson & Koivusaari, 1985). Low water temperatures and the concomitant low activity of glucuronidation enzyme activities may enable the fish to maintain high levels of steroid hormones necessary for reproductive development (Andersson & Koivusaari, 1985; Scott & Sumpter, 1983; Kime & Saksema, 1980).

In achieving these new levels of enzyme activity (Phase I or Phase II reactions), a range of factors could be regulating the response, from simple direct effects of temperature on rates of synthesis/degradation to more complex, hormonally mediated responses. Enzyme activities often show a relatively high variability within an acclimation group, indicating that other factors may play a role in modifying enzyme activities. This wide variability is also observed in other 'wild fish populations' from both polluted and 'non-polluted' sites, which is a major difficulty in using such enzymes as a biomonitoring tool for indices of aquatic pollution (Jimenez & Burtis, 1989). This variability may mask any effects of temperature acclimation and can be reduced if measurements of xenobiotic transformation enzymes are made on cells obtained from the same fish acclimated to different temperatures in a paired design. In this regard, cultured catfish hepatocytes were used to show that the regulation of energy-transforming enzyme response to temperature changes rested with the cell itself, independent of organismal-level response (Koban, 1986). In an analogous study of the effects of temperature on cytochrome P450 levels and aryl hydrocarbon hydroxylase (AHH), epoxide hydrolase (EH), glutathione S-transferase (GST), UDP-glucuronosyl transferase (UDPGT) and sulphotransferase (SFT) activities in cultured toadfish hepatocytes, it was found that regulation of much of the response to temperature also rested with the cell itself (Fig. 4; Kennedy et al., 1991). Acclimation to colder temperatures increased the cytochrome P450 levels and the activities of EH and AHH. UDPGT activity showed 'inverse compensation' in

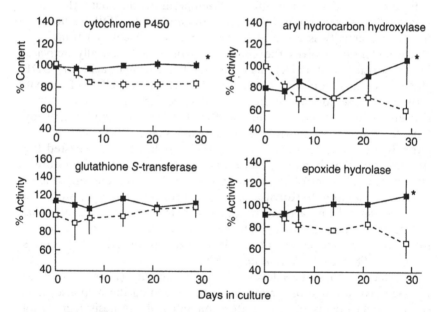

Fig. 4. Effects of culture temperature on cytochrome P450 levels and the activities of the microsomal enzymes aryl hydrocarbon hydroxylase, epoxide hydrolase and glutathione S-transferase in cultured hepatocytes. Hepatocytes from 18 °C-acclimated fish were incubated at 28 °C (□) and cells from 28 °C-acclimated fish were cultured at 18 °C (■). Enzyme activities (nmol.min⁻¹.10⁶ cells⁻¹) are expressed as percentage changes from day 0 which was normalized to 100% for the average value of 18 °C-acclimated cells. ∗ denotes a significant difference at $p < 0.05$ level at day 29. Bars represent SE. (From Kennedy, Gill & Walsh, 1991.)

toadfish hepatocyte cultures. Similar effects were seen in trout *in vivo* (Koivusaari, 1983). Levels of GST activity were similar in cultured cells from both cold- and warm-acclimated fish and it is speculated that the out of vogue 'positive thermal modulation' hypothesis (Hochachka & Somero, 1984), i.e. an increase in K_m with temperature, is responsible for its apparent temperature insensitivity.

Benzo[a]pyrene metabolism in isolated fish hepatocytes appears to be somewhat temperature insensitive at low benzo[a]pyrene concentrations. However, at concentrations approaching saturation of biotransformation pathways, effects were substantial (Gill & Walsh, 1990; Andersson & Koivusaari, 1986). Increasing rates of metabolism were seen with increasing temperature, and the results fit a pattern of 'rotation' (Prosser, 1973) whereby the warm-acclimated curve is rotated downward such that, following acclimation, the rate at the higher temperature is nearly identical to the rate at the lower temperature (Gill & Walsh, 1990). It is noteworthy that for enzymes of intermediary metabolism, where regulatory responsiveness is critical, K_m – temperature curves are tailored by evolution to be flat. In contrast, we believe that xenobiotic enzymes are designed for high throughput with on/off regulation achieved largely via changes in enzyme quantity. Thus, in an evolutionary sense, the regulatory responsiveness of a flat K_m–temperature curve may be sacrificed in favour of increases in catalytic efficiency.

Effects on enzyme microenvironment

Changes in environmental temperature experienced by aquatic ectotherms such as fish often lead to changes in biochemical structures which may be significant in the metabolism of xenobiotics. Adjustments in enzyme concentration may not be adequate for compensating the effects of changes in environmental factors, and alternative mechanisms which include a modification in catalytic efficiencies are needed for adjusting enzymatic rates in ectotherms (Somero, 1979; Koivusaari, 1983). Such changes may be realized through modifications in the hydrophobic microenvironment of these enzymes. Cold acclimation leads to an increase in the proportions of polyunsaturated fatty acids in membranes of several fish tissues which partly compensates for the effects of temperature on membrane fluidity (Hazel, 1984; Miller, Hill & Smith, 1976).

Cytochrome P450s and NADPH-cytochrome P450 reductase are associated with the membranes of the endoplasmic reticulum. Purified trout and rat P450 reductases were optimally functional (as measured as P450-linked *O*-deethylation of 7-ethoxycoumarin) at 26 and 37 °C, respectively (Gurumurthy & Mannering, 1985). When trout reductases were then added to rat microsomes and vice versa, the optimal functional temperatures of the reductases were reversed, indicating that the phospholipid annulus surrounding the active site of membrane-bound P450 may determine the optimal temperature of P450 systems (Fig. 5). It has been shown that the amount and the type of lipid present in the endoplasmic reticulum can influence aryl hydrocarbon monooxygenase activity. Wills (1980) reported that a higher rate of benzo[a]pyrene oxidation is associated with a higher proportion of polyunsaturated fatty acids in microsomal membranes. The Phase II enzyme UDP-glucuronosyl transferase is located in the same intracellular membrane as the P450 monooxygenase system. However, as mentioned earlier, this enzyme typically exhibits 'inverse compensation', a fact which is not easily explained by the data available. It should be noted that Blank *et al.* (1989) found that there were only small changes in the fatty acid composition of hepatic microsomes isolated from rainbow trout in August and December, but did not rule out the possibility that enzyme activities are modified by small temperature-induced changes in membrane lipids.

Global warming

Most global warming scenarios include the prediction of small long-term increases of 2–4 °C in water temperature, along with more abrupt acute

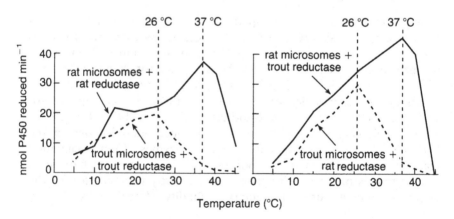

Fig. 5. Optimal temperatures of purified rat and trout NADPH-cytochrome P450 reductase combined with either rat or trout hepatic microsomes. (From Gurumurthy & Mannering, 1985.)

changes in temperature, for example during seasonal transitions in temperate zones. The responses of fish to aquatic contaminants will surely be modified by alterations in habitat temperature; however, the complex processes involved in the common paradigm in toxicology as described by Stegeman and Hahn (1994) make predictions of toxic endpoints difficult. Temperature may act on most elements of this paradigm, including environmental transport, partitioning, degradation and bioavailability which determine what the fish will actually be exposed to, what is available for uptake, the absorption of the chemical, its distribution within the fish, its metabolic transformation, and eventually its excretion. One of the most important determinants of toxicity is the biotransformation of the xenobiotic. The amount of a compound that is metabolized depends upon the delivery of the xenobiotic to the sites of metabolism, the quantities and activities at these sites and the removal of the metabolites from the system.

The effects of temperature change on the respiration of fishes are well documented. It is suggested that the branchial chemical uptake route is modulated by temperature through changes in fish respiration rate or convection volume. With an increase in global temperatures, chemical uptake rates by the branchial route are likely to increase even after a suitable acclimation period. Evidence also suggests that temperature effects on cell membrane fluidity and possibly fatty acid composition will affect the passive diffusion of xenobiotics through membranes, although these effects will probably contribute to enhanced uptake to a limited extent.

Most of the enzyme systems involved in both Phase I and Phase II reactions show temperature compensation. Fish acclimated to higher environmental temperatures typically have lower biotransformation enzyme activities compared with fish acclimated to colder temperatures. In addition to this fact, levels of biotransformation in fishes are typically lower than those found in mammals. Higher rates of chemical uptake with increasing global temperature are likely to lead to increased levels of these chemicals at the sites of biotransformation. In many cases, an increased uptake of xenobiotics may saturate metabolic pathways leading to the accumulation of xenobiotics to toxic levels.

Many xenobiotics undergo sequential reactions in the biotransformation process. Limited information shows that differing compensatory patterns to temperature changes exist for the enzyme systems, ranging from partial to perfect to inverse compensation. Increasing environmental temperature may lead to the buildup of intermediary metabolites and alterations in the types of metabolites produced. Since biotransformation reactions can result in products which are more toxic than the parent molecule, the buildup of reactive intermediates is possible at higher temperatures. Thus, the effects of increases in global temperatures on chemical toxicity will be due in part to toxicant-specific metabolism. Some compounds will be metabolized to non-toxic compounds and eliminated, whereas the bioactivation of some chemicals will increase the toxicity of others. Indeed, existing information indicates that increasing temperature can reduce, increase, or result in no change in the toxicity of xenobiotics.

The above alterations in xenobiotic uptake and biotransformation may result in the saturation of important metabolic pathways, alterations in parent compound and metabolite levels and metabolite profiles. As well, increases in temperature may alter organism susceptability to a given dose of a compound or a toxic metabolite. Other potential effects of altered metabolism by global temperature change may include alterations in the metabolism of endogenous molecules such as sex steroids which are also acted upon by these same enzyme systems. It must be made clear that temperature may act at many different points in the continuum between the absorption of a compound and its ultimate excretion or interaction with a receptor initiating a toxic response. The outcome of the interaction between the organism and chemical, being very toxin-specific, cannot be easily predicted.

Summary and suggestions for future research

Global warming scenarios will almost certainly be superimposed on a backdrop of substantial chemical pollution. However, the study of the effects of temperature on xenobiotic toxicity, and on the mechanisms of toxicity and detoxification, is still in its early stages. Temperature effects on toxicity and detoxification pathways are not as uniform as is typically seen for routine metabolism. This variation may derive partly from the variation in the chemistry of xenobiotics. However, we believe that the typical all-or-none inducible nature of enzymatic detoxification pathways contributes to this variation. Indeed, very recent studies of the effects of acclimation temperature on the induction process itself reveal potentially complex post-transcriptional regulatory mechanisms and species specificity (Stegeman, 1993). Therefore, we believe that to reduce this variability, future studies of temperature effects on xenobiotic toxicity and metabolism should not be piecemeal. Instead, laboratories should systematically focus on a given compound (or class of compounds) and follow the mechanism of toxicity and detoxification fully from the whole organism to the molecular level in a single test species of interest. The scant data available on substrate concentration–temperature interactions suggest that the older 'positive thermal modulation' hypothesis might be applicable to temperature–K_m interactions for xenobiotic detoxification enzymes. Therefore, we recommend studies of purified enzymes in this regard. Lastly, since the study of enzyme 'microenvironment' has proved so fruitful for enzymes of routine metabolism, and since many xenobiotic enzymes are membrane-bound, studies of temperature–enzyme–microenvironment interactions are strongly indicated.

References

Andersson, T. (1984). Oxidative and conjugative metabolism in isolated rainbow trout liver cells. *Marine Environmental Research*, **14**, 442–3.

Andersson, T. & Koivusaari, U. (1985). Influence of environmental temperature on the induction of xenobiotic metabolism by β-naphthoflavone in rainbow trout, *Salmo gairdneri*. *Toxicology and Applied Pharmacology*, **80**, 43–50.

Andersson, T. & Koivusaari, U. (1986). Oxidative and conjugative metabolism of xenobiotics in isolated liver cells from thermally acclimated rainbow trout. *Aquatic Toxicology*, **8**, 85–92.

Ankley, G. T., Reinert, R. E., Wade, A. E. & White, R. A. (1985). Temperature compensation in the hepatic mixed-function oxidase system of Bluegill. *Comparative Biochemistry and Physiology*, **81C**, 125–9.

Barron, M. G. (1990). Bioconcentration. *Environmental Science and Technology*, **24**, 1612–18.

Bartell, S. M., Gardner, R. H. & O'Neill, R. V. (1992). *Ecological Risk Estimation*. Chelsea, Mich: Lewis Publishers.

Baumann, P. C. (1989). PAH, metabolites and neoplasia in feral fish populations. In *Metabolism of Polycyclic Aromatic Hydrocarbons in the Aquatic Environment*, ed. U. Varanasi, pp. 269–89. Boca Raton, Fla: CRC Press.

Blank, J., Lindstrom-Seppa, P., Agren, J. J., Hanninen, O., Rein, H. & Ruckpaul, K. (1989). Temperature compensation of hepatic microsomal cytochrome P-450 activity in rainbow trout. I. Thermodynamic regulation during water cooling in autumn. *Comparative Biochemistry and Physiology*, **93C**, 55–60.

Brodie, B. B. & Maickel, R. P. (1962). Comparative biochemistry of drug metabolism. *Proceedings of the First International Pharmacology Meeting*, **6**, 299–324.

Brookes, P. & Lawley, P. D. (1964). Evidence for the binding of polynuclear aromatic hydrocarbons to the nucleic acids of mouse skin: relation between carcinogenic power of hydrocarbons and their binding to deoxyribonucleic acid. *Nature*, **202**, 781–4.

Brown, V. M. (1968). The calculation of the acute toxicity of mixtures of poisons to rainbow trout. *Water Research*, **2**, 723–33.

Brown, V. M., Jordan, D. H. M. & Tiller, B. A. (1967). The effect of temperature on the acute toxicity of phenol to rainbow trout in hard water. *Water Research*, **1**, 587–94.

Cairns, J. Jr, Heath, A. G. & Parker, B. C. (1975). The effects of temperature upon the toxicity of chemicals to aquatic organisms. *Hydrobiologia* **47**, 135–71.

Carpenter, J. M., Fredrickson, L. S., Williams, D. E., Buhler, D. B. & Curtis, L. R. (1990). The effect of thermal acclimation on the activity of aryl hydrocarbon hydroxylase in rainbow trout (*Oncorhynchus mykiss*). *Comparative Biochemistry and Physiology*, **97**C, 127–32.

Curtis, L. R. (1983). Glucuronidation and biliary excretion of phenol-phthalein in temperature-acclimated steelhead trout (*Salmo gairdneri*). *Comparative Biochemistry and Physiology*, **76**C, 107–11.

Curtis, L. R., Fredrickson, L. K. & Carpenter, H. M. (1990). Biliary excretion appears rate limiting for hepatic elimination of benzo[a]-pyrene by temperature-acclimated rainbow trout. *Fundamental and Applied Toxicology*, **15**, 420–8.

Dewaide, J. H. (1971). Alterations in microsomal drug-oxidation in thermally acclimated fish. *Arch. int. Pharmacodyn.* **189**, 377–9.

Egaas, E. & Varanasi, U. (1982). Effects of polychlorinated biphenyls and environmental temperature on *in vitro* formation of benzo(a)pyrene metabolites by liver of trout (*Salmo gairdneri*). *Biochemical Pharmacology*, **31**, 561–6.

Gelboin, H. V. (1980). Benzo[a]pyrene metabolism, activation, and carcinogenesis: role and regulation of mixed-function oxidases and related enzymes. *Physiological Reviews*, **60**, 1107–66.

Gill, K. A. & Walsh, P. J. (1990). Effects of temperature on metabolism of benzo[a]pyrene by toadfish (*Opsanus beta*) hepatocytes. *Canadian Journal of Fisheries and Aquatic Sciences*, **47**, 831–7.

Goksoyr, A. (1985). Purification of hepatic microsomal cytochromes P-450 from β-naphthoflavone-treated Atlantic cod (*Gadus morhua*), a marine teleost fish. *Biochimica et Biophysica Acta*, **840**, 409–17.

Gurumurthy, P. & Mannering, G. J. (1985). Membrane bound cytochrome P-450 determines the optimal temperatures of NADPH-cytochrome P-450 reductase and cytochrome P-450-linked mono-oxygenase reactions in rat and trout hepatic microsomes. *Biochemical and Biophysical Research Communications*, **127**, 571–7.

Hazel, J. R. (1979). Influence of thermal acclimation on membrane lipid composition of rainbow trout liver. *American Journal of Physiology*, **236**, R91–101.

Hazel, J. R. (1984). Effects of temperature on the structure and metabolism of cell membranes in fish. *American Journal of Physiology*, **246**, R460–70.

Hazel, J. R. (1993). Thermal biology. In *The Physiology of Fishes*, ed. D. H. Evans, pp. 427–67. Boca Raton, Fla: CRC Press.

Hazel, J. R. & Prosser, C. L. (1974). Molecular mechanisms of temperature compensation in poikilotherms. *Physiological Reviews*, **54**, 620–77.

Hendricks, J. D., Meyers, T. R., Casteel, J. L., Nixon, J. E., Loveland, P. M. & Bailey, G. S. (1984). Rainbow trout embryos: advantages and limitations for carcinogenesis research. *National Cancer Institute Monograph*, **65**, 129–37. NIH publication #84–2653, Bethesda, Md.

Hochachka, P. W. & Somero, G. N. (1984). *Biochemical Adaptation*. Princeton, NJ: Princeton University Press.

Hoover K. L. (1984). The use of small fish species in carcinogenicity testing. *National Cancer Institute Monograph*, **65**. NIH publication #84–2653, Bethesda, Md.

Jimenez, B. D. & Burtis, L. S. (1989). Influence of environmental variables on the hepatic mixed-function oxidase system in bluegill sunfish, *Lepomis macrochirus*. *Comparative Biochemistry and Physiology*, **93C**, 11–21.

Jimenez, B. D., Cirmo, C. P. & McCarthy, J. F. (1987). Effects of feeding and temperature on uptake, elimination and metabolism of benzo(a)pyrene in the bluegill sunfish (*Lepomis macrochirus*). *Aquatic Toxicology*, **10**, 41–57.

Kapoor, I. P., Metcalf, R. L., Hirwe, A. S., Coats, J. R. & Khalsa, M. S. (1973). Structure–activity correlations of biodegradability of DDT analogs. *Journal of Agriculture and Food Chemistry*, **21**, 310–15.

Karr, S. W., Reinert, R. E. & Wade, A. E. (1985). The effects of temperature on the cytochrome P-450 system of thermally acclimated bluegill. *Comparative Biochemistry and Physiology*, **80C**, 135–9.

Kennedy, C. J., Gill, K. A. & Walsh, P. J. (1989a). Thermal modulation of benzo[a]pyrene uptake in the gulf toadfish, *Opsanus beta*. *Environmental Toxicology and Chemistry*, **8**, 863–9.

Kennedy, C. J., Gill, K. A. & Walsh, P. J. (1989b). Thermal modulation of benzo[a]pyrene metabolism by the gulf toadfish, *Opsanus beta*. *Aquatic Toxicology*, **15**, 331–44.

Kennedy, C. J., Gill, K. A. & Walsh, P. J. (1991). Temperature acclimation of xenobiotic metabolizing enzymes in cultured hepatocytes and whole liver of the gulf toadfish, *Opsanus beta*. *Canadian Journal of Fisheries and Aquatic Sciences*, **48**, 1212–19.

Kennedy, C. J. & Walsh, P. J. (1991). The effects of temperature on benzo[a]pyrene metabolism and adduct formation in the gulf toadfish. *Opsanus beta. Fish Physiology and Biochemistry*, **9**, 179–87.

Kennedy, C. J. & Walsh, P. J. (1994). The effects of temperature on the uptake and metabolism of benzo[a]pyrene in isolated gill cells of the gulf toadfish, *Opsanus beta*. *Fish Physiology and Biochemistry*, **13**. 93–103.

Kime, D. E. & Saksema, D. N. (1980). The effect of temperature on the hepatic catabolism of testosterone in the rainbow trout (*Salmo gairdneri*) and the goldfish (*Carassius auratus*). *General and Comparative Endocrinology*, **42**, 228–4.

Khudoley, V. V. (1984). Use of aquarium fish, *Danio rerio* and *Poecilin reticulata*, as test species for evaluation of nitrosamine carcinogenicity. *National Cancer Institute Monograph*, **65**. 65–70. NIH publication #84–2653, Bethesda, Md.

Koban, M. (1986). Can cultured teleost hepatocytes show temperature acclimation? *American Journal of Physiology*, **250**, R211–20.

Koivusaari, U. (1983). Thermal acclimatization of hepatic polysubstrate monooxygenase and UDP-glucuronosyl transferase of mature rainbow trout (*Salmo gairdneri*). *Journal of Experimental Zoology*, **227**, 35–42.

Koivusaari, U. & Andersson, T. (1984). Partial temperature compensation of hepatic biotransformation enzymes in juvenile rainbow trout (*Salmo gairdneri*) during the warming of water in spring. *Comparative Biochemistry and Physiology*, **78B**, 223–6.

Koivusaari, U., Harri, M. & Hanninen, O. (1981). Seasonal variation of hepatic biotransformation in female and male rainbow trout (*Salmo gairdneri*). *Comparative Biochemistry and Physiology*, **70C**, 149–157.

Kyono-Hamaguchi, Y. (1984). Effects of temperature and partial hepatectomy on the induction of liver tumors in *Oryzias latipes*. *National Cancer Institute Monograph*, **65**, 337–44. NIH publication #84–2653, Bethesda, Md.

Law, F. C. P., Abedini, S. & Kennedy, C. J. (1991). A biologically based toxicokinetic model for pyrene in rainbow trout. *Toxicology and Applied Pharmacology*, **110**, 390–402.

Lech, J. J. (1974). Glucuronide formation in rainbow trout – effect of salicylamide on the acute toxicity, conjugation and excretion of 3-trifluoromethyl-4-nitrophenol. *Biochemical Pharmacology*, **23**, 2403–10.

Lee, A. G., Birdsall, N. J. M., Metcalfe, J. C., Toon, P. A. & Warren, G. B. (1974). Clusters in lipid bilayers and the interpretation of thermal effects in biological membranes. *Biochemistry*, **13**, 3699–705.

Lemly, A. D. (1993). Metabolic stress during winter increases the toxicity of selenium to fish. *Aquatic Toxicology*, **27**, 133–58.

Macek, K. J., Hutchinson, C. & Cope, O. B. (1969). The effects of temperature on the susceptibility of bluegills and rainbow trout to selected pesticides. *Bulletin of Environmental Contamination and Toxicology*, **4**, 174–83.

McEwen, F. L. & Stephenson, G. R. (1979). *The Use and Significance of Pesticides in the Environment.* New York: John Wiley.

Malins, D. C., Krahn, M. M., Myers, M. S., Rhodes, L. D., Brown, D. W., Krone, C. A., McCain, B. B. & Chan, S.-L. (1985). Toxic chemicals in sediments and biota from a creosote-polluted harbor: relationships with hepatic neoplasms and other hepatic lesions in English sole (*Parophrys vetulus*). *Carcinogenesis*, **6**, 1463–9.

Miller, N. G. A., Hill, M. W. & Smith, M. W. (1976). Positional and species analysis of membrane phospholipids extracted from goldfish adapted to different environmental temperatures. *Biochimica et Biophysica Acta*, **455**, 644–54.

Munkittrick, K. R., Portt, C. B., Van Der Kraak, G. J., Smith, I. R. & Rokosh, D. (1991). Impact of bleached kraft mill effluent on population characteristics, liver MFO activity and serum steroid levels of a Lake Superior white sucker (*Catastomus commersoni*) population. *Canadian Journal of Fisheries and Aquatic Sciences*, **48**, 1371–80.

Payne, J. F., Fancey, L. L., Rahimtula, A. D. & Porter, E. L. (1987). Review and perspective on the use of mixed-function oxygenase enzymes in biological monitoring. *Comparative Biochemistry and Physiology*, **86C**, 233–45.

Peters, G. & Peters, N. (1977). Temperature-dependent growth and regression of epidermal tumors in the European eel (*Anguilla anguilla* L.). *Annals of the New York Academy of Sciences*, **298**, 245–60.

Powers, D. A. (1989). Fish as model systems. *Science*, **246**, 352–8.

Prosser, C. L. (1973). *Comparative Animal Physiology.* Philadelphia, Pa: W. B. Saunders.

Schmidt-Nielsen, K. (1983). *Animal Physiology: Adaptation and Environment.* Cambridge: Cambridge University Press.

Scott, A. P. & Sumpter, J. P. (1983). The control of trout reproduction: basic and applied research on hormones. In *Control Processes in Fish Physiology*, ed. J. C. Rankin, T. J. Pricher & R. Duggan, pp. 200–20. London: Croom Helm.

Shugart, L., McCarthy, J., Jimenez, B. & Daniels, J. (1987). Analysis of adduct formation in the bluegill sunfish (*Lepomis macrochirus*) between benzo[a]pyrene and DNA of the liver and hemoglobin of the erythrocyte. *Aquatic Toxicology*, **9**, 319–25.

Sidell, B. D. & Hazel, J. R. (1987). Temperature affects the diffusion of small molecules through cytosol of fish muscle. *Journal of Experimental Biology*, **129**, 191–203.

Somero, G. N. (1979). Interacting effects of temperature and pressure on enzyme function and evolution in marine organisms. In *Biochemical and Biophysical Perspectives in Marine Biology*, Vol. 4, ed. D. C. Malins & J. R. Sargent, pp.1–27. New York: Academic Press.

Stegeman, J. J. (1979). Temperature influence on basal activity and induction of mixed-function oxygenase activity in *Fundulus heteroclitus*. *Journal of the Fisheries Research Board of Canada*, **36**, 1400–5.

Stegeman, J. J. (1993). The cytochrome P450s in fish. In *Biochemistry and Molecular Biology of Fishes*, Vol. 2. *Molecular Biology Frontiers*, ed. P. W. Hochachka & T. P. Mommsen, pp.137–58. Amsterdam: Elsevier.

Stegeman, J. & Hahn, M. (1994). Biochemistry and molecular biology of monooxygenases: current perspectives on forms, functions, and regulation of cytochrome P450 in aquatic species. In *Aquatic Toxicology: Molecular, Biochemical, and Cellular Perspectives*, ed. D. C. Malins & G. K. Ostrander. Boca Raton, Fla: Lewis Publishers.

Steidinger, K. A., Burklew, M. A. & Ingle, R. M. (1972). The effects of *Gymnodinium breve* toxin on estuarine animals. In *Marine Phycognosy*, ed. D. F. Martin & G. M. Padilla, pp. 179–202. New York: Academic Press.

Thilo, L., Trauble, H. & Overath, P. (1977). Mechanistic interpretation of the influence of lipid phase transitions on transport functions. *Biochemistry*, **16**, 1283–9.

Varanasi, U., Uhler, M. & Stranahan, S. I. (1978). Uptake and release of naphthalene and its metabolites in skin and epidermal mucous of salmonids. *Toxicology and Applied Pharmacology*, **44**, 277–89.

Wills, E. D. (1980). The role of polyunsaturated fatty acid composition of the endoplasmic reticulum in the regulation of the rate of oxidative drug and carcinogen metabolism. In *Microsomes, Drug Oxidations, and Chemical Carcinogenesis*, Vol. 1, ed. M. J. Coon, A. H. Conney, R. W. Estabrook, H. V. Gelboin, J. R. Gillette & P. J. O'Brien, pp. 545–79. New York: Academic Press.

Wunderlick, F., Ronai, A., Speth, V., Seelig, J. & Blume, A. (1975). Thermotropic lipid clustering in *Tetrahymena* membranes. *Biochemistry*, **14**, 3730–5.

S.D. REID, D.G. McDONALD and
C.M. WOOD

Interactive effects of temperature and pollutant stress

Introduction

The prediction of the global warning scenario is a 1–5 °C increase in
the mean global temperature as a result of a doubling of the so-called
greenhouse gases; methane, carbon dioxide and nitrous oxide
(Schneider, 1990; Mohnen & Wang, 1992). A variety of physiological
parameters of poikiliothermic fish are directly and indirectly impacted
by changes in environmental temperature, including metabolism (O_2
consumption), growth, cardiac output, ventilation and excretory processes. Specifically, environmental temperature determines the rate of
chemical reactions such that, in general, a 10 °C increase in temperature
enhances reaction rates by 2–3-fold (Q_{10} = 2–3). It is well established
that fish do have some capacity to compensate for changes in environmental temperature (see Hazel, 1993). However, in many natural situations the predicted change in the temperature will not be the only
environmental stressor with which the fish must cope because many
environments are no longer pristine. The metabolic cost of living in
polluted environments has yet to be clearly established, though it is
likely to be substantial (Calow, 1991). Therefore, the anticipated alterations in fish physiology associated with global warming have the
potential to increase the burden of stress already experienced by fish
living in marginalized environments.

The study of the relationship between environmental temperature
and pollutant toxicity in fish is not a new endeavour, but the literature
describing temperature effects on toxicity affords little in terms of
reliable generalization. Clearly, the relationship between temperature
and toxicity of aquatic pollutants can be extremely complex, depending
on the impact of temperature on specific physiological processes, the
chemistry of the toxicant, the chemistry of the environment and the
toxic mechanism of the pollutant (Sprague, 1985). For example, increasing temperature generally increases the rate of clearance of most aquatic

Society for Experimental Biology Seminar Series 61: *Global Warming: Implications for freshwater and marine fish*, ed. C. M. Wood & D. G. McDonald. © Cambridge University Press 1996, pp. 325–349.

toxicants via a stimulation of metabolic and excretory processes; therefore increases in environmental temperature may reduce pollutant toxicity. On the other hand, the requirement of aquatic organisms for oxygen is positively correlated with temperature, while oxygen solubility is inversely correlated with environmental temperature. The resulting increase in ventilation and gill permeability may increase pollutant uptake, thereby increasing toxicity. In addition, for a toxic mechanism involving an impairment of branchial gas exchange, increased environmental temperature may increase toxicity. Furthermore, the properties of the pollutant itself may be directly influenced by temperature through, for example, the effect of temperature on the equilibrium between molecular and ionized forms. Finally, temperature is itself an important limiting factor for aquatic organisms.

In this review, we will discuss the impact of temperature on the physiology of fish and the chemistry of a variety of aquatic toxicants to demonstrate the possibilities for interaction between temperature and pollutant stress in fish. Furthermore, we will address the method by which interactions between temperature and pollutant stress have been described and compare it with a method which addresses this issue from a global warming perspective.

Physiological implications of an elevated mean annual temperature

Fish occupy waters of extremely variable temperature. For example, in Canadian waters, the eurythermal rainbow trout (*Oncorhynchus mykiss*) is active at temperatures ranging from 0 to 26 °C (Kaya, 1978). However, when given the opportunity to select an environmental temperature, the preferred temperature for this species is approximately 13 °C (Coutant, 1977). When unable to select an environment of relatively constant temperature for such reasons as predation pressure or habitat availability, then the physiology of these ectotherms is altered in accordance with fluctuations in temperature. The temperature sensitivity results from insignificant metabolic heat production relative to the high thermal conductivity and specific heat of water.

Traditionally, the physiological impact of elevated temperature has been assessed by determining the temperature coefficient (Q_{10}) of a specific process at two constant temperatures to which the fish have been acclimated. The Q_{10} is calculated by using the van't Hoff equation:

$$Q_{10} = (k_2/k_1)^{10/t_2 - t_1}$$

where k_1 and k_2 are velocity constants or metabolic rates at temperatures t_1 and t_2 respectively. A temperature difference of 10 °C has become a standard span over which to determine the temperature sensitivity of biological functions. A Q_{10} of 2 indicates that for every 10 °C increase in temperature, the reaction rate or metabolic rate doubles. The Q_{10} for physical processes, such as diffusion, is about 1. However, that for biochemical reactions and many physiological rates is typically 2 to 3 (see Table 1). For example, muscle mechanics and maximum aerobic swimming performance (Randall & Brauner, 1991) in fish are similarly sensitive to temperature. Fauconneau and Arnal (1985) demonstrated that whole fish total protein synthesis has a Q_{10} of 2.0. Cardiac output has been shown to increase with increased temperature, with a temperature coefficient of approximately 2.6 (Graham & Farrell, 1989; Farrell & Jones, 1992). More importantly, from an aquatic toxicology perspective, active and standard metabolic rates (Evans 1990) increase, as does gill ventilation rate, with increased temperature (Randall & Cameron, 1973, Black *et al.*, 1991). The temperature sensitivity of ventilation is critical in a consideration of temperature and pollutant stress as it has a significant impact on the effective exposure of fish to environmental stressors. The gills are the

Table 1. *Temperature sensitivity of various physiological rates*

Physiological rate	Temperature coefficient (Q_{10})	Reference
Muscle shortening velocity	1.5–3	Randall & Brauner, 1991
Max. aerobic swimming performance	1.5	Randall & Brauner, 1991
White muscle protein synthesis	3.6	Loughna & Goldspink, 1985
Whole body protein synthesis	2.0	Fauconneau & Arnal, 1985
Gill and liver protein synthesis	2.1	Reid *et al.*, 1995
Standard metabolism	1.5	Evans, 1990
	2.0	Black *et al.*, 1991
	2.3	Randall & Cameron, 1973
Active metabolism	2.0	Evans, 1990
Cardiac output	2.6	Farrell & Jones, 1992
		Graham & Farrell, 1989
Gill ventilation	1.5	Black *et al.*, 1991
	1.3	Randall & Cameron, 1973
Ventilation volume	2.0	Black *et al.*, 1991
	2.0	Randall & Cameron, 1973

initial site of contact and uptake of water-borne environmental stressors owing to their physiological function and unique design. Gas exchange, electrolyte and acid–base homeostasis, regulated at the gills, require an organ with a large, permeable surface area in intimate contact with the surrounding environment. The gills of a 250 g rainbow trout, for example, represent 60% of the total surface area exposed to the environment (Reid, 1990) which is irrigated with water at a rate of *c.* 40–100 ml min^{-1} (Randall & Cameron, 1973). Therefore, any temperature-induced alteration in ventilation will directly alter the effective load of pollutant brought to the gills.

The influence of temperature on the environment

Water quality

Temperature has a direct impact on water quality, based on the temperature dependence of the pH of neutrality, oxygen solubility and water viscosity. Under ideal environmental conditions, temperature-dependent changes in pH or dissolved oxygen are usually not a major concern. However, these two factors may play a significant role in modifying pollutant toxicity in less than ideal environmental conditions. Temperature affects the physicochemical equilibrium between H_2O and H^+/OH^- such that the pH of water decreases by about 0.017 units per degree Celsius. Changes in water pH would be experienced only by fish living in softwater and likely to be a serious consideration only for fish living in waters already marginalized through dry or wet acid deposition.

The oxygen concentration of water is temperature dependent due to the influence of temperature on oxygen solubility (αO_2, ml O_2 l^{-1} torr^{-1}). The higher the temperature, the lower the oxygen solubility in water and other physiological solutions. In distilled water, the Q_{10} for αO_2 is approximately 0.79 over the temperature range 1 to 24 °C (Graham, 1987) which represents a nearly 40% drop in oxygen content at any given oxygen partial pressure. The impact of temperature-induced reductions in dissolved oxygen would be dependent on metabolic rate, and therefore oxygen demands. In 1957, Downing and Merkens conducted a study on the influence of temperature on the survival of adult rainbow trout (*Oncorhynchus mykiss*), perch (*Perca fluviatilis*), roach (*Rutilus rutilus*) and mirror carp (*Cyprinus carpio*) during short- to long-term exposure to low dissolved oxygen. They showed that at each temperature the period of survival decreased with reduction in water oxygen tension. More importantly, Downing and Merkens (1957) reported that a rise in temperature between 10 and

16 °C caused a reduction in the resistance to hypoxia of all species tested. In other words, 100% mortality was achieved at a significantly higher oxygen tension at 16 °C than at 10 °C.

Within the context of global warming, it is unlikely that either of these two temperature-dependent parameters of water quality would be limiting in the absence of additional stressors. For example, a 2 °C rise in mean temperature from 21 to 23 °C would result in only a 4.5% reduction in dissolved O_2 and 0.034 unit drop in pH. However, in combination with other aquatic pollutants or at temperatures where heat-induced mortality may occur, or if the gill oxygen diffusion distance has been increased due to toxin exposure, the influence of temperature on water quality may have devastating consequences.

Pollutant toxicity

The chemistry and toxicity of aquatic pollutants can be influenced by changes in temperature. For those environmental stressors which are toxic to fish in an ionized form, such as certain metals and ammonia, alterations in environmental temperature can shift the equilibrium between toxic and non-toxic forms of the stressor. Furthermore, the shift in neutrality of water, particularly previously acidified water, may have an additional influence on the physicochemical equilibrium of ionizable environmental stressors. Ammonia provides a useful example.

There is a wealth of literature on ammonia toxicity, and it is generally accepted that toxicity to fish is due to NH_3 (Thurston, Russo & Vinogradov, 1981, Broderius *et al.*, 1985, Sheehan & Lewis, 1986). However, in aqueous solutions, ammonia exists as an equilibrium between the ionized (ammonium ion, NH_4^+) and un-ionized (ammonia, NH_3) species. This equilibrium depends strongly upon the pH and, to a lesser extent, upon temperature and ionic composition. As either pH or temperature increases, the equilibrium shifts towards the NH_3 species based on the temperature dependency of the ammonia pK_a (Emerson *et al.*, 1975). In the context of the global warming scenario, according to Emerson *et al.* (1975), a 2 °C rise in mean annual temperature from both 2 to 4 °C and 20 to 22 °C at pH 8.0 would correspond to 1.18- and 1.15-fold increases in the percentage of NH_3. These shifts in the ammonia equilibrium would be reduced slightly by temperature-dependent reductions in water pH.

Despite the demonstrated increase in NH_3 associated with increased temperature, it may be noted that other studies have shown that the toxicity of ammonia is unaltered (Arthur *et al.*, 1987) or decreases (Thurston *et al.*, 1981, Jeney *et al.*, 1992) with increases in temperature.

This suggests that ammonia toxicity results from the joint toxicity of NH_3 and NH_4^+. Erickson (1985) attempted to model the effects of pH and temperature on ammonia toxicity assuming that both un-ionized ammonia and ammonium ion contribute equally to ammonia toxicity: the joint toxicity model. Although the model was effective in demonstrating that both species of ammonia were involved in the relationship between ammonia toxicity and pH, joint toxicity failed to explain the observed influence of temperature on ammonia toxicity. These findings suggest that the reduction in ammonia toxicity at higher temperatures, despite the relative increase in the more toxic species (NH_3), is due to a temperature-dependent increase in ammonia metabolism or detoxification.

Physiological impact of toxicants

Common aquatic pollutants are metals, such as copper (Cu), cadmium (Cd) and aluminium (Al), organic compounds such as detergents and pesticides, and environmental acidification. Many of these pollutants result in disturbances in electrolyte homeostasis and gas exchange, owing to the physiological role of the gills, the primary contact point for these stressors. Those pollutants that influence electrolyte balance do so by inhibiting normal uptake mechanisms, by enhancing gill epithelium permeability, or by both. Such toxicants are often termed 'surface-active'. For example, Cu has been shown to compete with sodium for the sodium binding site of the basolaterally located Na^+/K^+ ATPase (Lauren & McDonald, 1985), while environmental acidification induces both an increase in diffusive ion loss and a reduction in active uptake (McDonald, 1983). Aluminium toxicity is more complicated, reflecting the complex aqueous chemistry of aluminium. Aluminium toxicity can result from an increase in diffusive electrolyte efflux (Neville, 1985; Playle, Goss & Wood, 1989). However, hypoxia is also a mechanism by which fish mortality is induced in the presence of aluminium. According to Neville (1985) and Playle *et al.* (1989), fish death in aluminum rich water is due primarily to hypoxia at pH 6.1 and to electrolyte loss at pH 4.5 and 4.0. Between pH 5.0 and 5.5, it appears that both mechanisms of toxicity contribute to fish mortality. In addition, water hardness can shift the balance between ionoregulatory and respiratory toxicity (Wood *et al.*, 1988).

Other aquatic pollutants accumulate within the body and the mechanism of toxicity involves a critical body burden. Once reached, the toxins influence cellular processes. Insecticides, such as DDT, some metals (mercury and lead) and ammonia exhibit potent effects on the nervous system (Narahashi, 1987). With these environmental stressors, there is typically some degree of detoxification that occurs within the animal, such as the conjugation of phenol (Brown, Jordan & Tiller, 1967), sequestration of metals by soluble metal-binding proteins (Fowler, 1987) or oxidation of benzo[a]pyrene (Kennedy, Gill & Walsh, 1989). It is only when the rate of accumulation exceeds the rate at which clearance can be maintained that altered physiology and cell injury results.

Whether the toxin is predominantly surface active or internally active, alteration in gill morphology is a common response to acute exposure, although not always the cause of mortality. Common changes in gill structure as a result of exposure to aquatic pollutants are changes in mucous cell numbers and gill mucus content, pavement cell hypertrophy, degeneration of chloride cells, intercellular oedema, fusion of respiratory lamellae, or lifting or sloughing of the epithelium (see Mallatt, 1985). The main concern here is the direct impact these changes may have on gill diffusion distance and, ultimately, oxygen transport. In cases of extensive yet non-lethal damage, the gill can be successfully repaired and as a result, increased tolerance acquired (see McDonald & Wood, 1993). In others, however, the damage may be minimal and no repair initiated. If acclimation requires the occurrence of a repair event, then in such exposures, increased tolerance to further exposure would not result. This hypothesis has been used to explain why rainbow trout do not acclimate to low levels of environmental acidification (Mueller *et al.*, 1991; Audet & Wood, 1993).

Interactive effects of temperature and pollutant stress

The traditional laboratory approach

Much of what is known about temperature-dependent alterations in pollutant toxicity is based on the acclimation and exposure of fish to two or more constant temperatures and the monitoring of physiological parameters or toxicological indicators. These studies were generally of short duration (acute) and employed a 10 °C or greater separation in acclimation temperature. Typically, fish were acclimated to a specific temperature for a period of approximately 2 weeks, particularly when the test species was rainbow trout (*Oncorhynchus mykiss*). The rationale for this frequently used acclimation period is based on a commonly cited study by Peterson and Anderson (1969), in which they reported that the metabolic rate of salmonids stabilizes within 2 weeks following a temperature change from 18 to 6 °C or from 6 to 18 °C. With such an acclimation protocol, cadmium toxicity increased with increased temperature (Fig. 1; Roch & Maly, 1979). In other words, fish acclimated and tested at 6 °C survived longer (ET_{50}) than fish acclimated and tested at 12 °C, which in turn survived longer than fish acclimated and tested at 18 °C. Furthermore, the 10 day lethal threshold cadmium concentrations decreased with increased temperature. This is an example of a study in which the temperature sensitivity of a metal resembled the effects of temperature on metabolic rate following a rapid temperature change. Similar findings have been reported for acute exposure of fish to acidified water (Fig. 2; Kwain, 1975; Korwin-Kossakowski & Jezierska, 1985). With cadmium toxicity and other similar toxicants, it is thought that change in toxicity is due to temperature-dependent change in physiology (i.e. ventilation, metabolism) and water quality (i.e. dissolved oxygen) rather than changes in the chemistry of the toxin. Therefore, without a concurrent increase in the rate of detoxification, a lethal body burden can be reached more quickly at higher temperatures due simply to an increased rate of accumulation.

The relationship between metal toxicity and environmental temperature, using fish acclimated to fixed temperatures prior to exposure, does not always yield findings consistent with the temperature-dependence of metabolism. Although the survival time (ET_{50}) of Atlantic salmon exposed to zinc decreased with increased temperature ($Q_{10} \approx 0.31$),

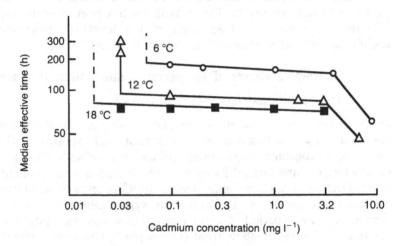

Fig. 1. Median mortality times of juvenile rainbow trout exposed to cadmium and acclimated at 6, 12 and 18 °C. Dashed lines represent approximate 10 day lethal thresholds (50% mortality) at 6 and 18 °C. (From Roch & Maly, 1979.)

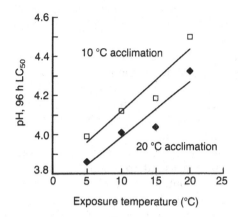

Fig. 2. Relationship between median lethal concentration (pH, 96 h LC_{50}) and testing water temperature of fingerling rainbow trout. (From Kwain, 1975.)

Hodson and Sprague (1975) noted that temperature-acclimated fish were about 50% more tolerant of zinc at 19 °C than at 3 °C (14 days LC_{50}). In other words, while the reduction in survival time for Atlantic salmon at higher temperatures corresponds with a temperature-dependent increase in metabolism, the influence of temperature on zinc toxicity does not. These results suggest that the internal detoxification (metabolism, sequestering) of zinc may be enhanced by elevated temperature, thereby increasing zinc tolerance. However, when the lethal internal concentration of zinc is accumulated, death occurs more rapidly. Similar findings and interpretations have been reported for chronic exposure of rainbow trout to arsenate (McGeachy and Dixon, 1990) and antimony (Doe *et al.*, 1987).

Similar findings have also been reported for organic toxins that can be detoxified internally through a complex series of enzymatic reactions, each influenced by temperature to varying degrees. For example, Brown *et al.* (1967) tested the acute toxicity of phenol to juvenile rainbow trout acclimated and tested at temperatures ranging from 6 to 18 °C. They reported that the resistance of trout to phenol increased with increases in temperature such that when comparing phenol toxicity at the extremes, at 18 °C the 48 hr LC_{50} was approximately 1.8 times that at 6 °C. In contrast to the apparent increase in phenol tolerance at higher temperatures, Brown *et al.* (1967) clearly demonstrated that survival time decreased with increased temperature. In order to explain their findings, it was suggested that the thermal coefficient for phenol conjugation or detoxification (Williams, 1959) is greater than that of the rate for phenol absorption.

The difference in the temperature-dependent modification in toxicity as assessed by chronic (threshold or incipient LC_{50}) or acute (24, 48 or 96 h LC_{50}; median survival time, ET_{50}) illustrates how critical it is to select an appropriate duration for any future studies (Fig. 3) This was demonstrated by an earlier study of McGeachy and Dixon (1990), in which the effect of temperature on acute arsenate toxicity was investigated. In the acute study, these authors reported that rainbow trout tested at 5 °C were more resistant to arsenate than those tested at 15 °C. These findings were contrary to their more recent chronic study (McGeachy & Dixon, 1991). According to Sprague (1970), this problem is most evident when one considers conclusions based on LC_{50} values at arbitrarily selected observation times (e.g. 24, 48 or 96 h) against those based on a concentration at which 50% of the test population can live for an indefinite time (incipient lethal level or lethal threshold concentration). This point is shown in Fig. 4, which illustrates

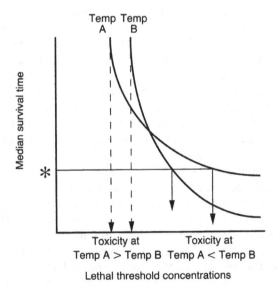

Fig. 3. Two hypothetical toxicity curves for an aquatic toxicant tested at two different temperatures. Asterisk indicates the possible difference in relative toxicity that can be obtained when the test is discontinued prior to the establishment of the lethal threshold concentration or incipient lethal level. (Simplified from Brown *et al.*, 1967.)

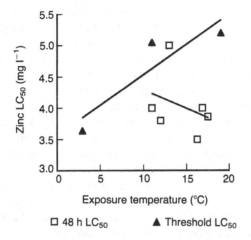

Fig. 4. Temperature-induced changes in toxicity of zinc to salmonids. Data were taken from six independent acute zinc exposures conducted at different temperatures (□: Lloyd, 1961; Herbert & Shurben, 1963; 1964, Herbert & Wakeford, 1964; Brown *et al.*, 1969; Brown & Dalton, 1970) and presented in contrast to data generated from a single chronic exposure experiment in which salmonids were acclimated and exposed to three different water temperatures (▲: Hodson & Sprague, 1975). Regressions were fitted to data using least squares.

the relationship between temperature and zinc toxicity to salmonids in acute (48 h) versus chronic exposures. Toxicity data obtained from a variety of acute zinc exposures shows no significant impact of exposure temperature on zinc toxicity despite the implication of a direct relationship between temperature and zinc toxicity. However, when data from a single study, in which zinc toxicity in a salmonid was assessed with the threshold LC_{50}, are presented in an identical manner, it is apparent that a contrary interpretation of the relationship between temperature and zinc toxicity results. In the case of Brown *et al.* (1967) and other studies in which the toxicity curves intersect, any conclusions based on short-term LC_{50} values would also lead to conclusions contrary to those based on lethal threshold concentration. Therefore it is difficult to evaluate the interactive effects of temperature and pollutant stress based on arbitrarily terminated toxicity tests (i.e. 48 or 96 h LC_{50}), unless it is clear that true incipient lethal levels have been determined (i.e. acute toxicity has long since ceased). Chronic toxicity tests conducted at sublethal concentrations would help eliminate such difficulties in interpretation.

Another difficulty in assessing the interactive effects of temperature and pollutant stress is that rarely do such studies look at the physiological implications of temperature either alone, or in combination with the sublethal pollutant, at the upper end of the temperature range of the animal. This is not to suggest that studies in which fish are exposed to near lethal temperature and pollutants have not been carried out. There are several studies in which the critical thermal maximum technique, CTM (Cowles & Bogert, 1944) has been used to delineate the effects of a chemical on the thermal tolerance of fish. In this method, fish are exposed to gradually increasing temperature, typically 0.3 °C min^{-1} (Becker & Genoway, 1979), until some pre-established endpoint criterion such as death or final loss of equilibrium is reached. The CTM of fish exposed to various concentrations of pollutants is then compared against the CTM of those not having prior pollutant exposure. The results of such studies tend to show either no impact of toxicants on the critical thermal maxima or a reduction in the CTM. For example, Paladino and Spotila (1978) determined with CTM methodology that arsenic depresses the thermal tolerance of young muskellunge (*Esox masquinongy*). Becker and Wolford (1980) reported similar results regarding the effect of nickel on rainbow trout, as did Watenpaugh and Beitinger (1985) when channel catfish (*Ictalurus punctatus*) were exposed to nitrite. However, an aquatic herbicide, aquathol-K, had no significant effect on the CTM of red shiner (*Notropis lutrensis*), even at a concentration many times its recommended use (Takle, Beitinger & Dickson, 1983). Unfortunately, the CTM is another acute lethality test, which may be useful in the determination of acquired tolerance or acclimation but which may be limited in its application to the questions of global warming and its cumulative effects on fish in marginalized environments.

The global warming scenario and pollutant stress

Relationships between temperature and toxicity that have been determined to date may not provide the proper basis for modelling the impact of global warming/pollution scenarios on fish experiencing a natural thermal exposure. Certainly, there are periods during the year when the water temperature is relatively constant (Fig. 6). The inshore temperature of Lake Ontario was *c.* 4–4.5 °C from 15 January to 15 March 1994. However, over the entire year there were periods of variable temperature. During the summer of 1993, water temperature rose from *c.* 13 to 22 °C in only 18 days (Reid *et al.*, 1995). To further complicate matters, temperature profiles are not consistent from year to year, even for the same body of water. The peak temperature during the summer of 1993 was *c.* 22 °C, while the temperature during the subsequent summer reached a maximum of only 19 °C. There is no expectation for seasonal or daily temperature profiles to be less variable. In fact, one of the predictions of global warming is that, although mean annual temperature may rise, temperate regions will experience an increase in the seasonal difference between winter minimum and summer maximum temperatures (Mohnen & Wang, 1992). This suggests an increase in rate at which seasonal temperature change will occur, strengthening the notion that the assessment of toxicity at fixed temperature may not be the predictive tool that is required for addressing this issue. Therefore, the data which will prove most useful are likely to be obtained from field studies in which seasonal variation in temperature occurs, and is variable from year to year, and from laboratory studies in which control over a number of parameters is maintained yet fish experience the natural variability in daily and seasonal temperature.

Recently we undertook a laboratory-based investigation into the interactive effects of temperature and pollutant stress which incorporates a natural thermal exposure with seasonal change, chronic sublethal pollutant exposure, juvenile fish, and the global warming scenario of an increase in ambient temperature of 2 °C. This ongoing study attempts to replicate the conditions to be experienced by fish living in marginalized environments under the global warming scenario. The objective is to determine the metabolic costs of increased thermal stress and increased anthropogenic toxicants, 'cost' being equated with an increase in animal energy expenditure or a re-partitioning of the energy budget (Calow, 1991). Similar approaches have been used to demonstrate the metabolic costs associated with zinc-induced tissue damage and subsequent repair in rainbow trout (Hogstrand, Reid & Wood, 1995), the re-partitioning of the energy budget in rainbow trout (Wilson, Bergman & Wood, 1994a,b; Wilson, Wood & Houlihan, 1996) and largemouth bass (*Mictopterus salmoides*; Leino & McCormick, 1993) exposed to aluminium, and the marine fish, dab (*Limanda limanda*), exposed to sewage sludge (Houlihan *et al.*, 1994). However these studies lack a temperature component to their design and are therefore of limited usefulness for modelling the impact of global warming on fish living in marginalized environments.

Our study monitors a number of physiological parameters (Fig. 5), ranging from integrative measures such as growth and metabolic rate, to tissue-specific rates of protein synthesis. These basic bioenergetic measurements are made throughout the exposures to produce comprehensive bioenergetic, toxicological and physiological data sets of value

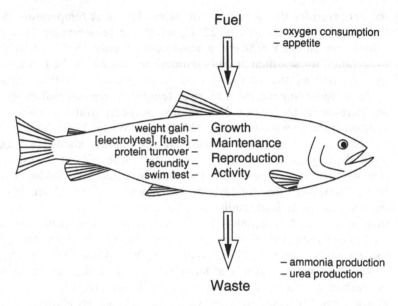

Fig. 5. Fish bioenergetics model used to assess the impact of the global warming scenario (2 °C elevation in mean temperature) on rainbow trout during chronic sublethal exposure to elevated ammonia or environmental acidification. Parameters in smaller type are examples of measurements that indicate the parameters in bold type.

for predictive and modelling purposes. One key component is our ability to take a natural seasonal temperature profile, in this case that of inshore Lake Ontario, and superimpose a 2 °C elevation throughout the year: the global warming scenario (Fig. 6). Using countercurrent heat exchangers, both hardwater and synthetic softwater are generated continuously at ambient and ambient +2 °C for exposure of juvenile fish to sublethal environmental acidification (pH 5.2, softwater) or sublethal elevated ammonia (hardwater) on a flow-through, single pass basis. Thus the fish were exposed to one of four conditions based on the water temperature and absence or addition of sublethal levels of two priority pollutants (ammonia and low pH) in test waters of relevant hardness.

Between June 1993 and September 1994, we completed a series of three 3 month exposures, at different times of the year and with different feeding regimes. The results of one such experiment will be highlighted here. A 3 month chronic summer exposure (June–September 1993) was characterized as a period of rising temperature, with a peak temperature at or near the upper lethal temperature for rainbow trout (Fig. 6). Most indicators of metabolic cost increased as temperature rose (i.e. appetite, protein synthesis, growth and oxygen consumption), with the addition of 2 °C to the ambient profile resulting in further increases in metabolic costs; the overall Q_{10} for these processes was about 2.1 over the first 60 days (Reid *et al.*, 1995). However, our most significant finding was the profound inhibitory influence of water temperature on all measured parameters during the period of peak temperatures ($Q_{10} = 0.47$). For example, appetite (Fig. 7), nitrogenous waste excretion and gill protein synthesis were reduced, not

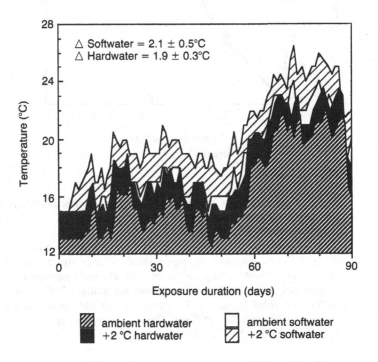

Fig. 6. Temperature profile for the synthetic softwater and hardwater exposure system during a 90 day experiment, which ran over the summer from 18 June to 15 September 1993. Juvenile rainbow trout were held at temperatures representative of inshore Lake Ontario (ambient temp., softwater = open area, ambient temp., hardwater = heavily cross-hatched area) or at a temperature 2 °C higher (lightly cross-hatched area and dark area for softwater and hardwater, respectively). (From Reid *et al.*, 1995.)

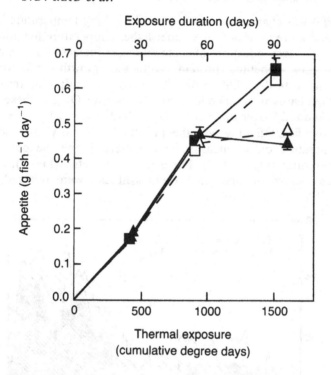

Fig. 7. The effect of water temperature and total ammonia level on appetite in hardwater-acclimated trout. Fish were fed to satiation twice daily with direct measurement of the amount of food consumed. Data from fish exposed to ambient water temperature and ambient ammonia (6.0 μM total ammonia) are represented by open squares and dotted lines, ambient water temperature and 69 μM total ammonia by filled squares and solid lines, ambient water ammonia +2 °C by solid triangles and solid lines, +2 °C water and 69 μM total ammonia by open triangles and dotted lines. Data presented as means ($N = 10$) ± SE. (From Reid *et al.*, 1995.)

stimulated, by the addition of 2 °C to the ambient temperature profile during the last 30 days of the exposure. We also confirmed the higher cost associated with life in softwater. Trout acclimated to and held in artificial softwater exhibited reduced appetites and growth rates, a compensatory increase in gross food conversion efficiency and higher gill protein synthesis rates, compared to hardwater-acclimated rainbow trout (Reid *et al.*, 1995; Dockray, Reid & Wood, 1996).

Exposure to sublethal acid also tended to increase metabolic cost, satisfied largely by increasing food intake. In contrast, hardwater fish exposed to elevated total ammonia did not consume more food, but instead re-partitioned their energy budgets to compensate for significantly greater rates of liver protein synthesis and nitrogenous waste excretion. The increase in nitrogenous waste excretion may act as an ammonia detoxification mechanism, as this increase occurred in the presence of elevated water total ammonia and without any alteration in food consumption. In contrast, nitrogenous waste excretion was greater in softwater fish at low pH, largely due to the enhanced appetite. Alterations in nitrogenous waste excretion, in relation to fuel intake (appetite) and ammonia exposure, appear to be of great importance.

Despite increased food intake in softwater fish exposed to pH 5.2, gill protein synthesis was significantly reduced while muscle and liver protein synthesis rates were little affected. We suggest that alterations in gill protein synthesis reflect changes in branchial electrolyte transport and permeability that are known to occur during sublethal acid exposure (see Wood, 1989).

Following each 3 month exposure, challenge experiments are conducted to evaluate whether the chronic, sublethal exposure induced increased temperature and/or pollutant tolerance. In general, sensitization rather than increased tolerance has been observed, underlying the synergistic nature of the stressors.

Due to the complex nature of experiments of this design, it is difficult to quantify the exact interaction between pollutant-induced changes in physiology and the impact of 2 °C superimposed on a natural temperature profile. However, if we look at the impact of elevated ammonia and temperature on rates of liver protein synthesis (Fig. 8a), we can demonstrate the deleterious impact that high water ammonia and only a 2 °C rise in ambient temperature might have on fish inhabiting such waters. Figure 8b illustrates the changes in liver protein synthesis that occur separately in response to the addition of 2 °C to the ambient temperature profile, and in response to elevated water ammonia during an ambient temperature profile, and the interactive effects of these two influences. As stated previously, after 60 and 90 days of exposure, the addition of either temperature or ammonia resulted in a stimulation in liver protein synthesis. The influence of ammonia was greater than that of the additional 2 °C and is thought to be associated with the stimulation of ammonia detoxification by this organ. However, the combined effect of added temperature and ammonia is clearly different

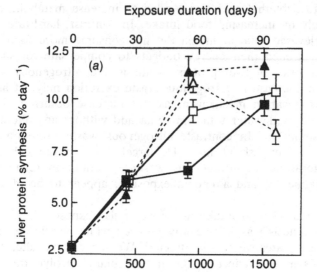

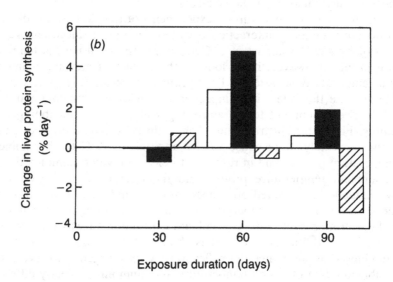

from the impact of the individual stressors. After both 60 and 90 days of exposure, the addition of 2 °C to the ambient temperature profile in the presence of elevated water ammonia is clearly inhibitory. Therefore, if we can assume that there is a direct correlation between liver protein synthesis and ammonia detoxification, the impact of global warming on fish living in waters of elevated ammonia will be to impair significantly the ability of these fish to deal with a warmer, polluted environment.

Summary and suggestions for future research

The study of the relationship between temperature and the toxicity of an environmental stressor is not a new research endeavour. Temperature has long been known to modify the chemistry of a number of aquatic pollutants resulting in significant alterations in their toxicities to fish. With the increased awareness and concern about the impact of global warming on inhabitants of aquatic environments, it is essential that data be generated that can be used to develop models for predicting the future status of these animals. Until recently, the interactive effects of temperature and pollutant stress have not been approached from a global warming perspective. The traditional approach involved acute, single-species toxicity tests completed at different, yet constant, environ-

---

Fig. 8. The effect of water temperature and total ammonia level on (a) liver protein synthesis (K_s) and (b) the change in liver protein synthesis rates ($\triangle K_s$) in hardwater-acclimated trout. All rates were calculated on a % day^{-1} basis. In (a) data from fish exposed to ambient water temperature and total ammonia (6.0 μM) are represented by ■, ambient water temperature with 69 μM total ammonia by □, ambient total ammonia at +2 °C by ▲ and +2 °C water with 69 μM total ammonia by △. Data presented as means ($n = 10$) ± 1 SE (Reid *et al.*, 1995). In (b) the change in liver K_s due to temperature alone was calculated as the difference between ambient rates and +2 °C rates, both in fish at ambient ammonia levels (white bars). Change in liver K_s due to ammonia alone was calculated as the difference between rates from fish exposed to 6.0 μM and 69 μM total ammonia, both at ambient temperature (black bars). The interactive influence was calculated as the difference between rates from fish exposed to ambient temperature and water ammonia and those from fish exposed to 69 μM total ammonia, +2 °C water temperature (hatched bars). Data presented as means ($n = 10$) ± 1 SE. (From Reid *et al.*, 1995.)

mental temperatures. Considerable information regarding the interaction on temperature and common aquatic pollutants is available, but for numerous reasons, the findings of many of these studies may not be directly applicable to the unique nature of the global warming scenario. In order to assess the impact of a chronic temperature elevation of marginalized environments in the context of global warming, the traditional experimental approach must be reconsidered. From a global warming perspective, investigations into the interactive effects of temperature and pollutant stress should take advantage of realistic temperature profiles, should examine life stages most likely to be exposed to that temperature profile or pollutant, and should utilize more than a single species. Experiments that mimic the natural temperature profile of a region onto which is superimposed an increase in temperature, although complicated and difficult to interpret, are possible and may represent the best method for predicting the impact of global warming on fish in pristine and polluted environments.

References

Arthur, J. W., West, C. W., Allen, K. N. & Hedtke, S. F. (1987). Seasonal toxicity of ammonia to five fish and nine invertebrate species. *Bulletin of Environmental Contamination and Toxicology*, **38**, 324–31.

Audet, C. & Wood, C. M. (1993). Branchial morphological and endocrine responses of rainbow trout (*Oncorhynchus mykiss*) during a long-term sublethal acid exposure in which acclimation did not occur. *Canadian Journal of Fisheries and Aquatic Sciences*, **50**, 198–209.

Becker, C. D. & Genoway, G. (1979). Evaluation of the critical thermal maximum for determining thermal tolerance of fresh water fish. *Environmental Biology of Fishes*, **4**, 245–56.

Becker, C. D. & Wolford, M. G. (1980). Thermal resistance of juvenile salmonids sublethally exposed to nickel, determined by the critical thermal maximum method. *Environmental Pollution Series A: Ecological and Biological*, **21**, 181–9.

Black, M. C., Millsap, D. S. & McCarthy, J. F. (1991). Effects of acute temperature change on respiration and toxicant uptake by rainbow trout, *Salmo gairdneri* (Richardson). *Physiological Zoology*, **64**, 145–68.

Broderius, S., Drummond, R., Fiandt, J. & Russom, C. (1985). Toxicity of ammonia to early life stages of the smallmouth bass at four pH values. *Environmental Toxicology and Chemistry*, **4**, 87–96.

Brown, V. M. & Dalton, R. A. (1970). The acute lethal toxicity to rainbow trout of mixtures of copper, phenol, zinc and nickel. *Journal of Fish Biology*, **2**, 211–16.

Brown, V. M., Jordan, D. H. M. & Tiller, B. A. (1967). The effect of temperature on the acute toxicity of phenol to rainbow trout in hard water. *Water Research*, 1, 587–94.

Brown, V. M., Jordan, D. H. M. & Tiller, B. A. (1969). The acute toxicity of fluctuating concentrations and mixtures of ammonia, phenol and zinc. *Journal of Fish Biology*, 1, 1–9.

Calow, P. (1991). Physiological costs of combating chemical toxicants: ecological implications. *Comparative Biochemistry and Physiology*, **100** C, 3–6.

Coutant, C. C. (1977). Compilation of temperature preference data. *Journal of the Fisheries Research Board of Canada*, **34**, 739–45.

Cowles, R. B. & Bogert, C. M. (1944). A preliminary study of the thermal requirements of desert reptiles. *Bulletin of the American Museum of Natural History*, **83**, 265–96.

Dockray, J. J., Reid, S. D. & Wood, C. M. (1996). Effects of elevated summer temperatures and reduced pH on metabolism and growth of juvenile rainbow trout *(Oncorhynchus mykiss)* on unlimited ration. *Canadian Journal of Fisheries and Aquatic Sciences* (in press).

Doe, K. G., Parker, W. R., Ponsford, S. J. & Vaughan, J. D. A. (1987). The acute and chronic toxicity of antimony to *Daphnia magna* and rainbow trout. In *Proceedings of the Thirteenth Annual Aquatic Toxicity Workshop, 12–24 November 1986, Moncton, New Brunswick*, ed. J. S. S. Lakshminarayana, pp. 40–3. Canadian Technical Report of Fisheries and Aquatic Science, No. 1575.

Downing, K. M. & Merkens, J. C. (1957). The influence of temperature on the survival of several species of fish in low tensions of dissolved oxygen. *Annals of Applied Biology*, **45**, 261–7.

Emerson, K., Russo, R. C., Lund, R. E. & Thurston, R. V. (1975). Aqueous ammonia equilibrium calculations: effect of pH and temperature. *Journal of the Fisheries Research Board of Canada*, **32**, 2379–82.

Erickson, R. J. (1985). An evaluation of mathematical models for the effects of pH and temperature on ammonia toxicity to aquatic organisms. *Water Research*, **8**, 1047–58.

Evans, D. O. (1990). Metabolic thermal compensation by rainbow trout: effects on standard metabolic rate and potential usable power. *Transactions of the American Fisheries Society*, **119**, 585–600.

Farrell, A. P. & Jones, D. R. (1992). The heart. In *Fish Physiology; vol. 12A. Cardiovascular Systems*, ed. W. S. Hoar, D. J. Randall & A. P. Farrell, pp. 1–88. New York: Academic Press.

Fauconneau, B. & Arnal, M. (1985). *In vivo* protein synthesis in different tissues and the whole body of rainbow trout (*Salmo gairdneri* R.). Influence of environmental temperature. *Comparative Biochemistry and Physiology*, **82**A, 179–87.

346   S.D. REID *et al.*

Fowler, B. A. (1987). Intracellular compartmentation of metals in aquatic organisms: roles in mechanisms of cell injury. *Environmental Health Perspectives*, **71**, 121–8.

Graham, M. (1987). The solubility of oxygen in physiological salines. *Fish Physiology and Biochemistry*, **4**, 1–4.

Graham, M. S. & Farrell, A. P. (1989). The effect of temperature acclimation and adrenaline on the performance of a perfused trout heart. *Physiological Zoology*, **62**, 38–61.

Hazel, J. R. (1993). Thermal biology. In *The Physiology of Fishes*, ed. D. H. Evans, pp. 427–67. Boca Raton, Fla: CRC Press.

Herbert, D. W. M. & Shurben, D. S. (1963). A preliminary study of the effects of physical activity on the resistance of rainbow trout (*Salmo gairdneri* Richardson) to two poisons. *Annals of Applied Biology*, **52**, 321–6.

Herbert, D. W. M. & Shurben, D. S. (1964). The toxicity to fish of mixtures of poisons. I. Salts of ammonia and zinc. *Annals of Applied Biology*, **53**, 33–41.

Herbert, D. W. M. & Wakeford, A. C. (1964). The susceptibility of salmonid fishes to poisons under estuarine conditions. I. Zinc sulphate. *Journal of Air and Water Pollution*, **8**, 251–6.

Hodson, P. A. & Sprague, J. B. (1975). Temperature-induced changes in acute toxicity of zinc to Atlantic salmon (*Salmo salar*). *Journal of the Fisheries Research Board of Canada*, **32**, 1–10.

Hogstrand, C., Reid, S. D. & Wood, C. M. (1995). Calcium versus zinc transport in the gills of fresh water rainbow trout, and the cost of adaptation to waterborne zinc. *Journal of Experimental Biology*, **198**, 337–418.

Houlihan, D. F., Costello, M. J., Secombes, C. J., Stagg, R. & Brechin, J. (1994). Effects of sewage sludge exposure on growth, feeding and protein synthesis of Dab (*Limanda limanda* L.). *Marine Environmental Research*, **37**, 331–53.

Jeney, Zs, Nemcsók, J., Jeney, G. & Oláh, J. (1992). Acute effect of sublethal ammonia concentrations on common carp (*Cyprinus carpio* L.). I. Effect of ammonia on adrenaline and noradrenaline levels in different organs. *Aquaculture*. **104**, 139–48.

Kaya, C. M. (1978). Thermal resistance of rainbow trout from a permanently heated stream, and of two hatchery strains. *Progressive Fish-Culturist*, **40**, 138–42.

Kennedy, C. J., Gill, K. A. & Walsh, P. J. (1989). Thermal modulation of benzo[a]pyrene metabolism by the gulf toadfish, *Opsanus beta*. *Aquatic Toxicology*, **15**, 331–44.

Korwin-Kossakowski, M. & Jezierska, B. (1985). The effect of temperature on survival of carp fry, *Cyprinus carpio* L., in acid water. *Journal of Fisheries Biology*, **26**, 43–7.

Kwain, W. H. (1975). Effects of temperature on development and survival of rainbow trout, *Salmo gairdneri*, in acid water. *Journal of the Fisheries Research Board of Cananda*, **32**, 493–7.

Lauren, D. J. & McDonald, D. G. (1985). Effects of copper on branchial ionoregulation in the rainbow trout, *Salmo gairdneri* Richardson. *Journal of Comparative Physiology B*, **155**, 635–44.

Leino, R. L. & McCormick, J. H. (1993). Responses of juvenile largemouth bass to different pH and aluminum levels at overwintering temperatures: effects on gill morphology, electrolyte balance, scale calcium, liver glycogen, and depot fat. *Canadian Journal of Zoology*, **71**, 531–43.

Lloyd, R. (1961). The toxicity of mixtures of zinc and copper sulphates to rainbow trout (*Salmo gairdneri* Richardson). *Annals of Applied Biology*, **49**, 535–8.

Loughna, P. T. & Goldspink, G. (1985). Muscle protein synthesis rates during temperature acclimation in a eurythermal (*Cyprinus carpio*) and a stenothermal (*Salmo gairdneri*) species of teleost. *Journal of Experimental Biology*, **118**, 267–76.

McDonald, D. G. (1983). The effects of H^+ upon the gills of fresh water fish. *Canadian Journal of Zoology*, **61**, 691–703.

McDonald, D. G. & Wood, C. M. (1993). Branchial mechanisms of acclimation to metals in fresh water fish. In *Fish Ecophysiology*, ed. J. C. Rankin & F. B. Jensen, pp. 297–321. London: Chapman & Hall.

McGeachy, S. M. & Dixon, D. G. (1990). The impact of temperature on the acute toxicity of arsenate and arsenite to rainbow trout (*Salmo gairdneri*). *Ecotoxicology and Environmental Safety*, **17**, 86–93.

McGeachy, S. M. & Dixon, D. G. (1991). Effect of temperature on the chronic toxicity of arsenate to rainbow trout (*Oncorhynchus mykiss*). *Canadian Journal of Fisheries and Aquatic Sciences*, **47**, 2228–34.

Mallatt, J. (1985). Fish gill structural changes induced by toxicants and other irritants: a statistical review. *Canadian Journal of Fisheries and Aquatic Science*, **42**, 630–48.

Mohnen, V. A. & Wang, W. -C. (1992). An overview of global warming. *Environmental Toxicology and Chemistry*, **11**, 1051–9.

Mueller, M. E., Sanchez, D. A., Bergman, H. L., McDonald, D. G., Rhem, R. G. & Wood, C. M. (1991). Nature and time course of acclimation to aluminum in juvenile brook trout (*Salvelinus fontinalis*). II. Histology. *Canadian Journal of Fisheries and Aquatic Sciences*, **48**, 2016–27.

Narahashi, T. (1987). Nerve membrane ion channels as the target site of environmental toxicants. *Environmental Health Perspectives*, **71**, 25–9.

Neville, C. M. (1985). Physiological response of juvenile rainbow trout, *Salmo gairdneri*, to acid and aluminum – prediction of field responses from laboratory data. *Canadian Journal of Fisheries and Aquatic Sciences*, **42**, 2004–19.

Paladino, F. V. & Spotila, J. R. (1978). The effect of arsenic on the thermal tolerance of newly hatched muskellunge fry (*Esox masquinongy*). *Journal of Thermal Biology*, **3**, 223–7.

Peterson, R. H. & Anderson, J. M. (1969). Influence of temperature change on spontaneous locomotor activity and oxygen consumption of Atlantic salmon, *Salmo salar*, acclimated to two temperatures. *Journal of the Fisheries Research Board of Canada*, **26**, 93–109.

Playle, R. C., Goss, G. G. & Wood, C. M. (1989). Physiological disturbances in rainbow trout (*Salmo gairdneri*) during acid and aluminum exposures in soft water of two calcium concentrations. *Canadian Journal of Zoology*, **67**, 314–24.

Randall, D. & Brauner, C. (1991). Effects of environmental factors on exercise in fish. *Journal of Experimental Biology*, **160**, 113–26.

Randall, D. J. & Cameron, J. N. (1973). Respiratory control of arterial pH as temperature changes in rainbow trout, *Salmo gairdneri* R. *Americain Journal of Physiology*, **225**, 997–1002.

Reid, S. D. (1990). Metal–gill surface interactions in rainbow trout (*Oncorhynchus mykiss*). Ph.D. thesis, McMaster University, pp. 178.

Reid, S. D., Dockray, J. J., Linton, T. K., McDonald, D. G. & Wood, C. M. (1995). Effects of a summer temperature regime representative of a global warming scenario on growth and protein synthesis in hardwater- and softwater-acclimated juvenile rainbow trout (*Oncorhynchus mykiss*). *Journal of Thermal Biology*, **20**, 231–44.

Roch, M. & Maly, J. (1979). Relationship of cadmium-induced hypocalcemia with mortality in rainbow trout (*Salmo gairdneri*) and the influence of temperature on toxicity. *Journal of the Fisheries Research Board of Canada*, **36**, 1297–303.

Schneider, S. H. (1990). The global warming debate: science or politics? *Environmental Science and Technology*, **24**, 432–5.

Sheehan, R. J. & Lewis, W. M. (1986). Influence of pH and ammonia salts on ammonia toxicity and water balance in young channel catfish. *Transactions of the American Fisheries Society*, **115**, 891–9.

Sprague, J. B. (1970). Measurement of pollutant toxicity to fish. II. Utilizing and applying bioassay results. *Water Research*, **4**, 3–32.

Sprague, J. B. (1985). Factors that modify toxicity. In *Fundamentals of Aquatic Toxicology: Methods and Applications*, ed. G. M. Rand & S. R. Petrocelli, pp. 124–63. New York: Hemisphere Publishing Corporation.

Takle, J. C. C., Beitinger, T. L. & Dickson, K. L. (1983). Effect of the aquatic herbicide endothal on the critical thermal maximum of red shiner, *Notropis lutrensis*. *Bulletin of Environmental Contamination and Toxicology*, **31**, 512–17.

Thurston, R. V., Russo, R. C. & Vinogradov, G. A. (1981). Ammonia toxicity to fishes. Effect of pH on the toxicity of the un-ionized ammonia species. *Environmental Science and Technology*, **15**, 837–40.

Watenpaugh, D. E. & Beitinger, T. L. (1985). Temperature tolerance of nitrite-exposed channel catfish. *Transactions of the American Fisheries Society*, **114**, 274–8.

Williams, R. T. (1959). *Detoxification Mechanisms: The Metabolism and detoxification of Drugs, Toxic Substances and other Organic Compounds*, 2nd edn London: Chapman & Hall.

Wilson, R. W., Bergman, H. L. & Wood, C. M. (1994a). Metabolic costs and physiological consequences of acclimation to aluminum in juvenile rainbow trout (*Oncorhynus mykiss*). 1: acclimation specificity, resting physiology, feeding, and growth. *Canadian Journal of Fisheries and Aquatic Sciences*, **51**, 527–35.

Wilson, R. W., Bergman, H. L. & Wood, C. M. (1994b). Metabolic costs and physiological consequences of acclimation to aluminum in juvenile rainbow trout (*Oncorhynus mykiss*). 2: gill morphology, swimming performance, and aerobic scope. *Canadian Journal of Fisheries and Aquatic Sciences*, **51**, 536–44.

Wilson, R., Wood, C. M. & Houlihan, D. F. (1996). Growth, feeding and protein turnover during acclimation to acid/aluminum in juvenile rainbow trout (*Oncorhychus mykiss*) *Canadian Journal of Fisheries and Aquatic Sciences*, **53**, 802–11.

Wood, C. M. (1989). The physiological problems of fish in acid waters. In *Acid Toxicity and Aquatic Animals*, ed. R. Morris, E. W. Taylor, D. J. A. Brown & J. A. Brown. *Society for Experimental Biology Seminar Series* **34**, pp. 125–48. Cambridge: Cambridge University Press.

Wood, C. M., Playle, R. C., Simons, B. P., Goss, G. G. & McDonald, D. G. (1988). Blood gases, acid–base status, ions, and hematology in adult brook trout (*Salvelinus fontinalis*) under acid/aluminum exposure. *Canadian Journal of Fisheries and Aquatic Sciences*, **45**, 1575–86.

LARRY I. CRAWSHAW and
CANDACE S. O'CONNOR

Behavioural compensation for long-term thermal change

We found the main stream so low that the teeter-snipe pattered about in what last year were trout riffles, and so warm that we could duck in its deepest pool without a shout . . . There were no trout . . . But . . . up near the headwaters we had once seen a fork . . . fed by cold springs that gurgled out under its close-hemmed walls of alder. What would a self-respecting trout do in such weather? Just what we did: go up. I climbed down the dewy bank and stepped into the Alder Fork. A trout was rising just upstream . . .

(Aldo Leopold, 1949, p. 40)

Introduction and overview

A 1 °C alteration in a heterothermal aquatic environment produces behavioural adjustments in many fishes; a 4 °C shift leads to major changes in fish distribution. This is expected to occur for a majority of species and is likely to be important both ecologically and physiologically. Temperature changes of this magnitude are predicted during the next century, owing to the greenhouse effects of anthropogenically generated gases. These climate warming scenarios differ by region and latitude and would be influenced by other human activities such as logging, use of dams, power generation, and urbanization (Kemp, 1994).

In this chapter thermoregulatory behaviour is discussed as a short-term, local adjustment to the posited increases in global temperature. However, thermoregulation is only one of many behaviours that fish utilize to survive and reproduce; behavioural adjustments to other physical aspects of the environment are also important and are considered. Next, the nature of the thermoregulatory system of fishes will

Society for Experimental Biology Seminar Series 61: *Global Warming: Implications for freshwater and marine fish*, ed. C. M. Wood & D. G. McDonald. © Cambridge University Press 1996, pp. 351–376.

be reviewed and circumstances that lead to changes in the regulated temperature will be noted.

The thermoregulatory capabilities of fishes will then be related to particular species in specific geographic areas. The habitats focused on will include those areas in central North America, where the largest changes in temperature and decreases in soil moisture are anticipated; coastal areas of the North Atlantic, where some of the largest marine changes are anticipated; and areas important to pelagic or anadromous fish migrations. A limited number of species will serve as examples to develop a general understanding of potential outcomes; there will be no attempt to cover exhaustively all affected species nor to include every investigation.

Temperature and species distribution

Each species of fish is present in some habitats and absent from others. Light, dissolved oxygen, current or salinity often contribute to habitat selection, but the role of temperature is the easiest to document. Temperature exerts its influence in all environments, directly affects all physiological systems, usually changes periodically, and is distributed heterogeneously in space.

In the ocean, species distributions may change rapidly and dramatically when the thermal environment changes. For example, water temperatures along the North American west coast rose to unusually high levels in 1958, and 10 fish species were recorded at their maximal northward extension (Norris, 1963). This did not result from advection as the coastal current continued to flow southward.

Vertical movements of fish are important in determining the distribution of less active species. Temperature often dictates the depth at which a fish is found. In the cold North Pacific, the chimaera (*Hydrolagus colliei*) occurs in shallow water, while further south it is found at similar temperatures, which only occur at depths exceeding 90 m (Norris, 1963).

Different species can respond in dramatically different ways to seasonal and annual variations in water temperature. Murawski (1993) analysed the results of 24 years of standardized bottom-trawl and water temperature surveys for 34 species in the North Atlantic, an area of unusually great latitudinal and seasonal thermal heterogeneity. Four distinct groups of fishes could be discerned. Shallow-water, sedentary species (including the yellowtail flounder, *Pleuronectes ferrugineus*, and the longhorn sculpin, *Myoxocephalus octodecemspinosus*) exhibited wide variations in body temperature and little directed movement.

Deepwater, sedentary species (pollock, *Pollachius virens*, and Atlantic plaice, *Hippoglossoides platessoides*) also showed little movement, but were exposed to less thermal variation because deep waters in the Gulf of Maine have more uniform temperatures. Mobile warmwater species (bluefish, *Pomatomus saltatrix*, and black sea bass, *Centropristis striata*) were found at highly variable depths and geographical locations, but at relatively constant temperatures. The greatest interannual variation in latitude was found in the coolwater migratory species (Atlantic herring, *Clupea harengus*, and Atlantic mackerel, *Scomber scombrus*). For these fish, a 1 °C increase in average water temperature led to poleward distributional shifts of about 110 km. Murawski (1993) noted that widespread thermal changes could decrease the distributional overlap between relatively sedentary predators such as the Atlantic cod (*Gadus morhua*) and highly mobile prey such as mackerel. Less severe consequences would be expected for more mobile predators like bluefish.

Responses to environmental stimuli

While broad-scale oceanic distributions offer meaningful ecological information, understanding the orientation of individual fish to particular environmental stimuli is particularly informative. Responses to these stimuli include simple taxes, sophisticated regulatory systems, and orientation involving complex innate and learned behaviours. All sensed environmental stimuli interact with temperature in the behavioural response hierarchy of a fish. It is important to consider the competing cues.

Matthews and Hill (1979) concluded that 'The influence of any variable upon habitat selection depends on the entire milieu of conditions at a particular time, and organisms seldom select habitat in response to single factors' (p. 79). They did note a definite heirarchy: the most important factors for red shiner (*Notropis lutrensis*) in midwestern North America were temperature, lack of current, and water deeper than 20 cm. Salinity and pH exerted effects, but only with substantial gradients. Dissolved oxygen was of little importance, except at low levels. Turbidity, shelter, shade and substrate were also less important.

Avoidance of intense light by juvenile walleye pollock in the North Pacific was interpreted as a behavioural adaptation to elude avian predators working the ocean surface (Olla & Davis, 1990). In a laboratory tank with a thermocline of temperatures similar to those in their usual environment, the pollock stayed in 10 °C water above the thermocline. Introduction of a predator (an adult walleye pollock) caused the juveniles to dive to the bottom of the tank if there was no thermocline but only to mid-depths with a thermocline present. When responses were arranged hierarchically, the greatest effect was from thermal stratification, followed by the presence of a predator and, finally, light intensity (Sogard & Olla, 1993).

Recognition of and response to potential predators or intricate habitat attributes involves more complex neural processing than that required by a physical gradient. Sophisticated choices by many animal species are critical to survival and reproductive success (Alcock, 1993). Juvenile lemon sharks (*Negaprion brevirostris*) in the North Sound of Bimini, Bahamas provide a good example (Morrissey & Gruber, 1993). Although many temperatures, depths and substrate types were available, the sharks spent the majority of their time in water in the low 30s °C at depths less than 50 cm and over a mixed rock and sand substrate. This habitat presumably provided optimal physiological temperatures, food availability, and protection from predators. Juvenile sharks were tracked with implanted ultrasonic transmitters. Observations indicated that predation was an important aspect of the sharks' environment. In one case, the authors 'observed a large splash near the estimated location of the telemetered shark, immediately followed by the signal travelling at an unusually high rate of speed toward deeper water' (p. 317).

Temperature and season were important in habitat selection by Atlantic salmon (*Salmo salar*) parr in a stream tank (Gibson, 1978). In late summer, parr occupied most locations but favoured deeper (50 cm) over shallower water (30 cm). When in shallow water they were attracted to shade, but showed a higher preference for surface turbulence. In autumn, when temperatures dropped to 10 °C, fish began taking shelter in rocky rubble; by the time 9 °C was reached most fish had disappeared. Habitat occupation is also affected by social interactions. The deepest areas of stream pools are preferred locations for many salmonids. In a coastal stream in British Columbia, such areas were occupied by the largest, presumably dominant, cutthroat trout (*Oncorhynchus clarki*) (Heggenes, Northcote & Armin, 1991). Social dominance also can affect temperature selection; in a bichambered shuttle aquarium, dominant bluegills (*Lepomis macrochirus*) occupied the normally preferred 31 °C side, while the subordinates were displaced to 26 °C (Medvick, Magnuson & Sharr, 1981).

So temperature is important, but only one of many physical factors such as light level and water depth that determine the location of an individual fish. Complex habitat characteristics as well as social and anti-predator interactions are also important. The relative influence of these elements may also change daily, seasonally and developmentally.

Importance of preferred temperature

Preferred temperature has been related to both ecological and physiological optima, and even utilized to indicate the areas better for commercial fishing. Magnuson, Crowder and Medvick (1979) suggested that thermal habitat should be considered as an ecological resource in the same manner as food. Animals compete for the optimal thermal niche and, when successful, are able to maximize growth and other aspects of fitness. Physiological thermal optima were characterized in a series of studies on the sockeye salmon (*Oncorhynchus nerka*), summarized by Brett (1971). The usual preferred temperature of 15 °C was also the temperature at which metabolic scope, cardiac scope, overall performance, and growth rate under excess ration all were maximum. For young rainbow trout (*O. mykiss*) with a slightly higher preferred temperature (Jobling, 1981), important performance variables and response speeds related to prey capture are maximal at slightly above 15 °C (Webb, 1978).

Jobling (1981) reviewed growth optima and preferred temperatures for many species, and concluded that each species' preferred temperature provided a rapid, practical indication of the best temperature to utilize for fish culture. Similar findings pertain to the production of fish in natural waters. The realized commercial yield, under varying conditions, of lake trout (*Salvelinus namaycush*), lake whitefish (*Coregonus clupeaformis*), walleye (*Stizostedion vitreum vitreum*) and northern pike (*Esox lucius*) was assessed in respect to thermal habitat for a set of 21 large, north temperate lakes (Christie & Regier, 1988). Sustained yields of the four species had strong positive correlations with optimal thermal habitat sizes, defined by slightly modifying the 4 °C optimal range delimited by Magnuson *et al.* (1979).

Thermal regulation by fish

Each fish species has a repertoire of responses used to regulate body temperature. In some cases simple taxes help fish avoid lethal temperatures; such innate responses function best when a behaviour bears a fixed relation to a particular outcome. Young fish are particularly susceptible to thermal stress and often swim straight down when confronted with a hot environment; under most conditions this leads them into cooler water (Norris, 1963).

Fish are particularly adept at learning the thermal characteristics of an environment, and utilizing this information to anticipate and avoid potential thermal stresses. In (admittedly naive) initial experiments performed by one of the authors (LIC), thermal choice tanks were constructed in a fixed configuration such that a particular tank was always colder or warmer. The fish quickly detected this condition and, when placed in the apparatus, avoided inappropriate chambers without even sampling them. This necessitated completely rebuilding the equipment in such a way that the thermal configuration could be altered for each trial. In a more rigorous demonstration of learning (Rozin & Mayer, 1961), goldfish were readily trained to perform an arbitrary task (poking a lever) to obtain an injection of cold water that transiently lowered the temperature of a 41 °C tank by 0.3 °C.

The motivation for responding to stressful thermal conditions derives from the reduction of an error signal emanating from the thermoregulatory network of the central nervous system. This homeostatic system is anatomically and functionally similar in Chondrichthyes (Crawshaw & Hammel, 1973) and Osteichthyes (Crawshaw & Hammel, 1974), and thus may have kept vertebrate species within optimal temperatures for over 400 million years (Kardong, 1995). These similarities also extend to Mammalia (Fig. 1). Because many other similarities have been noted (Hammel, 1968; Crawshaw, 1980), judicious use of evidence from studies on mammals will be used to fill in missing information and thus provide a more complete description of the thermoregulatory system relevant to fish.

A functional thermoregulatory system must sense the controlled variable, compare it with an ideal value, and produce the correct behavioural response. Sensing occurs both centrally and peripherally. Temperature can be sensed at many spots over the entire surface of the body, including the fins (Bardach, 1956). This thermosensitivity is likely to be subserved by highly responsive, rate-sensitive, free nerve endings, as documented for mammals (Brück & Hinckel, 1990). When the entire body surface is warmed, fish can discriminate changes of 0.05 °C (Bardach & Bjorklund, 1957). Although changes this small usually do not elicit thermoregulatory responses, the evidence indicates that fish can be quite precise in their response to the thermal environment.

The nucleus preopticus periventricularis (NPP) of the teleost anterior brainstem integrates thermal inputs. The NPP is within the area of functional thermosensitivity and it possesses thermosensitive neurons. When this area is lesioned, thermoregulatory behaviour is lost (Crawshaw et al., 1990). Studies of neurons in the mammalian homologue to this area, the preoptic nucleus/anterior hypothalamic area,

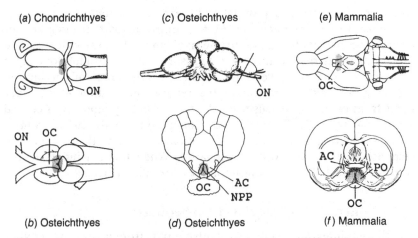

Fig. 1. Similarities in the anatomical substrate for the regulation of
body temperature in various classes of the subphylum Vertebrata.
Shaded areas of (*a*), (*b*) and (*e*) represent areas of the ventral brain
surface that overlie interior portions of the brain that possess functional
thermosensitivity and thermo-integrative properties. The straight line
in (*c*) represents the location and plane of the cross-section in (*d*);
(*f*) is a cross-section through the shaded portion of the brain depicted
in (*e*). The shaded portions of the cross-sections depicted in (*d*) and (*f*)
are areas wherein bioactive chemicals produce major thermoregulatory
effects. AC, anterior commissure; OC, optic chiasma; ON, optic
nerve; NPP, nucleus preopticus periventricularis; PO, preoptic nucleus.
(From Crawshaw, 1980; Crawshaw & Hammel, 1973, 1974; Crawshaw,
O'Connor & Wollmuth, 1992.)

have confirmed the presence of cells sensitive to both warm and cold.
Also, only warm sensitivity appears to be an inherent property of the
cells; cold sensitivity most likely derives from synaptic input via warm
cells. Cells in the hypothalamus respond to absolute temperature,
but are not rate sensitive. These cells thus provide input when body
temperature shifts significantly or when the regulated temperature is
altered. Some of the interplay between environmental and physiological
factors may be mediated by this area of the brain. About half of the
temperature-sensitive cells also respond to non-thermal stimuli such as
osmotic pressure, glucose concentration, or steroid hormone concentration (Boulant, Curras & Dean, 1989).

Various models, based largely on mammalian data, have been proposed to explain the generation of a regulated temperature by anterior
brainstem neurons (Boulant *et al.*, 1989). Inputs from peripheral warm

and cold receptors are postulated to excite the central warm- and cold-sensitive cells respectively. Motor control areas that integrate co-ordinated movements leading to the avoidance of warm water receive excitatory input from central warm-sensitive cells and inhibitory input from central cold-sensitive cells. Central nervous areas subserving the avoidance of cold water receive the reverse innervation – excitatory input from central cold-sensitive cells and inhibitory input from central warm-sensitive cells. The reciprocal innervation of the motor control areas subserving warm and cold avoidance ensure that either one response or the other will be activated. When thermal inputs to the two areas are balanced, non-thermal factors direct behaviour.

Alterations of regulated temperature

The thermo-integrative cells also receive inputs from many other physiological systems and neuronal networks. Thus, the regulated temperature is actually altered, rather than being overridden by competing inputs, as was discussed earlier. Such changes in the regulated temperature may occur as diel rhythms, during ontogenetic development or following acclimation to different temperatures. Regulatory changes also occur in response to bacterial infections, hypoxia, salinity shifts, and exposure to toxins.

Diel rhythms of regulated body temperature are characteristic of most vertebrates (Reinberg & Smolensky, 1990), and have been documented in many fishes, including goldfish (*Carassius auratus*), smallmouth bass (*Micropterus dolomieui*) and polar cod (*Boreogadus saida*). Such temperature rhythms may help cue vertical migrations or could be involved in energy conservation during inactive periods. Other fish such as bluegills, brown bullheads (*Ictalurus nebulosus*), and navaga (*Eleginus navaga*) clearly do not exhibit such rhythms (Crawshaw, 1975a; Reynolds & Casterlin, 1976; Reynolds *et al.*, 1978; Schurmann & Christiansen, 1994).

For a number of species, juveniles prefer warmer temperatures than adults. This is believed to aid foraging and growth for developing fish, and could serve to segregate sizes for species where the adults are cannibalistic (Coutant, 1987a).

For fish, residence at a particular temperature leads to biochemical and structural adjustments that permit efficient operation. These alterations, termed thermal acclimation, are documented amply in other chapters of this book. The net consequence of such acclimation alters lethal limits and many other temperature-related functions; prominent among these is preferred temperature. Thus, brown bullheads acclimated to 7 °C choose 16 °C water, while those acclimated to 25 °C choose water of about 29 °C. If fish remain in the temperature gradient (or other choice device), they will all eventually gravitate to the same temperature, termed the final thermal preferendum (Fry, 1947). Even for acclimation to extreme temperatures, this usually occurs in 1–2 days (Crawshaw, 1975b).

Like other vertebrates, fish respond to bacterial infections with a fever – a 2 or 3 °C increase in regulated body temperature. This response provides a major stimulus to the immune system: neutrophil migration, the secretion of antibacterial chemicals, and interferon production are all augmented. Survival rate is increased if fish are maintained at (or can select) the higher temperatures (Reynolds, Casterlin & Covert, 1976; Kluger, 1986).

The stress of hypoxia can lead to a dramatic decrease in preferred temperature (Wood, 1991). For most fish species, ambient P_{O_2} has a minimal effect on thermoregulatory behaviour until a rather low threshold value is reached (Fig. 2). Physiological advantages that accrue to hypoxic fish at lower temperatures include a decreased metabolic rate, plus increases in both gill extraction efficiency and haemoglobin–oxygen affinity (Schurmann & Steffensen, 1992).

Although the physiological benefits of a low body temperature are not so clear as for hypoxia, osmotic stress, like hypoxia, appears to lower the preferred temperature. The mummichog (*Fundulus heteroclitus*) prefers temperatures 3–6 °C cooler in fresh water than in its normal saltwater habitat. In contrast, the banded killifish (*F. diaphanus*), prefers temperatures 5–8 °C cooler in saltwater than it does in its normal freshwater habitat (Garside & Morrison, 1977).

The rate at which fish accumulate heavy metals and pesticide residues is higher at increased water temperatures (Reinert, Stone & Willford, 1974). For salmonids, exposure to low concentrations of various chlorinated hydrocarbons often leads to decreases in the preferred temperature. Higher concentrations of these and other toxins can increase, decrease or not affect the preferred temperature. Substances that increase the preferred temperature are more toxic at lower temperatures, while those that decrease the preferred temperature are more toxic at higher temperatures. Availability of a range of temperatures is thus important for the short-term survival of fish exposed to toxins (Peterson, 1973, 1976).

Genetic change and temperature selection

When fish are faced with a sustained increase in the thermal environment, one way for the species to survive could involve genetic selection. Fish exhibit intraspecific variability in preferred temperature, on which

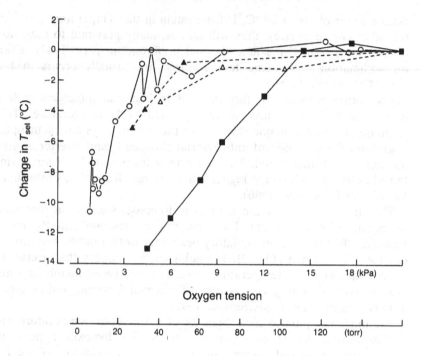

Fig. 2. Effects of oxygen tension on the temperatures selected by four different species of fish. O, Goldfish; ▲, Atlantic cod; △, rainbow trout; ■, Plains minnow. (Data from Bryan, Hill & Neill, 1984; Rausch and Crawshaw, 1990; Schurmann & Steffensen, 1992; Schurmann, Steffensen & Lomholt, 1991.)

selection could act. To the extent that such variation is heritable, the possibility of altering gene frequencies in a population under the pressure of selection exists.

Water temperature preferred by fish is a species characteristic; that is, similar-aged fish of the same species, with the same thermal history, will prefer similar temperatures in a temperature gradient. However, grouped around any species-specific mean preferred temperature, a range of individual differences in preference exists. A tendency to focus our attention on the mean value for a species and to address variation in a population only as 'noise' may predispose us to ignore the genetic richness reflected in the natural variability that exists among individuals (Bennett, 1987). Figure 3 shows the preferred temperature of three individual channel catfish (*Ictalurus punctatus*) whose simultaneous behaviour in individual temperature gradients was recorded

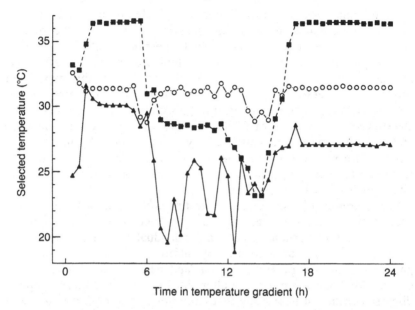

Fig. 3. The temperatures selected, over the same 24 hour period, for three channel catfish (*Ictalurus punctatus*) of the same age, kept under similar conditions.

for 24 hours. These fish were of similar age and were kept under identical thermal conditions. All three exhibit stable behaviour, yet their thermal preferences differ markedly. For these channel catfish the observed variability cannot be partitioned into genetic and environmental components. But if preferred temperature were partly heritable, some fishes may possess sufficient genetic variability to respond to selection pressure with a change in preferred temperature.

Variation in critical and chronic thermal maxima has occurred in genetically distinct populations of largemouth bass (*Micropterus salmoides*) (Fields *et al.*, 1987). However, these populations do not have different thermal preferenda (Koppelman, Whitt & Philipp, 1988). This raises interesting questions about the possibility of rapid genetic selection for altered thermal tolerances or preferenda which might be adaptive under sustained thermal increases. Selection experiments that bear directly on this question are lacking in fish. However, given the similarity in the thermoregulatory centres of the central nervous system in fish and mammals, experiments using the mouse might shed some light on this question.

Response by mice to selection for a maximal decrease (COLD mice) or a minimal decrease (HOT mice) in body temperature after an injection of ethanol suggests that genetic selection may be possible (Fig. 4). Crabbe *et al.* (1987) applied selection pressure to a single phenotypic characteristic only – rectal temperature change after drug injection. The change in phenotype selected for by this experiment produced an alteration in the response to ethanol by the central nervous system regulator of body temperature (O'Connor, Crawshaw & Crabbe, 1993). Over the course of 17 generations (about 3 years) the HOT and COLD mouse lines diverged by more than 3 °C in the body temperature response to the selection stimulus. The regulator of body temperature was quite plastic, and thus may be amenable to rapid alteration from selection pressure.

For a mouse, a lower body temperature after ethanol is an adaptive characteristic that prolongs survival (Gordon, 1993). The questions remain: whether variability between individual fish in preferred temperature (Fig. 3) is accompanied by differences in other physiological parameters and temperature optima; and whether the differences are genetic. Without accompanying alterations in the other thermal optima discussed earlier, it is unlikely that alteration of the preferred tempera-

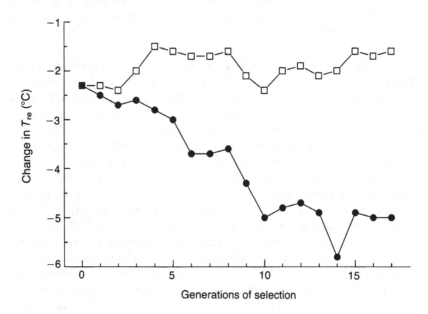

Fig. 4. The change in rectal temperature (T_{re}) of mice following 17 generations of selection for minimal (HOT, □) and maximal (COLD, ●) changes in body temperature following an injection of ethanol. (From Crabbe *et al.*, 1987.)

ture would prove beneficial. The time course for a simultaneous adjustment of all the various physiological and ecological thermal optima is not likely to occur in the time span allotted by global warming trends.

Striped bass and warming

The striped bass (*Morone saxatilis*) off the eastern coastline of the United States and in the inland waters of the southeast provide a useful field example of thermal behaviour related to climate change. This species has failed to thrive in some habitats into which it has been introduced, and many of its natural habitats have been degraded. The ensuing life history is based on summaries by Dudley, Mullis and Terrell (1977) and Coutant (1985, 1990).

Striped bass are native to the Atlantic coast of North America from the St Lawrence River to Florida, and also to the eastern half of the Gulf of Mexico. The geographic groups form separate races, although stocking has weakened the distinction. In southern rivers, the Gulf stocks fare better, and are presumably adapted to higher temperatures. Nevertheless, the temperatures preferred by the Atlantic and Gulf stocks are similar (Van Den Avyle & Evans, 1990). The Gulf striped bass is almost entirely confined to rivers.

Striped bass from the Atlantic Coast are anadromous, and northerly portions of the population migrate north along the coast in summer and south in winter. Spawning generally occurs in spring when water temperatures reach 15–19 °C. The pelagic larvae move downriver and shoreward as summer progresses. In the first year of life, striped bass remain in shallow fresh waters or estuaries. As development continues, young fish move into deeper portions of the estuaries and then to adjacent coastal areas over a few years' time. Adults enter particular areas within a river to spawn and have a weak homing tendency.

Temperature optima and preferences are evident throughout the life of the striped bass. These relationships (Fig. 5) have been summarized by Coutant (1987a). Growth during earlier life stages is optimal at the lower temperatures characteristic of spawning areas. Juveniles prefer 25 °C, which approximates their growth optima. This temperature also places them in shallow areas which provide food resources and separates them from the predatory adults. As they mature, preferred temperatures steadily decline, eventually to about 20 °C for large adults. Temperature preferences are presumably a major factor in directing the annual migrations up and down the Atlantic coast.

Striped bass have been introduced to southeastern reservoirs; in some cases they thrived, in others they fared poorly. Remote tracking of subadults in well oxygenated lakes confirmed that thermal preference

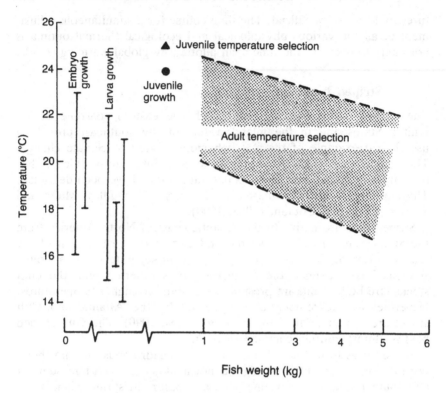

Fig. 5. Temperatures for optimal growth at various stages of development and temperature selection by juvenile and adult striped bass
(*Morone saxatilis*). (From Coutant, 1987a.)

is paramount in nature. Through spring and summer, fish adjusted
their depth so that 75% of the time they were in water at 20–24 °C.
When water temperatures were below 22 °C, they remained near the
surface (though below 1.5 m) in the warmest water (Coutant, 1980).
A major difficulty facing striped bass in certain reservoir habitats
involves locating temperatures below 25 °C and dissolved oxygen levels
above 2 mg l^{-1} throughout the year (Coutant, 1985). As summer
progresses, deoxygenation moves upward and excessively warm temperatures move downward. This greatly reduces, or in some cases
eliminates, tolerable habitat. Larger individuals are more susceptible
to warm, deoxygenated water, and summer die-offs typically involve
fish greater than 5 kg. When the interaction between thermal prefer-

ences and habitat degradation aggregates large numbers of adults, the fish become stressed and exhibit decreased reproductive success (Coutant, 1987b).

Adult striped bass often survive apparently untenable conditions, as occurred during the summer period illustrated in Fig. 6. Body temperatures lower the than the water temperature at the reservoir surface or bottom were maintained by remaining around cooler spring-fed areas (Van Den Avyle & Evans, 1990). During the remainder of the year the striped bass ranged widely throughout the entire reservoir.

A careful, thorough scenario describing the likely effects of continued warming on striped bass populations of the Atlantic coast was crafted by Coutant (1990). Historically, the centre of abundance for striped bass has been Chesapeake Bay, where habitat degradation has already occurred and is likely to intensify. Human activities already have greatly expanded anoxic areas and with continued heating, areas with little or no oxygen will begin to overlap regions warmer than 25 °C. This premier habitat for the striped bass could be lost. Annual northerly migrations coupled with feeble homing instincts might facilitate a northward shift in the population. Optimal temperatures, and population centres, would by then occur perhaps in mid New England. Areas

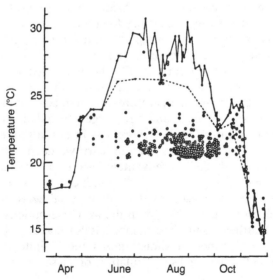

Fig. 6. Temperatures of a reservoir surface (solid line), the reservoir bottom (broken line), and the body cavity of striped bass (circles) at different times of the year. (From Van Den Avyle & Evans, 1990.)

around the Gulf of St Lawrence should fall within the productive range of striped bass, although juvenile development could be deficient because a projected increase in the cold Labrador current could decrease areas with water temperatures optimal for juveniles.

Northern pike and warming

The northern pike inhabits portions of North America predicted to sustain the most severe consequences from global warming. In the more southerly portions of its range, this fish already encounters problems similar to those described for striped bass. The northern pike avoids surface temperatures above 25 °C, which will become more frequent with continued warming. In waters that are or have become eutrophic, cooler bottom waters often become anoxic. This restricts the pike to the coolest water available that still meets minimum oxygen needs of about 3 mg l^{-1}. During summer, thermal habitat was constricted; the pike lost weight and showed little development. (Headrick & Carline, 1993).

Great plains fishes and warming

For native fishes in the southern Great Plains and southwestern portions of North America, northward migration to cooler waters is not an option (Matthews & Zimmerman, 1990). Wide, slow streams flow east, and slightly south; distances to satisfactory long-term retreats are great, and behavioural cues of cooler water point away from possible long-term emigration routes. The fishes use numerous small-scale thermoregulatory adjustments just to survive. Of the many species, mostly minnows, found in this area, none has a critical thermal maximum above 40 °C. Summer stream temperatures often reach 37 or 38 °C; these temperatures are stressful and are largely avoided by native species. Cooler temperatures (32–35 °C) are sought out in shaded pools on the outside curve of meanders and in narrow, deeper tributary creeks which usually are shaded. Disadvantages accrue as large numbers of fish aggregate into small areas; crowding-induced stress, increased parasitism, disease and decreased food availability all serve to make the minnows much more vulnerable. With the available options already fully subscribed, Matthews and Zimmerman (1990) make it clear that a 3–4 °C increase in summer maximum poses the definite threat of extinction for the entire native fish population of these areas.

Salmonids and warming

At higher latitudes and altitudes cooler water supports permanent salmonid populations. While all salmonids have been placed in the cold water guild (Magnuson *et al.*, 1979), differences in temperature preference occur among them. As with many families, age affects preference. For rainbow trout (*Oncorhynchus mykiss*) the preferred temperature decreases from 18 °C for 0.2 kg individuals to 13 °C for those of 5 kg (Spigarelli & Thommes, 1979). The rainbow trout is at the warmer end of salmonid thermal preferences. In natural waters, rainbow trout occasionally move into water above 20 °C (Horak & Tanner, 1964; Spigarelli & Thommes, 1979). In southern portions of the Columbia River system a subspecies, the redband trout, has adapted to extremes in stream flow, temperature and dissolved oxygen (Vinson & Levesque, 1994).

In a lake occupied by both rainbow and cutthroat (*Oncorhynchus clarki*) trout in western Washington (USA), cutthroats were generally found in cooler water (15 °C) than rainbows (18 °C), while a member of the warm water guild, the brown bullhead (*Ictalurus nebulosus*), occurred in the shallowest and warmest water (up to 23 °C). Trout distributions were not affected by regional differences in dissolved oxygen concentration from 3.9 to 8.6 mg l^{-1} (Stables & Thomas, 1992).

Where species overlap in streams, there are clear altitudinal effects which are based largely on temperature differences. The highest, coolest sites are most favourable for cutthroat trout. This species gives way to brook trout at lower elevations. Where brook trout and rainbow trout overlap, the brook trout (*Salvelinus fontinalis*) occupy the higher elevations (see De Staso & Rahel, 1994). Laboratory observations support this hierarchy: At 10 °C brook trout and cutthroat trout are fairly equal competitors, while at 20 °C brook trout are clearly dominant (De Staso & Rahel, 1994). All three species are dependent upon and utilize groundwater discharges, riparian shading and cool tributaries as refuges from stressful summer temperatures. Effects of global warming on stream trout habitats have been modelled, and discussed by Meissner (1990); it is probable that the predicted temperature changes will seriously decrease the lower altitude habitat for salmonids. Initial effects would decrease the zones where high altitude species such as the cutthroat are dominant. Meissner (1990) documented upstream migrations of brook trout during summer as warming of the lower reaches progressed. The importance of unimpeded movement through large extents of stream habitat is thus critical for short-term survival. The decrease in available habitat, however, can still be expected to produce the stresses discussed earlier.

Brown trout (*Salmo trutta*) prefer cooler temperatures of about 12 °C in the laboratory (Reynolds & Casterlin, 1979). Ultrasonic tracking in Lake Constance, Germany from spring to autumn revealed that the brown trout were generally found from 11 to 15 °C. This typically involved residence above the thermocline at a depth of 8–16 m. Under the homothermic conditions of winter, the brown trout swam near the surface, often at a depth of about 10 cm (Schulz & Berg, 1992). If levels of dissolved oxygen remain adequate, lacustrine salmonids are likely to respond to climatic warming by following the downward movement of preferred temperature zones. In some cases the downward migration will affect availability of food and shelter. Acceptable habitat will then become more scarce, and increased overlap between presently separated species and growth stages seems inevitable.

Unfettered stream movements may be particularly important for large brown trout; these fish are of singular importance for perpetuating the species. Clapp, Clark and Diana (1990) used telemetry to track such trout in a Michigan (USA) stream. The range of movement was considerable, with one fish covering over 33 km. Almost all fish had different summer and winter ranges, with an average spring upstream movement of 10 km to slower, deeper and cooler water. The ability to spread into acceptable thermal environments as the need arises is an important behavioural adaption for increasing effective habitat size.

Upstream spawning runs are decisive for the survival of anadromous salmonids. While olfactory stimuli provide the primary cues for locating spawning tributaries (Hasler & Scholz, 1983), temperature can be an important trigger for initiating key physiological and behavioural changes that make migration possible, and may be pivotal in determining whether a particular migration is successful. Increased fresh water temperatures usually reduce survival of adult salmon, through both an increased incidence of disease (Chatters et al., 1991) and an increased utilization of energy stores (Berman & Quinn, 1991).

Berman and Quinn (1991) documented complex behavioural thermoregulation by homing spring chinook salmon (Oncorhynchus tshawytscha) in the Yakima River of central Washington (USA). Transmitters placed in the stomach were utilized to locate the fish, and obtain a reading of deep body temperatures. Recording began in mid-June and was continued for up to 4 months. These salmon arrived in the vicinity of their spawning areas many weeks before spawning occurred. They travelled to holding areas in the Yakima River Canyon, where they were able to find cold water refuges. Lower temperatures occurred here owing to shading from canyon walls, outcroppings, and vegetation, groundwater seeps, and input from cooler tributaries. As spawning time approached, salmon began their final migration to nearby spawning beds. In about half of the cases, this involved movement downstream.

Deep body temperature measurements highlighted the thermal heterogeneity in the river and the ability of the salmon to find pockets of cooler water. Body temperature was lower than mid-river temperature in all 197 measurements on 19 different salmon. The mean difference between ambient river temperatures and deep body temperatures was a constant 2.5 °C for mid-river temperatures from 9 to 19 °C. If the preferred temperature of the salmon were above 9 °C, mid-river temperature and body temperature should be similar at and below that temperature. This indicates that either the preferred temperature of these salmon was below 9 °C or that, during upstream migration, and while holding in thermal refuges, the salmon maintained the lowest body temperature possible. The salmon already were making maximum use of their behavioural strategies to avoid high body temperatures. Any increase in stream temperature probably would be added to the mean body temperature of the fish, thus decreasing reproductive success.

During their offshore migrations, salmon also avoid warmer water when possible. As mature chum salmon (*Oncorhynchus keta*) approach natal spawning streams by moving along the coast of Honshu, Japan, they encounter warmer (12–20 °C) water of the Tsugaru current. The water of this current is well above the normally frequented sea surface temperatures of 3–11 °C. Analysis of trawl catches where the currents intersect indicates that the fish descend to 200–350 m where the water is 11 °C or cooler. The salmon travel in this cold water until they turn shoreward and move through the warm surface layer during their final approach to the streams (Ueno, 1992). Telemetry data indicate that sockeye salmon (*Oncorhynchus nerka*) also avoid warm surface water during travels adjacent to the mainland of British Columbia, Canada (Quinn, Terhart & Groot, 1989). Behavioural adjustments during offshore migrations are definitely important, and unlike the riverine phase, there is still latitude for further adjustment by moving deeper or by remaining for less time in the warmer surface waters.

Suggestions for future research

Approaches to the effects of global warming on fishes would do well to take into account how a combination of habitat alteration and continued warming will affect the myriad behaviours that evolved to ensure occupancy of an appropriate environment. The warming may render these behaviours ineffective, or even deleterious. Coutant (1987a) has made this point quite clearly in regard to temperature, and his article is an excellent point of departure. As he notes, attention should be focused on the evaluation of thermal preference in determining habitat suitability, on the maintenance and creation of optimal temperatures and thermal structures, and on how to best avoid further degradation of preferred thermal habitat.

Further investigation of how fishes orient in natural environments, and utilize thermal heterogeneity to increase physiological fitness is warranted. Long-term, remote tracking studies in both natural and perturbed environments are of clear value and should be emphasized in future work. In coastal areas, the rate of migration in response to thermal shifts, the modes of migration utilized, and thermal tolerances of key species should be documented. Special attention should be given to species which move slowly or are subject to potential thermal 'traps' which might bar movement to more suitable areas.

Developing a greater understanding of the cues that compete in determining the behaviour of a fish in a particular environment is important. Shelter, dissolved oxygen, depth, light levels, and many other factors compete with temperature in determining where a fish is likely to be. Understanding the nature of these determinants of behaviour will increase our ability to avoid situations where fish are drawn away from viable thermal habitats.

The importance of a readily available heterothermal environment needs further evaluation and documentation. Temperatures both above and below those normally selected can be critical for survival. Fish infected with bacteria select warmer water; this increases the activity and effectiveness of the immune system. Some pesticides are apparently also better survived in warmer water. Cooler water, on the other hand, increases fishes' ability to survive short-term exposures to many physiological stressors such as water containing low oxygen levels or toxic chemicals. Information on alterations in selected temperature under stressful conditions is minimal; more work in this area would greatly aid in making wise choices about developing and protecting thermal resources.

Work to date has overwhelmingly documented the importance of thermal refuges for escaping excessively warm environments. This information could well be expanded and extended to further chronicle the presence of refuges and the extent of their utilization by fishes. A deeper understanding of the stress that accompanies refuge utilization is needed. The nature and extent of stress is partially dependent upon the species involved. Developing an understanding, for a number of species, of how to decrease the adverse side effects of refuge utilization as well as how to increase the thermal effectiveness of refuges would be invaluable. Methods for protecting large aggregations of adult fish at thermal refuges should also be developed.

Finally, the genetic basis for thermal preferences and their importance in the evolution of thermal adaptation to environmental conditions should be investigated. Why do thermal preferences appear to remain stable in isolated races of fish that have developed different thermal maxima? Are the individual differences in thermal preference within a particular species correlated with differences in thermal optima for physiological system function or with thermal maxima?

Acknowledgments

Mr Richard Rausch and Dr Nancy Bowers are thanked for their useful suggestions regarding the manuscript. Ms Helen Wallace and Ms Robin Jensen are thanked for their astute editing, as well as for keeping things in the lab going while the authors dropped out to write this chapter.

References

Alcock, J. (1993). *Animal Behavior: An Evolutionary Approach*, 5th edn. Sunderland, Mass. Sinauer Associates.

Bardach, J. E. (1956). The sensitivity of the goldfish (*Carassius auratus*) to point heat stimulation. *American Naturalist*, **90**, 309–17.

Bardach, J. E. & Bjorklund, R. G. (1957). The temperature sensitivity of some American freshwater fishes. *American Naturalist*, **91**, 233–51.

Bennett, A. F. (1987). Interindividual variability: an underutilized resource. In *New Directions in Ecological Physiology*, ed. M. E. Feder, A. F. Bennett, W. W. Burggren & R. B. Huey, pp. 147–69. Cambridge: Cambridge University Press.

Berman, C. H. & Quinn, T. P. (1991). Behavioral thermoregulation and homing by spring chinook salmon, *Oncorhynchus tshawytscha* (Walbaum), in the Yakima River. *Journal of Fish Biology*, **39**, 301–12.

Boulant, J. A., Curras, M. C. & Dean, J. B. (1989). Neurophysiological aspects of thermoregulation. In *Advances in Comparative and Environmental Physiology*, Vol. 4, ed. L. C. H. Wang, pp. 117–60. Berlin: Springer-Verlag.

Brett, J. R. (1971) Energetic responses of salmon to temperature. A study of some thermal relations in the physiology and freshwater ecology of sockeye salmon (*Oncorhynchus nerka*). *American Zoologist*, **11**, 99–113.

Brück, K. & Hinckel, P. (1990). Thermoafferent networks and their adaptive modifications. In *Thermoregulation: Physiology and Biochemistry*, ed. E. Schönbaum & P. Lomax, pp. 129–52. New York: Pergamon Press.

Bryan, J. D., Hill, L. G. & Neill, W. H. (1984). Interdependence of acute temperature preference and respiration in the plains minnow. *Transactions of the American Fisheries Society*, 113, 557–62.

Chatters, J. C., Neitzel, D. A., Scott, M. J. & Shankle, S. A. (1991). Potential impacts of global climate change on Pacific Northwest spring chinook salmon (*Oncorhynchus tshawytscha*): an exploratory case study. *The Northwest Environmental Journal*, 7, 71–92.

Christie, G. C. & Regier, H. A. (1988). Measures of optimal thermal habitat and their relationship to yields for four commercial fish species. *Canadian Journal of Fisheries and Aquatic Sciences*, 45, 301–14.

Clapp, D. F., Clark, R. D. Jr & Diana, J. S. (1990). Range, activity, and habitat of large, free-ranging brown trout in a Michigan stream. *Transactions of the American Fisheries Society*, 119, 1022–34.

Coutant, C. C. (1980). Temperatures occupied by ten ultrasonic-tagged striped bass in fresh water lakes. *Transactions of the American Fisheries Society*, 109, 195–202.

Coutant, C. C. (1985). Striped bass, temperature, and dissolved oxygen: a speculative hypothesis for environmental risk. *Transactions of the American Fisheries Society*, 114, 31–61.

Coutant, C. C. (1987a). Thermal preference: when does an asset become a liability? *Environmental Biology of Fishes*, 18, 161–72.

Coutant, C. C. (1987b). Poor reproductive success of striped bass from a reservoir with reduced summer habitat. *Transactions of the American Fisheries Society*, 116, 154–60.

Coutant, C. C. (1990). Temperature–oxygen habitat for fresh water and coastal striped bass in a changing climate. *Transactions of the American Fisheries Society*, 119, 240–53.

Crabbe, J. C., Kosobud, A., Tam, B. R., Young, E. R. & Deutsch, C. M. (1987). Genetic selection of mouse lines sensitive (COLD) and resistant (HOT) to acute ethanol hypothermia. *Alcohol and Drug Research*, 7, 163–74.

Crawshaw, L. I. (1975a). Twenty-four hour records of body temperature and activity in bluegill sunfish (*Lepomis macrochirus*) and brown bullheads (*Ictalurus nebulosus*). *Comparative Biochemistry and Physiology*, 51A, 11–14.

Crawshaw, L. I. (1975b). Attainment of the final thermal preferendum in brown bullheads acclimated to different temperatures. *Comparative Biochemistry and Physiology*, 52A, 171–3.

Crawshaw, L. I. (1980). Temperature regulation in vertebrates. *Annual Review of Physiology*, 42, 473–91.

Crawshaw, L. I. & Hammel, H. T. (1973). Behavioral temperature regulation in the California horn shark, *Heterodontus francisci. Brain, Behavior and Evolution*, **7**, 447–52.

Crawshaw, L. I. & Hammel, H. T. (1974). Behavioral temperature regulation of internal temperature in the brown bullhead, *Ictalurus nebulosus. Comparative Biochemistry and Physiology*, **47A**, 51–60.

Crawshaw, L. I. O'Connor, C. S. & Wollmuth, L. P. (1992). Ethanol and the neurobiology of temperature regulation. In *Alcohol and Neurobiology: Brain Development and Hormone Regulation*, ed. R. R. Watson, pp. 341–60. Boca Raton, Fla: CRC Press.

Crawshaw, L. I. Wollmuth, L. P., O'Connor, C. S., Rausch, R. N. & Simpson, L. (1990). Body temperature regulation in vertebrates: comparative aspects and neuronal elements. In *Thermoregulation: Physiology and Biochemistry*, ed. E. Schönbaum and P. Lomax, pp. 209–20. New York: Pergamon Press.

De Staso J. III, & Rahel, F. J. (1994). Influence of water temperature on interactions between juvenile Colorado River cutthroat trout and brook trout in a laboratory stream. *Transactions of the American Fisheries Society*, **123**, 289–97.

Dudley, R. G., Mullis, A. W. & Terrell, J. W. (1977). Movements of adult striped bass (*Morone saxatilis*) in the Savannah River, Georgia. *Transactions of the American Fisheries Society*, **106**, 314–22.

Fields, R., Lowe, S. S., Kaminski, C., Whitt, G. S & Philipp, D. P. (1987). Critical and chronic thermal maxima of northern and Florida largemouth bass and their reciprocal F_1 and F_2 hybrids. *Transactions of the American Fisheries Society*, **116**, 856–63.

Fry, F. E. J. (1947). Effects of the environment on animal activity. *Publications of the Ontario Fisheries Research Laboratory*, **68**, 1–62.

Garside, E. T. & Morrison, G. C. (1977). Thermal preferences of mummichog, *Fundulus heteroclitus* L., and banded killifish, *F. diaphanus* (LeSueur), (Cyprinodontidae) in relation to thermal acclimation and salinity. *Canadian Journal of Zoology*, **55**, 1190–4.

Gibson, J. R. (1978). The behavior of juvenile Atlantic salmon (*Salmo salar*) and brook trout (*Salvelinus fontinalis*) with regard to temperature and to water velocity. *Transactions of the American Fisheries Society*, **107**, 703–12.

Gordon, C. J. (1993). *Temperature Regulation in Laboratory Rodents*. New York: Cambridge University Press.

Hammel, H. T. (1968). Regulation of internal body temperature. *Annual Review of Physiology*, **30**, 641–710.

Hasler, A. D. & Scholz, A. D. (1983). *Olfactory Imprinting and Homing in Salmon*. Berlin: Springer–Verlag.

Headrick, M. R. & Carline, R. F. (1993). Restricted summer habitat and growth of northern pike in two southern Ohio impoundments. *Transactions of the American Fisheries Society*, **122**, 228–36.

Heggenes, J., Northcote, T. G. & Armin, P. (1991) Seasonal habitat selection and preferences by cutthroat trout (*Oncorhynchus clarki*) in a small, coastal stream. *Canadian Journal of Fisheries and Aquatic Sciences*, **48**, 1364–70.

Horak, D. L. & Tanner, H. A. (1964). The use of vertical gillnets in studying fish depth distribution in Horsetooth Reservoir, Colorado. *Transactions of the American Fisheries Society*, **93**, 137–45.

Jobling, M. (1981). Temperature tolerance and the final preferendum – rapid methods for the assessment of optimum growth temperatures. *Journal of Fish Biology*, **19**, 439–55.

Kardong, K. V. (1995). *Vertebrates – Comparative Anatomy, Function, Evolution*. Dubuque, Iowa: William C. Brown.

Kemp, D. D. (1994). *Global Environmental Issues: A Climatological Approach*, 2nd edn. New York: Routledge.

Kluger, M. J. (1986). Is fever beneficial? *Yale Journal of Biology and Medicine*, **59**, 89–95.

Koppelman, J. B., Whitt, G. S., & Philipp, D. P. (1988). Thermal preferenda of northern, Florida, and reciprocal F_1 hybrid largemouth bass. *Transactions of the American Fisheries Society*, **117**, 238–44.

Leopold, A. (1949). *A Sand County Almanac*. New York: Oxford University Press (Ballantine Paperback edition).

Magnuson, J. J., Crowder, L. B. & Medvick, P. A. (1979). Temperature as an ecological resource. *American Zoologist*, **19**, 331–43.

Matthews, W. J. & Hill, L. G. (1979). Influence of physico-chemical factors on habitat selection by red shiners, *Notropis lutrensis*, (Pisces: Cyprinidae). *Copeia, 1979*, 70–81.

Matthews, W. J. & Zimmerman, E. G. (1990). Potential effects of global warming on native fishes of the southern Great Plains and the Southwest. *Fisheries*, **15**, 26–32.

Medvick, P. A., Magnuson, J. J. & Sharr, S. (1981). Behavioral thermoregulation and social interactions of bluegills, *Lepomis macrochirus*. *Copeia, 1981*, 9–13.

Meissner, J. K. (1990). Potential loss of thermal habitat for brook trout, due to climatic warming, in two southern Ontario streams. *Transactions of the American Fisheries Society*, **119**, 282–91.

Morrissey, J. F. & Gruber, S. H. (1993). Habitat selection by juvenile lemon sharks, *Negaprion brevirostris*. *Environmental Biology of Fishes*, **38**, 311–19.

Murawski, S. A. (1993). Climate change and marine fish distributions: forecasting from historical analogy. *Transactions of the American Fisheries Society*, **122**, 647–58.

Norris, K. S. (1963). The functions of temperature in the ecology of the percoid fish *Girella nigricans* (Ayres). *Ecological Monographs*, **33**, 23–62.

O'Connor, C. S., Crawshaw, L. I. & Crabbe, J. C. (1993). Genetic selection alters thermoregulatory response to ethanol. *Pharmacology, Biochemistry and Behaviour*, **44**, 501–8.

Olla, B. L. & Davis, M. W. (1990). Behavioral responses of juvenile walleye pollock *Theragra chalcogramma* Pallas to light, thermoclines and food: possible role in vertical distribution. *Journal of Experimental Marine Biology and Ecology*, **135**, 59–68.

Peterson, R. H. (1973). Temperature selection of Atlantic salmon (*Salmo salar*) and brook trout (*Salvelinus fontinalis*) as influenced by various chlorinated hydrocarbons. *Journal of the Fisheries Research Board of Canada*, **30**, 1091–7.

Peterson, R. H. (1976). Temperature selection of juvenile Atlantic salmon (*Salmo salar*) as influenced by various toxic substances. *Journal of the Fisheries Research Board of Canada*, **33**, 1722–30.

Quinn, T. P., Terhart, B. A. & Groot, C. (1989). Migratory orientation and vertical movements of homing adult sockeye salmon, *Oncorhynchus nerka*, in coastal waters. *Animal Behavior*, **37**, 587–99.

Rausch, R. N. & Crawshaw, L. I. (1990). Anoxia and hypoxia lower the selected temperature of goldfish. *The Physiologist*, **33**, A40.

Reinberg, A. & Smolensky, M. (1990). Chronobiology and thermoregulation. In *Thermoregulation: Physiology and Biochemistry*, ed. E. Schönbaum & P. Lomax, pp.61–100. New York: Pergamon Press.

Reinert, R. E., Stone, L. J. & Willford, W. A. (1974). Effect of temperature on accumulation of methylmercuric chloride and p,p' DDT by rainbow trout (*Salmo gairdneri*). (*Journal of the Fisheries Research Board of Canada*, **31**, 1649–52.

Reynolds, W. W. & Casterlin, M. E. (1976). Thermal preferenda and behavioral thermoregulation in three centrarchid fishes. In *Thermal Ecology II*, ed. G. W. Esch & R. W. McFarlane, pp.185–90. Springfield, Va: US National Technical Information Service.

Reynolds, W. W. & Casterlin, M. E. (1979). Thermoregulatory behavior of brown trout, *Salmo trutta*. *Hydrobiologica*, **62**, 79–80.

Reynolds, W. W., Casterlin, M. E. & Covert, J. B. (1976). Behavioural fever in teleost fishes. *Nature*, **259**, 412.

Reynolds, W. W., Casterlin, M. E., Matthey, J. K., Millington, S. T. & Ostrowski, A. C. (1978). Diel patterns of preferred temperature and locomotor activity in the goldfish, *Carassius auratus*. *Comparative Biochemistry and Physiology*, **59A**, 225–7.

Rozin, P. N. & Mayer, J. (1961). Thermal reinforcement and thermal behavior in the goldfish, *Carassius auratus*. *Science*, **134**, 942–3.

Schulz, U. & Berg, R. (1992). Movements of ultrasonically tagged brown trout (*Salmo trutta* L.) in Lake Constance. *Journal of Fish Biology*, **40**, 909–17.

Schurmann, H. & Christiansen, J. S. (1994). Behavioral thermoregulation and swimming activity of two arctic teleosts (subfamily Gadinae) – the polar cod (*Boreogadus saida*) and the navaga (*Eleqinus navaga*). *Journal of Thermal Biology*, **19**, 207–12.

Schurmann, H. & Steffensen, J. F. (1992). Lethal oxygen levels at different temperatures and the preferred temperature during hypoxia of the Atlantic cod, *Gadus morhua* L. *Journal of Fish Biology*, **41**, 927–34.

Schurmann, H., Steffensen, J. F. & Lomholt, J. P. (1991). The influence of hypoxia on the preferred temperature of rainbow trout *Oncorhynchus mykiss*. *Journal of Experimental Biology*, **157**, 75–86.

Sogard, S. M. & Olla, B. L. (1993). Effects of light, thermoclines and predator presence on vertical distribution and behavioral interactions of juvenile walleye pollock, *Theragra chalcogramma* Pallas. *Journal of Experimental Marine Biology and Ecology*, **167**, 179–95.

Spigarelli, S. A. & Thommes, M. M. (1979). Temperature selection and estimated thermal acclimation by rainbow trout in a thermal plume. *Journal of the Fisheries Research Board of Canada*, **36**, 366–76.

Stables, T. B. & Thomas, G. L. (1992). Acoustic measurement of trout distributions in Spada Lake, Washington, using stationary transducers. *Journal of Fish Biology*, **40**, 191–203.

Ueno, Y. (1992). Deepwater migrations of chum salmon (*Oncorhynchus keta*) along the Pacific Coast of Northern Japan. *Canadian Journal of Fisheries and Aquatic Sciences*, **49**, 2307–12.

Van Den Avyle, M. J. & Evans, J. W. (1990). Temperature selection by striped bass in a Gulf of Mexico coastal river system. *North American Journal of Fisheries Management*, **10**, 58–66.

Vinson, M. & Levesque, S. (1994). Redband trout response to hypoxia in a natural environment. *Great Basin Naturalist*, **54**, 150–5.

Webb, P. W. (1978). Temperature effects on acceleration of rainbow trout, *Salmo gairdneri*. *Journal of the Fisheries Research Board of Canada*, **35**, 1417–22.

Wood, S. C. (1991). Interactions between hypoxia and hypothermia. *Annual Review of Physiology*, **53**, 71–85.

J.J. MAGNUSON and B.T. DESTASIO

Thermal niche of fishes and global warming

Introduction

In 1979, we proposed the idea of a thermal niche for fishes as analogous to niches for other resources such as food type and size (Magnuson, Crowder & Medvick, 1979). In our view, competition occurs for space with the appropriate thermal properties in a manner similar to competition for places suitable for refuge or spawning. We concluded that 'fish do compete for thermal resources and that considerations of temperature as a resource are generally consistent with the characteristics of food as a resource.'

We defined the thermal niche of fishes as the preferred temperature ± 2 °C or ± 5 °C. This definition was derived from laboratory gradient studies which showed that fish spend two-thirds of their time within 2 °C and all of their time within 5 °C of their preferred temperature (Magnuson et al., 1979). A large number of, but not all, performance optima for individual species occur within these ranges (Magnuson et al., 1979; Jobling, 1981). In North America, freshwater fish have been grouped into three thermal guilds – coldwater, coolwater and warmwater – based on their temperature preference (Hokanson, 1977; Magnuson et al., 1979). Other criteria for the thermal niche, based on temperature acting as a controlling or a lethal factor, enrich the concept (see Fry paradigm below).

Global warming would be expected to alter the temperatures of lakes and streams and thus favour some species over others in relation to present conditions. This premise has led us to examine the potential effects of climate change on the thermal niche space of fishes for the Laurentian Great Lakes (Magnuson, Meisner & Hill, 1990) and small inland lakes of Wisconsin (DeStasio et al., 1996). Climate, lake thermal structure, and resulting size of habitat within the thermal niche were simulated for a doubling of greenhouse gases and compared with

Society for Experimental Biology Seminar Series 61: *Global Warming: Implications for freshwater and marine fish*, ed. C. M. Wood & D. G. McDonald. © Cambridge University Press 1996, pp. 377–408.

simulations for the present climate. Our general conclusion was somewhat counter-intuitive: global warming would increase the size of habitat
within the thermal niche not only for warmwater fish but also for
coolwater and coldwater fish for deeper lakes at temperate latitudes.

The purposes of this chapter are to: (i) briefly review the concept
of thermal niche with respect to Fry's paradigm of fish physiological
ecology and Hutchinson's N-dimensional niche; (ii) evaluate the influence of environmental constraints that limit the fundamental niche to
an actual or realized thermal niche; and (iii) consider the role of
thermal niche in the potential effects of climate change on fish in small
and large lakes.

Elements of the thermal niche

Fry paradigm

The late F.E.J. Fry at the University of Toronto, Canada, classified
physical and chemical features of fish habitat as lethal, controlling, and
directive factors on the basis of how they influenced fish (Fry, 1947,
1971; see also Kerr, 1990; Neill *et al.*, 1994). Temperature could, for
example, kill the fish, control physiological rates, and influence habitat
preference (direct its location in a gradient). Some factors he also
considered as limiting in supply, like oxygen, or capable of masking
the influence of other factors. The range of conditions specified by
these definitions is narrowest when defined as a directive factor, and
progressively broader when defined as a controlling factor or a lethal
factor (Fig. 1). These ranges can be used quantitatively to delimit the
niche of a species or species life stage along a physical/chemical gradient
or niche dimension.

Hutchinson's N-dimensional niche

The late G.E. Hutchinson of Yale University, USA, thought of the
niche as a multidimensional world in biotic and abiotic gradients that
described the environmental conditions in which the organism would
do better than other species (Hutchinson, 1978). His earliest example
was for temperature and food particle size as niche axes for zooplankton
(Hutchinson, 1957, 1967). He also described areas of niche overlap and
potential interaction between species in these N-dimensional worlds. In
his view, competition was more likely for biotic axes than for abiotic
axes. The non-interactive (fundamental) niche was without interaction
with other species or based on that factor acting alone; the interactive
(realized) niche was what the fish actually occupied in the field interacting with other species and factors.

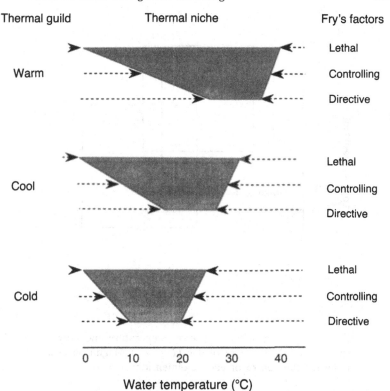

Fig. 1. Diagram of the thermal niches in terms of Fry's lethal, controlling, and directive factors for warmwater, coolwater, and coldwater fish. They approximate the niche of a largemouth bass (*Micropterus salmoides*), a yellow perch (*Perca flavescens*) and a cisco (*Coregonus artedii*), respectively.

Niches can be visualized graphically for 2 dimensions (Fig. 2) and 3 dimensions (Fig. 3). The 2-dimensional example (Fig. 2) quantifies the thermal axis as a directive factor and the prey size axis from optimum foraging data for three fishes. The 3-dimensional example is based on lethal effects of temperature, dissolved oxygen, and salinity on egg viability of Pacific cod (*Gadus macrocephalus*). Many major distribution patterns and differences in distribution patterns of fishes in time and space can be visualized well with only two or three niche axes.

Dealing with more than three environmental factors simultaneously soon becomes impossible graphically but successful attempts have been made with multivariate analyses of distributions of fishes across aquatic

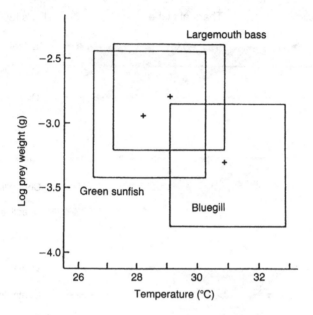

Fig. 2. Two-dimensional niche of temperature and prey size for three species of fish at a size of 4 g (Fig. 4 in Magnuson *et al.*, 1979). + indicates the centre of each 2-dimensional niche.

systems that differ in abiotic and biotic characteristics (e.g. Tonn & Magnuson, 1982; Rahel, 1984).

The realized thermal niche

The fundamental thermal niche can be used to describe optimum habitat with respect to temperature for particular fish during the growing season. It also can be used to indicate potential interactions with other species by explicitly evaluating overlap in thermal space. To the extent that organisms are able to occupy their fundamental thermal niche, they will be better able to escape predation or search for and capture prey, to grow faster, and to produce offspring sooner and in higher numbers than those that are less able to occupy their fundamental thermal niche.

However, fish live in a multivariate and changing world in which optimum conditions with respect to any single factor may not be achieved. The primary reasons that a fish may not occupy its funda-

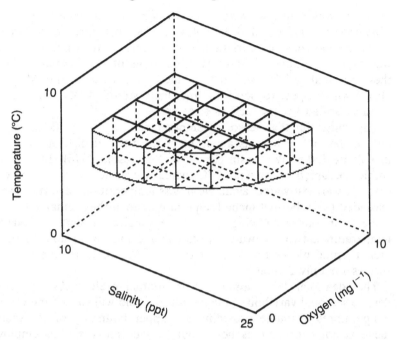

Fig. 3. Three-dimensional niche in temperature, salinity, and oxygen for Pacific cod (*Gadus macrocephalus*) based on the environmental space in which 90% of the eggs hatch. (Adapted from Alderdice & Forrester, 1971.)

mental thermal niche are to be found in (i) the actual physical availability or existence of those temperatures in the ecosystem it inhabits, (ii) interactions among temperature and other abiotic and biotic niche axes, and (iii) intra- and interspecific competition with other organisms for thermal resources. The realized thermal niche represents the more constrained temperature distributions of fishes after these realities are taken into account.

Seasonal constraints

Seasonal variation in the realized thermal niche results from an opposition between the fish's thermal behaviour and the constraints of the physical world. Recognizing the seasonal change in temperature occupation by temperate freshwater fish, Hokanson (1977) referred to the warmwater fish as *temperate eurytherms*, the coolwater fish as *temperate mesotherms*, and the coldwater fish as *temperate stenotherms*, based on the seasonal variation in temperatures to which each group was exposed. Expected seasonal realizations of the thermal niche for these thermal guilds were diagrammed by MacLean and Magnuson (1977) with respect to depth for thermally stratified lakes and latitude for unstratified waters such as streams.

The constraint that seasonal temperature structure of lakes imposes on the thermal niche was apparent in the vertical distribution of fish in Pallette Lake, Wisconsin (Engel & Magnuson, 1976). In winter all species occupied water near 4 °C, while in warmer seasons they occupied the warmest waters available unless temperatures of surface waters exceeded their thermal niche temperatures; then they remained at their niche temperatures. Owing to these seasonal constraints on available temperatures, two coldwater species in the lake were able to occupy their thermal niches for only 5 months of the year and the coolwater species for only 2 months.

The idea that fish maintain their thermal preferences year round despite seasonal variability in temperatures, and will choose the warmest temperatures available providing the upper boundary of the thermal niche is not exceeded, is borne out by laboratory studies employing thermal gradients. The bluegill (*Lepomis macrochirus*) preferred warm temperatures, near 30 °C, regardless of photoperiod or acclimation temperatures (Magnuson & Beitinger, 1978). Other species also prefer higher temperatures than occur in winter: bitterling (*Rhodeus sericeus*) (Zahn, 1963), yellow perch (*Perca flavescens*) (McCauley & Huggins, 1979) and larval sea lamprey (*Petromyzon marinus*) (Holmes & Lin, 1994). In winter, such fish would occupy the warmest waters available. A more precise behavioural mechanism seems unnecessary for closed water bodies where winter temperatures cannot be avoided by emigration. However, seasonal changes in temperature preference that can influence thermal distribution in winter do occur in some fish, for example juvenile Atlantic cod (*Gadus morhua*) (Clark & Green, 1991).

Interaction constraints among niche axes

The ability of fish to occupy their fundamental thermal niche depends very much on the suitability of other niche axes. Simply put, if a fish is to occupy its thermal niche, those temperatures must also co-occur within the dissolved oxygen niche (Kramer, 1987), the pH niche, etc. Such interactions can be viewed as environmental constraints in N-dimensional space that can be in opposition to the fishes' responses to temperature alone. For example, anoxic or low oxygen conditions in the deep, colder waters of eutrophic lakes and estuaries can prevent some fish from occupying their thermal niche (Rudstam & Magnuson, 1985; Coutant, 1987, 1990). It is the coincidence of suitable conditions along multiple niche axes that makes it possible for a species to occupy a particular habitat.

In the case of temperature and oxygen, and other pairs of axes such as conductivity and pH, or turbidity and prey density, direct physical/chemical relationships between the environmental factors will further influence the suitability of N-dimensional niches. In addition, fish response to one factor may be influenced by its response to the other. Warmer waters contain less oxygen than cooler waters and fish have higher metabolic demands at warmer temperatures. Thus, oxygen is more critical at warmer than at cooler temperatures. Not surprisingly, fish capable of aerial respiration are more likely to switch to air breathing in warmer than cooler waters (Johansen, Hanson & Lenfant, 1970). In an analogous case, pH is less buffered from change, and effects of low pH on fish eggs and larvae are more severe, in waters of lower ionic strength (Chulakasem, Nelson & Magnuson, 1989). Interaction between axes is also evident when zooplanktivores that locate their prey visually have a smaller search field in more turbid waters (O'Brien, 1987). Ability to locate prey in the warmer surface waters could be reduced by phytoplankton blooms. Thus, the realized availability of prey in the warmest surface waters could be less than in deeper, clearer waters for fish with good scotoptic vision. These examples help explain why a fish in an N-dimensional world may not occupy its thermal niche or may shift to a suboptimal temperature.

The distribution of prey also influences how a fish distributes itself with respect to its thermal niche (Crowder & Magnuson, 1983). Both prey abundance and occupied temperatures are linked closely to net energy gain and growth in fish. Behavioural thermoregulation maximizes growth through regulation of net energy gain given some level of ingestion. Optimal foraging predicts levels of energy intake given certain abundances and distributions of prey. Bioenergetics provides a mechanism for joint evaluating of food intake and food utilization. In simulations, the value of a food patch was determined not only by prey density, but also by the temperature of the patch. Warmer patches were, in a sense, more expensive and higher prey densities were required for a fish to achieve a positive energy budget.

An interesting question is whether the thermal niche changes systematically with prey abundance. At lower levels of prey abundance, net energy gain would be greater if the fish shifted to cooler temperatures;

at higher levels of prey abundance the fish could shift to more expensive, warmer habitats, and still achieve a greater net gain or growth (Spigarelli *et al.*, 1974). Brandt (1993) evaluated food and temperature simultaneously through spatially explicit modelling of fish growth across thermal fronts (or thermoclines). The modelling results were consistent with these ideas: when food was distributed uniformly across the front, growth was highest at the optimum temperatures; lower temperatures produced better growth when food was low in abundance, and high food availability increased growth rate potential at the front.

Observational and experimental evidence for a level of functional interaction between prey abundance and behavioural thermal regulation is equivocal, but several field and laboratory studies suggest that it may occur. Vertical migration of several species appears to exemplify a set of behaviours that can optimize net energy intake for some species by feeding at suboptimal temperatures (too warm or too cold) where prey are abundant, and then either minimizing energy loss from metabolism by spending the non-feeding hours at cooler temperatures (Brett, 1971; Clark & Green, 1991), or maximizing digestion, food processing and growth by spending the non-feeding hours at warmer temperatures (Wurtzbaugh & Neverman, 1988). Rudstam and Magnuson (1985) observed that in eutrophic Lake Mendota with anoxic deep water in summer, a coldwater fish, the cisco (*Coregonus artedii*), occupied the warm epilimnetic waters and achieved large body sizes. These waters were more than 10 °C warmer than the middle temperature of their thermal niche. In oligotrophic Wisconsin lakes with adequate oxygen in deep water, the same species occupied the deeper, colder waters. In these unproductive lakes cisco did not reach large body size.

Results of experimental studies are somewhat contradictory. Several studies (Javaid & Anderson, 1967; Stuntz & Magnuson, 1976; Reynolds & Casterlin, 1979; Mac, 1985; Boltz, Siemien & Stauffer, 1987; Wildhaber & Crowder, 1990; Morgan, 1993) suggest that when ration is reduced, fish choose only slightly cooler temperatures. Also, Wildhaber and Crowder (1990) found that temperature was more important in choice of food patches than was the bioenergetic optimum based on both prey density and temperature. They suggested that temperature was a macrohabitat choice for fish and food density was chosen at a smaller scale of patchiness within thermal macrohabitats. Other interesting discussions of temperature as a micro- or macrohabitat feature in resource partitioning are in the literature on lizards (Hertz, 1992; Wikramanayake & Dryden, 1993).

Fish do make brief feeding forays into waters that are suboptimal and would be lethal if prolonged (Hasler, 1945; Neil & Magnuson, 1974; Engel & Magnuson, 1976; Janssen & Giesy, 1984; Rahel & Nutzman, 1994). In these cases the conditions outside the thermal and oxygen niches are not ones in which the fish must or can dwell permanently. They are, or at least can be, lethal. Fish must simply resist the lethal or suboptimal conditions long enough to forage success-fully and then return to waters within their thermal and oxygen niche. Such behaviours may grade into vertical migrations for energetic optim-ization as discussed above.

Finally, predators of fish are distributed along the thermal niche axis and could alter the value of inhabiting the thermal niche for prey fishes. Magnuson and Beitinger (1978) were not able to find much direct evidence for this. The reason may well have been related more to the difficulty of collecting such information than to the absence of the phenomenon. The parallel issue of fish weighing predation risk against access to prey has been evaluated in some detail (see discussion in Wildhaber & Crowder, 1991).

Competition for thermal resources

Magnuson *et al.* (1979) presented evidence that competition for thermal resources does occur. Interactions among fish along the thermal niche axis do result in familiar distributional properties characteristic of competition between species, such as thermal resource partitioning (Brandt, Magnuson & Crowder, 1980; Olson *et al.*, 1988), thermal niche shifts (Crowder & Magnuson, 1982), niche complimentarity (Crowder, Magnuson & Brandt, 1981), and thermal niche compression (see Gehlbach, Bryan & Reno, 1978). Interspecific competition along the thermal niche axis alters their realized thermal niche. These interpret-ations are reviewed further by Crowder and Magnuson (1983) and Magnuson and Crowder (1984). Investigations of interspecific compe-tition for suitable thermal space are made more complex by the physical/ chemical constraints and interactions between niche axes discussed above.

Examples of realized thermal niches

Recent long-term and comparative regional studies have begun to provide a more complete view of the realized thermal niche of fresh-water fish in temperate waters. Below we present interyear variations in thermal niches for two lakes and compare the fundamental thermal niche and the realized thermal niche in lakes and streams.

Interyear variation in thermal niche of lakes

Interyear variation in climatic factors produced marked differences in observed thermal structure and thus habitat for coldwater (10 and 15 °C), coolwater (23 °C), and warmwater (27.5 °C) fish in Trout Lake and Lake Mendota, Wisconsin (Fig. 4). These graphs depict the fundamental thermal niche as constrained only by the existing temperatures in the lakes.

Several features are apparent and deserve comment. (i) The size of the habitat within the thermal niche was inversely related to the middle temperature of the niche; the warmwater habitat was the smallest and the 10 °C-coldwater fish had the largest habitat. Life at suboptimum temperatures was the rule, not the exception, for all guilds but especially for warm- and coolwater fish at these latitudes. (ii) The more southerly Lake Mendota had larger optimal thermal habitats for all guilds than the more northerly Trout Lake, with the possible exception of the 10 °C-coldwater niche. Small differences in latitude, about 3° here, significantly influenced the thermal values of a lake for all thermal guilds. Other differences between the lakes such as differences in water clarity and depth also may have influenced the differences. (iii) The warmwater 27.5 °C-niche was realized only in about one half of the years in Trout Lake. Trout Lake is near the northern boundary of many warmwater fish (see Becker, 1983). (iv) None of the thermal niches were realized in winter. Winter can be both a stressful season for fish (see Johnson & Evans, 1990; McCormick & Jensen, 1992) and important to reproductive biology (see Hokanson, 1977, Jones, Hokanson & McCormick, 1977). (v) Occasionally the 10 °C-coldwater niche occurred in the deepest waters which were certainly anoxic, for example 1985 in Lake Mendota. Extinction of coldwater fish without a flexible thermal niche would be likely to occur in such years. The only coldwater fish remaining in Lake Mendota is the cisco. (vi) Niche partitioning between the guilds was not complete. In these two lakes, the overlap in the thermal niche was quite apparent in the lakes' thermal structures between the 27.5- and 23 °C-niches, the 23- and 15 °C-niches, and the 15- and 10 °C-niches. (vii) Interyear differences did not appear to affect all thermal niches in any obvious directly or inversely proportional manner, either for within- or between-lake comparisons. A year with a small habitat for one thermal niche did not consistently have either a small or a large habitat for the other

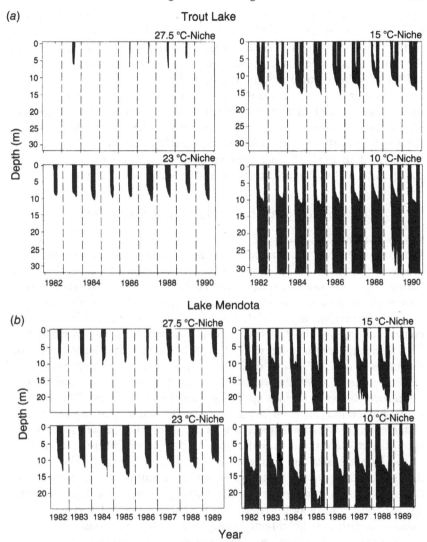

Fig. 4. Time series of fundamental thermal niche sizes plotted against water depth for a warmwater (27.5 °C), a coolwater (23 °C), and two coldwater (15 and 10 °C) fish, based on the observed seasonal isotherms for (a) mesotrophic Trout Lake in northern Wisconsin, 1982–90 and (b) eutrophic Lake Mendota in southern Wisconsin, 1982–9. The broad thermal niches (10 °C wide) are filled in black. Water temperature data are from the Long-Term Ecological Research (LTER) North Temperate Lakes and Lake Mendota databases at the Center for Limnology, University of Wisconsin–Madison, USA.

niches. (viii) Differences among years resulted perhaps more from the seasonal duration of the thermal habitats than from the vertical extent of the thermal habitats. This was especially apparent in the 23 °C-niche for Trout Lake where the season was longest in 1987 and shortest in 1983 while the vertical extent was near 9 m in all years.

A number of other patterns were apparent, but the obvious point to note is that sizes of habitat within the thermal niches differed greatly among lakes, years and thermal guilds, given the observed climate of the 1980s in north temperate lakes. These differences have the potential to be used to evaluate interyear variations and long-term changes in the quality of fish habitat in lakes and other aquatic systems.

Comparisons with the realized thermal niche

Lakes

We compared the vertical temperature distributions of fish to their fundamental thermal niches and to the amounts of water available at various temperatures in several temperate lakes. Fish thermal distribution in nature, that is the realized thermal niche, results from the thermal behaviour of the fish (the fundamental thermal niche) and constraints set by the thermal habitat available (habitat availability). The resulting temperature distributions of fish in the deepest part of the lakes (Figs. 5–7) were derived from data on vertical distributions of temperature and fish on dates in late summer when temperature differences between surface and deep waters were greatest. The amount of habitat (water) at each temperature was summed for metres of the water column at each temperature; also, using the known area of the lake at each depth and thus the area at each temperature, the volume of the lake at each temperature was calculated. Depth distributions of fish were transformed to temperature distributions by summing the catches at each temperature and plotting their vertical distribution against temperature rather than depth. These observed thermal distributions of fish can be directly compared with the amount of habitat existing at each temperature, either the number of metres in the water column at each temperature or the volume of water in the lake at each temperature.

Three general distribution patterns are apparent (Figs. 5–7). First, fish occupied their thermal niche when those temperatures were abundant in the lake; second, fish occupied the most available temperatures even if those temperatures were not within their thermal niche; and third, fish often exhibited bimodal distributions which we interpreted as age-related differences or forays outside their thermal niche to feed.

Compromises between inhabiting optimum thermal habitat and available thermal habitat resulted in the fish occupying different temperatures in different lakes, years, and seasons.

CISCO, A COLDWATER SPECIES

In Trout Lake, cisco distributions were centred at 9 °C in 1988, 7 °C in 1987, and 5–7 °C in 1989 (Fig. 5a), in direct response to differences in available temperatures in deep water. All of these temperatures were below the lower bound of their thermal niche by 2–6 °C. However, there is some uncertainty about the thermal niche for cisco. Here 15 °C was used for a midpoint as a general value for coregonid fish (Hewett & Johnson, 1987), but if 13 °C was used (Rudstam & Magnuson, 1985), the temperatures would be at or below the thermal niche by only 0–4 °C. Regardless, cisco were at or below the bounds of their thermal niche by either criterion. In Lake Mendota (Fig. 5b; Rudstam & Magnuson, 1985), cisco occupied temperatures 0–5 °C above the upper bounds of their thermal niche; they were unable to occupy permanently the deep and cold waters of this eutrophic lake because the hypolimnion was anoxic.

The fundamental thermal niche for cisco was located primarily at the thermocline which, owing to the sharp gradients of temperature, was a rather rare thermal habitat compared with the more thermally homogeneous epilimnion and hypolimnion. With their thermal flexibility, cisco in different lakes seemed able to use suboptimal habitats such as either the too-cold hypolimnion or the too-warm epilimnion in spite of a thermal preference for the rare temperatures between these two abundant habitats.

YELLOW PERCH, A COOLWATER SPECIES

In Crystal Lake and Lake Mendota, yellow perch did occur within the bounds of their thermal niche in the warmer upper waters (Fig. 6a, b). However, the temperature that they occupied within the niche depended on habitat availability. The centre of this realized niche ranged from 21 to 27 °C in the various years in Lake Mendota and was constant at 21 °C in Crystal Lake, entirely in response to differences in the most abundant waters within the thermal niche. Most realized thermal niches of yellow perch were bimodal, with a second cooler mode 1–9 °C below the bounds of the thermal niche (Lake Mendota in 1986 and 1989; Crystal Lake in 1987 and 1989). The deeper, cooler mode probably resulted from feeding forays to eat zooplankton in the thermocline or macroinvertebrates on the bottom. Crystal Lake is

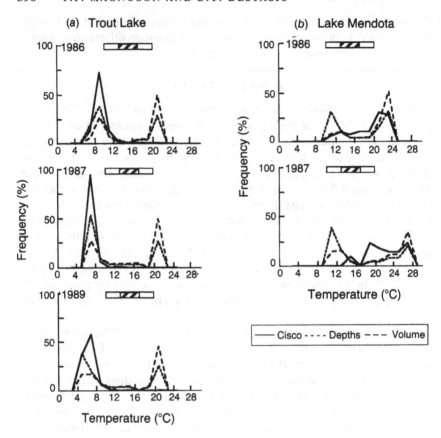

Fig. 5. Observed thermal distributions (the realized niche) of the coldwater fish (cisco) and distributions of the available thermal habitat measured as m (depth) of water and m³ (volume) of water at given temperatures in (a) Trout Lake in late July 1986, 1987 and 1989 and (b) Lake Mendota for late summer dates in August 1986 and late July 1987. The broad (10 °C-wide) and narrow (4 °C-wide) thermal niches are indicated by the horizontal bars above each distribution. Data are from the LTER project on North Temperate Lakes at the Center for Limnology, University of Wisconsin–Madison. Fish data were collected with vertical gill nets of various mesh sizes. Vertical gill nets were averaged from two successive nights; vertical temperatures were averaged from two profiles within c. 1 week before and after the gill net samples. See Rudstam and Magnuson (1985) for complete description of collection methods.

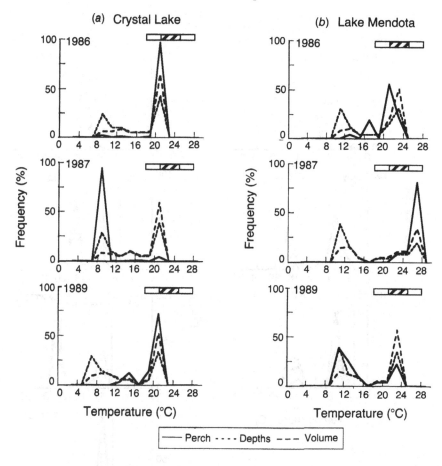

Fig. 6. Observed thermal distributions, i.e. the realized niche, of the coolwater fish (yellow perch) along with the thermal distributions of the available habitat measured as m (depth) or m³ (volume) of water at given temperatures in (*a*) Crystal Lake for August 1986, 1987 and 1989 and (*b*) Lake Mendota for August 1986, late July 1987 and August 1989. Definitions and methods as in Fig. 5.

oligotrophic and food limitations are apparent in the slow growth of perch.

Changes in seasonal constraints on the thermal niche were apparent for cisco in Trout Lake (Fig. 7*a*) and yellow perch in Lake Mendota (Fig. 7*b*). In spring both species occupied waters colder than their thermal niche simply because no waters within the niche were available.

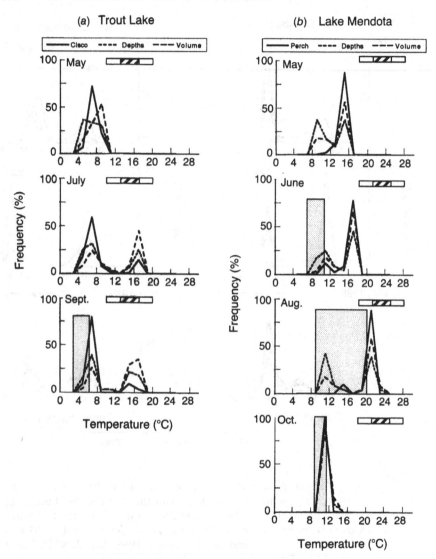

Fig. 7. Observed thermal distribution across seasons, i.e. the realized niche, from May to October 1992, of (*a*) the coldwater fish (cisco) in Trout Lake and (*b*) the coolwater fish (yellow perch) in Lake Mendota, along with the thermal distributions of the available habitat measured as m (depth) or m^3 (volume) at given temperatures. Stippled areas indicate temperatures with 5 mg l^{-1} of dissolved oxygen or less. Other definitions and methods are as in Fig. 5.

The same was true for yellow perch in June and October. In July and September for cisco, and in August for perch, temperatures within the thermal niche were available, and bimodal realized niches developed with one maximum within the thermal niche and the other in colder water. For cisco in Trout Lake the shallower, warmer mode was made up of young of the year and the deeper, colder mode of older fish. Resource partitioning between young and adults of the same species is reported for other species (Brandt, 1980). This appears to be general. Larger individuals occur at greater depths (Macpherson & Duarte, 1991) and in laboratory studies young fish of some species prefer warmer waters than do adults (McCauley & Read, 1973). Information on the thermal behaviour of cisco under laboratory conditions is too limited to provide estimates of preference for the various ages. For yellow perch in Lake Mendota the bimodality was not the result of age structure in the population; the small cooler mode in June and August probably represents fish in feeding forays to deeper, colder water.

Seasonal patterns in realized niche size did not appear to be constrained by low oxygen for cisco in Trout Lake but did for yellow perch in Lake Mendota (Fig. 7). Trout Lake had only a small zone of water below 5 mg l^{-1} which developed by September. A few cisco continued to occupy the low oxygen waters either permanently or in forays to feed near the lake bottom. Lake Mendota had a large zone of water below 5 mg l^{-1} which by August encompassed the entire cold mode of the yellow perch's realized niche. A few yellow perch continued to make forays down to feed despite the low oxygen, which was typically 0 mg l^{-1}. The deeper, colder mode of yellow perch distribution did not include the most abundant coldwater habitat even though the June mode did when oxygen was not as constraining. Thus, low oxygen appeared to constrain perch feeding forays more in August than in June.

Streams

The temperature distribution of fish in streams of the USA has been summarized from a massive database by John G. Eaton and colleagues of the US Environmental Protection Agency, Duluth, Minnesota (Biesinger *et al.*, 1979; Eaton *et al.*, 1995; Eaton & Scheller, 1996). From this database they produce seasonal distributions of the temperature in stream locations at which certain fish species have been found. Their methods and criteria for data selection are provided in their papers. These distributions are being used by Eaton and colleagues to establish thermal criteria for fish, but they also provide the most extensive presentation of the realized thermal niche of fishes across the USA.

The realized thermal niche of the coolwater yellow perch and the warmwater largemouth bass (*Micropterus salmoides*) are presented for Wisconsin and the entire USA in Fig. 8. The joint influences of constraints from available temperatures and preference of fish for certain temperatures are both apparent.

Seasonal constraints from available temperatures were apparent in that the temperatures inhabited by cool- and warmwater fish were colder than their thermal niche, except in summer when the thermal distributions of yellow perch were remarkably close to temperature preferences of fish exhibited in the laboratory. This similarity was in both the central tendency and variability. Most occurrences were within the 10 °C range of the broad fundamental thermal niche. The exception was the largemouth bass (Fig. 8*a*, *b*) where the warmest temperatures of occurrence were at the lower boundary of their fundamental niche for Wisconsin and below the fundamental thermal niche for many places in the national data set even in summer. Apparently largemouth bass live in suboptimal thermal habitats in many areas; many of these occurrences may have coincided with the warmest temperatures available across the northern part of the US database.

Temperatures occupied by yellow perch in summer were not constrained by available temperatures in either the Wisconsin or the national database. When compared with temperatures occupied by largemouth bass for the national database, it was clear that warmer waters were available than those at which yellow perch occurred (Fig. 8*a* vs 8*c*). Similarly, the comparison of yellow perch occurrences in the more restricted Wisconsin database with their occurrences in the larger database for the USA (Fig. 8*d* vs 8*c*) indicated that Wisconsin contained river habitats with the full range of water temperatures used by this coolwater species. The implication of these results is that during summer, some fish move in the rivers to locations within their thermal niche, that is, they thermoregulate behaviourally if possible.

Differences in temperatures among adjacent streams that differ in size, position in the flow system, riparian vegetation, groundwater input, and heated waste effluents, may provide the heterogeneity in thermal habitat for stream fish to occupy their thermal niche. In mountainous regions with substantial altitudinal relief (Wyoming), the zonation of river fish according to temperatures has been documented (Rahel & Hubert, 1991; Rahel, Keleher & Anderson, 1996). Similarly, Meisner (1990) documented the seasonal movement of a coldwater fish, brook trout (*Salvelinus fontinalis*), based on seasonal changes in temperature distributions in Ontario trout streams. This same pattern was apparent in Eaton's database, but at an even larger spatial scale.

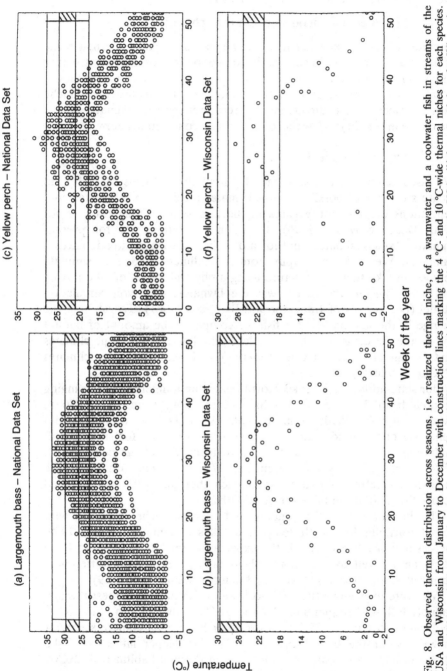

Fig. 8. Observed thermal distribution across seasons, i.e. realized thermal niche, of a warmwater and a coolwater fish in streams of the USA and Wisconsin from January to December with construction lines marking the 4 °C- and 10 °C-wide thermal niches for each species. Niche widths are indicated at the margins of each panel. Warmwater: (a) largemouth bass for the USA; (b) largemouth bass for Wisconsin. Coolwater: (c) yellow perch for the USA; (d) yellow perch for Wisconsin. Data from the 'Fish and Temperature Database Matching System' (Eaton et al., 1995), provided for presentation by J.G. Eaton, Environmental Research Laboratory of the US Environmental Protection Agency, Duluth, MN 55804, USA.

The thermal niche and climate change

Climate change is expected to alter the size of realized thermal niche of fishes because it would change the annual pattern of thermal structure and would occur simultaneously with other changes in precipitation, cloudiness, windiness, and ice cover that would influence other niche axes of fishes. These direct and indirect influences of climate change on the realized thermal niche of fishes are considered below.

Thermal niche in lakes

We used climate models, General Circulation Models (GCMs), to create scenarios of possible future climates resulting from greater concentrations of greenhouse gases in the atmosphere (Magnuson *et al.*, 1990; DeStasio *et al.*, 1996). Common simulations are for present climates (1xCO$_2$ climates) and for a doubling of carbon dioxide (2xCO$_2$ climates). We took the output from four climate models to simulate changes in lake thermal structure using a physical model of lake mixing and heat transfer; the model was the Dynamic Reservoir Simulation Model (DYRESM) of Imberger and Patterson (1981). To estimate the change in thermal habitat for fish we compared the amount of habitat within the fundamental thermal niche simulated for present climates (1xCO$_2$ climates, i.e. BASE scenarios) and possible future climates (2xCO$_2$ scenarios).

Our simulations were for six North American temperate lakes ranging in area from 37 to 5 800 000 ha and in depth from 20 to 280 m (Magnuson *et al.*, 1990; DeStasio *et al.*, 1996). The broad (10 °C-wide) definition of the fundamental thermal niche was used for species with midpoints of their niche at 10 °C (lake trout, *Salvelinus namaycush*), 15 °C (cisco), 23 °C (yellow perch) and 27.5 °C (largemouth bass). The number of metres of depth within each thermal niche was summed across the year in units of metre months.

The habitat within the fundamental thermal niche increases with global warming for north temperate lakes (Magnuson *et al.*, 1990; DeStasio *et al.*, 1996). This is the expectation for large lakes like Lake Michigan as well as for small inland lakes, but notable differences occurred among species, lakes, and climate scenarios (Fig. 9).

With rare exceptions the 10 °C-coldwater fish had the largest thermal habitat, followed sequentially by the 15 °C-coldwater fish, the 23 °C-coolwater fish and the 27.5 °C-warmwater fish for all climate scenarios as well as for the BASE case (Fig. 9). For deeper lakes like Lake Michigan and Trout Lake, the increases in thermal habitat with 2xCO$_2$ climates were greater for the 10 °C-coldwater fish than for all warmer

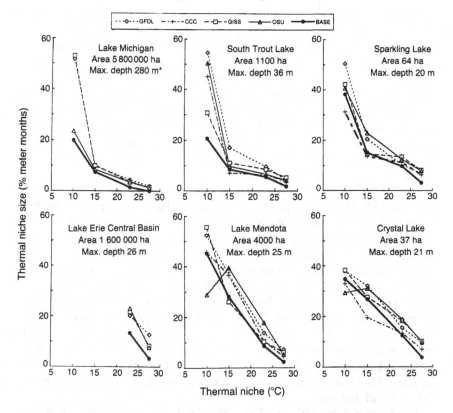

Fig. 9. Changes in habitat within fundamental thermal niches simulated with a doubling of atmospheric CO_2 for warm-, cool- and coldwater fish for six north temperate lakes (Magnuson *et al.*, 1990;
DeStasio *et al.*, 1996). Simulations are based on base climates for
Lake Michigan from 1981 to 1984, Lake Erie for 1970 and 1975, and
the other lakes for 1986, 1987 and 1989. Calculations for Lake Michigan included the upper 100 m which are the depths found in a transect
across the lake at Milwaukee. The four climate scenarios from General
Circulation Models were GFDL (General Fluid Dynamics Laboratory),
GISS (Goddard Institute of Space Science), OSU (Oregon State
University), and CCC (the Canadian Climate Centre). Methods and
information on the climate models are detailed in Magnuson *et al.*
(1990) and DeStasio *et al.* (1996).

water fish, while for the shallower lakes, 20–25 m deep, increases were greater for one or more of the groups. Increases in thermal habitat were similar for the 23 °C-coolwater fish and the 27.5 °C-warmwater fish in each lake; they approximately doubled in size on average. Both the 10 °C- and the 15 °C-coldwater fish demonstrated combinations of decreases and increases in thermal habitat with climate warming among lakes and scenarios; four out of 16 cases decreased for the 10 °C-fish, and five out of 16 cases decreased for the 15 °C-fish. Decreases were more likely for coldwater than for cool- and warmwater fish because in the shallower lakes, the bottom waters were too warm in some scenarios during summer.

Size of habitat within the fundamental thermal niche differed (Fig. 9). Lake Michigan, the largest and deepest lake, tended to have the smallest thermal habitat for all fishes both in the BASE case and in the most altered climate scenarios when measured as the percentage of water depths within each niche. That is, Lake Michigan has the largest fraction of total habitat that is suboptimal for each of the thermal guilds. In only two scenarios with the 10 °C-coldwater fish were the niche sizes larger than or comparable to the other lakes. The thermal niches in Lake Michigan would have appeared even smaller as a percentage of total water depths if a deeper part of the lake had been used for the calculations of size; calculations were for the upper 100 m while some places are as deep as 280 m. Water temperatures in Lake Michigan were below the thermal niche temperatures for all fish most of the year.

Trout Lake was similar to Lake Michigan for the coldwater fish but had more thermal habitat for cool- and warmwater fish than Lake Michigan, when measured as the percentage of water depth within each niche, because surface waters were warmer in this smaller lake. This was generally true for the shallower lakes. Thermal habitat size was also larger especially for the 15 °C-coldwater fish in the shallower lakes than in Michigan in the BASE case and most scenarios. This occurred because the deeper waters warmed more than they did in Lake Michigan and even in Trout Lake; the smaller lakes have smaller volumes per unit area and thus warm more with the same heat input than larger lakes.

Before leaving the differences in thermal habitat sizes among lakes, note should be taken of the measure used. We used % meter months, the percentage of depths within the thermal niche summed across the year. If an absolute measure of depths within the niche were used, deeper lakes would be expected to have larger thermal habitat than shallower lakes simply because they have more depths to be allocated to the niches than do shallower lakes. If bottom areas or volumes of water in each niche were used, large lakes like Lake Michigan would have immense thermal habitats compared with relatively tiny Crystal Lake. Volume and area measures also have been used to assess changes in thermal habitat (Christie and Regier, 1988; Stefan *et al.*, 1995). There are advantages to each measure depending on the objectives of the analysis.

If we had used relative volumes or relative bottom areas rather than the relative depths, the size of the warmer niches would have increased relative to the colder niches because lakes have larger areas at the shallower/warmer strata than at the deeper/colder strata. However, for the lakes we analysed, thermal habitats still would be greater for colder water fish than for warmer water fish. Much of this cold habitat is accumulated during spring and autumn periods of mixing.

The different climate models provide a range of scenarios for changes in water temperatures with climate warming. The general similarity in the thermal habitat simulated with the various models, suggests that the results are robust at least in terms of direction of change. A doubling of CO_2 would generally increase thermal habitat for all thermal guilds in north temperate regions, at least for lakes like those considered here. This conclusion, however, is limited by its thermal centric view where only water temperatures and changes in water temperatures are considered in the life of fish. Many other environmental features of the fish world are likely to be altered by climate change.

Influence on other niche axes

Climate warming sounds deceptively simple to evaluate for one trying to judge effects of warmer climates on ectothermic heterotherms like fish. However, it would be a mistake to stop with a single, non-interactive niche axis approach to such a complex issue as climate change. Factors likely to be altered by climate warming that should influence the realized thermal niche include changes in (i) the availability of dissolved oxygen resulting from altered productivities (Blumberg & Di Toro, 1990; Stefan, Hondzo & Fang, 1993; Stefan *et al.*, 1995), water levels (Croley, 1990), and durations of summer stratification (DeStasio *et al.*, 1996) and winter ice cover (Assel, 1991); (ii) littoral zone substrates, vegetation, and benthos resulting from altered water levels and flood regimes; (iii) the surface water connectivity between lakes and between rivers and their floodplains resulting from altered precipitation and evapotranspiration; (iv) lake chemistry from altered precipitation (Schindler *et al.*, 1996; Webster *et al.*, 1996), groundwater discharge (Meisner, Rosenfeld & Regier, 1988), and associated inputs of nutrients, suspended solids, dissolved organic matter and acidity (Fee *et al.*, 1996; Schindler *et al.*, 1996); (v) lake productivity owing to altered lake chemistry and solar radiation; and (vi) the rates of invasion of warmth-loving exotics (Mandrak, 1989; Johnson & Evans, 1990; Shuter & Post, 1990; Edsall *et al.*, 1993; Minns & Moore, 1995). Articles that touch on these and other factors that will interact with the thermal niche of fishes in a warming climate in north temperate waters are: Coutant (1981); Meisner *et al.* (1987), Magnuson *et al.* (1989), Regier *et al.* (1989), Smith (1991), Hill & Magnuson (1990), Schindler *et al.* (1990), Carpenter *et al.* (1992), McLain, Magnuson & Hill (1994), Stefan *et al.* (1995); Arnell *et al.* (1996); and Magnuson *et al.* (1996).

Temperature distributions and temperature dynamics are important features in the climate change ecology of fish. The dominant result from simulations is that thermal habitat for fish in north temperate lakes would increase. The signal of temperature in the success of fish populations is sufficiently strong that Christie and Regier (1988) were able to explain differences in yields of fish in large North American lakes from differences in the amount of optimum thermal habitat in the lakes. The temperature signal should be strong enough to project long-term changes in fish thermal habitat, but the fruits of such attempts await the future for confirmation or rejection. In the meantime, year-to-year variability and latitudinal differences in climate can provide a fertile testing ground for the ideas and simulations. The formalism of using temperature as a niche axis should help sort out the complex changes that may occur in a warming climate. Other properties in the environment of fish need to be incorporated into such a scheme. A useful beginning is provided by Stefan *et al.* (1995) who simulated realized niches for $2xCO_2$ climates using both temperature and oxygen. Particularly important are those that are both responsive to climate change and important to fish.

Acknowledgments

We thank John G. Eaton of the US Environmental Protection Agency for providing data on the temperature distribution of fishes in rivers and Linda D. Mortsch of the Atmospheric Environment Service Environment Canada for providing the CCC GCM II scenario data. We thank Larry I. Crawshaw, John G. Eaton, J. Howard McCormick and D. Gordon McDonald for their suggestions on the chapter. We thank Beth A. DeStasio, Alice Justice, Norma D. Magnuson, Robert

M. Scheller, and Jennifer R. Sharkey for facilitating the production of the manuscript and the literature search. This research was supported by the Midwestern Regional Center for the National Institute for Global Change, US Department of Energy (AGR.DTD 12/7/92) and the US National Science Foundation's Division of Environmental Biology (DEB 90110660, BSR 8514330, DEB 8012313).

References

Alderdice, D. F. & Forester, C. R. (1971). Effects of salinity, temperature, and dissolved oxygen on early development of the Pacific Cod (*Gadus macrocephalus*). *Journal of the Fisheries Research Board of Canada*, **28**, 883–902.

Assel, R. A. (1991). Implications of CO_2 global warming on Great Lakes ice cover. *Climate Change*, **18**, 377–95.

Arnell, N., Bates, B., Lang, H., Magnuson, J. J. & Mulholland, P. (1996). Hydrology and fresh water ecology. In *Climate Change 1995: Impacts, Adaptations and Mitigation of Climate Change: Scientific–Technical Analyses*. Contribution of Working Group II to the Second Assessment Report of the Intergovernmental Panel on Climate Change, ed. R. T. Watson, M. C. Zinyowera & R. H. Moss, ch. 10. Cambridge: Cambridge University Press.

Becker, G. C. (1983). *Fishes of Wisconsin*. Madison, Wis: University of Wisconsin Press.

Biesinger, K. E., Brown, R. P., Bernick, C. R., Flittner, G. A. & Hokanson, K. E. F. (1979). *A National Compendium of Fresh water Fish and Water Temperature Data*. Vol. 1. *Data Management Techniques, Output Examples and Limitations. Duluth, Minnesota: EPA, Report 600/3-79-056.*

Blumberg, A. F. & Di Toro, D. M. (1990). Effects of climate warming on dissolved oxygen concentrations in Lake Erie. *Transactions of the American Fisheries Society*, **119**, 210–23.

Boltz, J. M., Siemien, M. J. & Stauffer, J. R. Jr (1987). Influence of starvation on the preferred temperature of *Oreochromis mossambicus* (Peters). *Archives für Hydrobiology*, **110**, 143–6.

Brandt, S. B. (1980). Spatial segregation of adult and young-of-the-year alewives across a thermocline in Lake Michigan. *Transactions of the American Fisheries Society*, **109**, 469–78.

Brandt, S. B. (1993). The effect of thermal fronts on fish growth: a bioenergetics evaluation of food and temperature. *Estuaries*, **16**, 142–59.

Brandt, S. B., Magnuson, J. J. & Crowder, L. B. (1980). Thermal habitat partitioning by fishes in Lake Michigan. *Canadian Journal of Fisheries and Aquatic Sciences*, **37**, 1557–64.

Brett, J. R. (1971). Energetic responses of salmon to temperature. A study of some thermal relations in the physiology and fresh water ecology of sockeye salmon (*Oncorhynchus nerka*). *American Zoologist*, **11**, 99–113.

Carpenter, S. R. Fisher, S. G., Grimm, N. B. & Kitchell, J. F. (1992). Global change and fresh water ecosystems. *Annual Reviews of Ecology and Systematics*, **23**, 119–39.

Christie, G. C. & Regier, H. A. (1988). Measures of optimal thermal habitat and their relationship to yields for four commercial fish species. *Canadian Journal of Fisheries and Aquatic Sciences*, **45**, 301–14.

Chulakasem, W., Nelson, J. A. & Magnuson, J. J. (1989). Interaction between effects of low pH and low ion concentration on mortality during early development of medaka, *Oryzias latipes*. *Canadian Journal of Zoology*, **67**, 2158–68.

Clark, D. S. & Green, J. M. (1991). Seasonal variation in temperature preference of juvenile Atlantic cod (*Gadus morhua*), with evidence supporting an energetic basis for their diel vertical migration. *Canadian Journal of Zoology*, **69**, 1302–7.

Coutant, C. C. (1981). Foreseeable effects of CO_2-induced climate change. *Freshwater Concerns, Environmental Conservation*, **8**, 285–97.

Coutant, C. C. (1987). Thermal preference: when does an asset become a liability? *Environmental Biology of Fishes*, **18**, 161–72.

Coutant, C. C. (1990). Temperature-oxygen habitat for fresh water and coastal striped bass in a changing climate. *Transactions of the American Fisheries Society*, **119**, 240–53.

Croley, T. F. (1990). Laurentian Great Lakes double CO_2 climate change hydrologic impacts. *Journal of Climate Change*, **17**, 27–48.

Crowder, L. B. & Magnuson, J. J. (1982). Thermal habitat shifts by fishes at the thermocline in Lake Michigan. *Canadian Journal of Fisheries and Aquatic Sciences*, **39**, 1046–50.

Crowder, L. B. & Magnuson, J. J. (1983). Cost benefit analysis of temperature and food resource use: a synthesis with examples from the fishes. In *Behavioral Energetics: The Cost of Survival in Vertebrates*, ed. W. P. Aspey & S. I. Lustick, pp. 189–221. Columbus, Ohio: Ohio State University Press.

Crowder, L. B., Magnuson, J. J. & Brandt, S. B. (1981). Complementarity in the use of food and habitat resources by Lake Michigan fishes. *Canadian Journal of Fisheries and Aquatic Sciences*, **38**, 662–8.

DeStasio, B. T., Hill, D. K., Kleinhans, J. M., Nibbelink, N. P. & Magnuson, J. J. (1996). Potential effects of global climate change on small north temperate lakes: physics, fishes and plankton. *Limnology and Oceanography*, **41**, 1136–49.

Eaton, J. G., McCormick, J. H., Goodno, B. E., O'Brien, D. G., Stefan, H. G., Hondzo, M. & Scheller, R. M. (1995). A field information-based system for estimating fish temperature tolerances. *Fisheries*, **20**, 10–18.

Eaton, J. G. & Scheller, R. M. (1996). Effects of climate warming on fish thermal habitat in streams of the United States. *Limnology and Oceanography*, **41**, 1109–15.

Edsall, T. A., Selgeby, J. H., DeSorcie, T. J. & French, J. R. P. III (1993). Growth-temperature relation for young-of-the-year ruffe. *Journal of Great Lakes Research*, **19**, 530–3.

Engel, S. & Magnuson, J. J. (1976). Vertical and horizontal distributions of coho salmon (*Oncorhynchus kisutch*), yellow perch (*Perca flavescens*), and cisco (*Coregonus artedii*) in Pallette Lake, Wisconsin. *Journal of the Fisheries Research Board of Canada*, **33**, 2710–15.

Fee, E. J., Hecky, R. E., Kasian, S. W. & Cruikshank, D. (1996). Potential size-related effects of climate change on mixing depths in Canadian Shield Lakes. *Limnology and Oceanography*, **41**, 912–20.

Fry, F. E. J. (1947). Effects of the environment on animal activity. *University of Toronto Studies Biology Series*, **55**, 1–62.

Fry, F. E. J. (1971). The effect of environmental factors on the physiology of fish. In *Fish Physiology: Environmental Factors,* ed. W. S. Hoar & D. J. Randall, pp. 1–98. New York: Academic Press.

Gehlbach, F. R., Bryan, C. L. & Reno, H. A. (1978). Thermal ecological features of *Cyprinodon elegans* and *Gambusia affinis*, endangered Texas fishes. *The Texas Journal of Science*, **30**, 99–101.

Hasler, A. D. (1945). Observations on the winter perch population of Lake Mendota. *Ecology*, **26**, 90–4.

Hertz, P. E. (1992). Evaluating thermal resource partitioning. *Oecologia*, **90**, 127–36.

Hewett, S. W. & Johnson, B. L. (1987). A generalized bioenergetics model of fish growth for microcomputers. University of Wisconsin Sea Grant Institute, Technical Report WIS-SG-87–245, Madison, Wisconsin: Wisconsin Sea Grant Institute.

Hill, D. K. & Magnuson, J. J. (1990). Potential effects of global climate warming on the growth and prey consumption of Great Lakes fish. *Transactions of the American Fisheries Society*, **119**, 265–75.

Hokanson, K. E. F. (1977). Temperature requirements of some percids and adaptations to the seasonal temperature cycle. *Journal of the Fisheries Research Board of Canada*, **34**, 1524–50.

Holmes, J. A. & Lin, P. (1994). Thermal niche of larval sea lamprey, *Petromyzon marinus*. *Canadian Journal of Fisheries and Aquatic Sciences*, **51**, 253–62.

Hutchinson, G. E. (1957). Concluding remarks. *Cold Spring Harbor Symposium in Quantitative Biology*, **22**, 415–27.

404   J.J. MAGNUSON AND B.T. DESTASIO

Hutchinson, G. E. (1967). *A Treatise on Limnology*, Vol. 2. *Introduction to lake biology and the limnoplankton.* ed. G. Evelyn Hutchinson. New York: John Wiley.

Hutchinson, G. E. (1978). *An Introduction to Population Ecology.* New Haven, Conn: Yale University Press.

Imberger, J. & Patterson, J. C. (1981). A dynamic reservoir simulation model – DYRESM:5. In *Transport Models for Inland and Coastal Waters: Proceedings of a Symposium on Predictive Models,* ed. H. B. Fischer, pp. 310–61. New York: Academic Press.

Janssen J. & Giesy, J. P. (1984). A thermal effluent as a sporadic cornucopia. *Environmental Biology of Fishes,* **11**, 191–203.

Javaid, M. & Anderson, J. M. (1967). Influence of starvation on selected temperature of some salmonids. *Journal of the Fisheries Research Board of Canada,* **24**, 1515–19.

Jobling, M. (1981). Temperature tolerance and final preferendum – rapid methods for the assessment of optimum growth temperatures. *Journal of Fish Biology,* **19**, 439–55.

Johansen, K., Hanson, D. & Lenfant, C. (1970). Respiration in a primitive air breather, *Amia calva. Respiration Physiology,* **9**, 162–74.

Johnson, T. B. & Evans, D. O. (1990). Size-dependent winter mortality of young-of-the-year white perch: climate warming and the invasion of the Laurentian Great Lakes. *Transactions of the American Fisheries Society,* **119**, 301–13.

Jones, R. R., Hokanson, K. E. F. & McCormick, J. H. (1977). Winter temperature requirements for maturation and spawning of yellow perch *Perca flavescens* (Mitchill). In *Proceedings of the World Conference Towards a Plan of Actions for Mankind, Biological and Thermal Modifications,* Vol. 3, ed. M. Marois, pp. 189–92. New York: Pergamon Press.

Kerr, S. R. (1990). The Fry paradigm: its significance for contemporary ecology. *Transactions of the American Fisheries Society,* **119**, 779–85.

Kramer, D. L. (1987). Dissolved oxygen and fish behavior. *Environmental Biology of Fishes,* **18**, 81–92.

Mac, M. J. (1985). Effects of ration size on preferred temperature of lake charr (*Salvelinus namaycush*). *Environmental Biology of Fishes,* **14**, 227–31.

McCauley, R. W. & Huggins, N. W. (1979). Ontogenetic and nonthermal seasonal effects on thermal preferenda of fish. *American Zoologist,* **19**, 267–71.

McCauley, R. W. & Read, L. A. A. (1973). Temperature selection by juvenile and adult yellow perch (*Perca flavescens*) acclimated to 24 °C. *Journal of the Fisheries Research Board of Canada,* **30**, 1253–5.

McCormick, J. H. & Jensen, K. M. (1992). Osmoregulatory failure and death of first-year largemouth bass (*Micropterus salmonoides*) exposed to low pH and elevated aluminum, at low temperatures in soft water. *Canadian Journal of Fisheries and Aquatic Sciences*, **49**, 1189–97.

McLain, A. S., Magnuson, J. J. & Hill, D. K. (1994). Latitudinal and longitudinal differences in thermal habitat for fishes influenced by climate warming: expectations from simulations. *Verhandlungen der Internationalen Vereinigung für Theoretische und Angewandte Limnologie*, **25**, 2080–5.

MacLean, J. & Magnuson, J. J. (1977). Species interactions in percid communities. *Journal of the Fisheries Research Board of Canada*, **34**, 1941–51.

Macpherson, E. & Duarte, C. M. (1991). Bathymetric trends in demersal fish size: is there a general relationship. *Marine Ecology Progress Series*, **71**, 103–12.

Magnuson, J. J., Assel, R. A., Bowser, C. J., Dillon, P. J., Eaton, J. G., Evans, H. E., Fee, E. J., Hall, R. I., Mortsch, L. R., Schindler, D. W., Quinn, F. H. & Webster, K. E. (1996). Potential effects of climate change on aquatic systems: Laurentian Great Lakes and Precambrian Shield Region. *Hydrological Processes* (in press).

Magnuson, J. J. & Beitinger, T. L. (1978). Stability of temperatures preferred by centrarchid fishes and terrestrial reptiles. In *Contrasts in Behavior*, ed. E. S. Reese & F. J. Lighter, pp. 181–216. New York: John Wiley.

Magnuson, J. J. & Crowder, L. B. (1984). Species interactions in the dynamics of planktivorous fish assemblages in Lake Michigan. In *Proceedings of the Expert Consultation to Examine Changes in Abundance and Species Composition of Neritic Fish Stocks*, Food and Agriculture Organization of the United Nations, San Jose, Costa Rica, April 1983. FAO Fisheries Report 291(2): 779–88.

Magnuson, J. J., Crowder, L. B. & Medvick, P. A. (1979). Temperature as an ecological resource. *American Zoologist*, **19**, 331–43.

Magnuson, J. J., Hill, D. K., Regier, H. A., Holmes, J. A., Meisner, J. D. & Shuter, B. J. (1989). Potential responses of Great Lakes fishes and their habitat to global climate warming. In *The Potential Effects of Global Climate Change on the United States*, Appendix E. Aquatic Resources, ed. J. B. Smith & D. A. Tirpak, pp. 2-1–2-42. Washington, DC: US Environmental Protection Agency, EPA-230-05-89-055.

Magnuson, J. J., Meisner, J. D. & Hill, D. K. (1990). Potential changes in the thermal habitat of Great Lakes fish after global climate warming. *Transactions of the American Fisheries Society*, **119**, 254–64.

Mandrak, N. E. (1989). Potential invasion of the Great Lakes by fish species associated with climate warming. *Journal of Great Lakes Research*, **15**, 306–16.

Meisner, J. D. (1990). Effect of climatic warming on the southern margins of the native range of brook trout, *Salvelinus fontinalis*. *Canadian Journal of Fisheries and Aquatic Sciences*, **47**, 1065–70.

Meisner, J. D., Goodier, J. L., Regier, H. A., Shuter, B. M. & Christie, W. J. (1987). An assessment of the effects of climate warming on Great Lakes basin fishes. *Journal of Great Lakes Research*, **13**, 340–52.

Meisner, J. D., Rosenfeld, J. S. & Regier, H. A. (1988). The role of groundwater in the impact of climate warming on stream salmonines. *Fisheries*, **13**, 2–8.

Minns, C. K. & Moore, J. E. (1995). Factors limiting the distributions of Ontario's freshwater fishes: the role of climate and other variables, and the potential impacts of climate change. In *Climate Change and Northern Fish Populations*, ed. R. J. Beamish, pp. 137–60. *Canadian Journal of Fisheries and Aquatic Sciences, special publication* **121**.

Morgan, M. J. (1993). Ration level and temperature preference of American plaice. *Marine Behavior and Physiology*, **24**, 117–22.

Neill, W. H. & Magnuson, J. J. (1974). Distributional ecology and behavioral thermoregulation of fishes in relation to heated effluent from a power plant at Lake Monona, Wisconsin. *Transactions of the American Fisheries Society*, **103**, 663–710.

Neill, W. H., Miller, J. M., Van der Veer, H. W. & Winemiller, K. O. (1994). Ecophysiology of marine fish recruitment: a conceptual framework for understanding interannual variability. *Netherlands Journal of Sea Research*, **32**, 135–52.

O'Brien, J. J. (1987). Planktivory by freshwater fish: thrust and parry in the pelagia. In *Predation: Direct and Indirect Impacts on Aquatic Communities*, ed. W. C. Kerfoot & A. Sih, pp. 3–16. University Press of New England.

Olson, R. A., Winter, J. D., Nettles, D. C. & Hayes, J. M. (1988). Resource partitioning in summer by salmonids in south-central Lake Ontario. *Transactions of the American Fisheries Society*, **117**, 552–9.

Rahel, F. J. (1984). Factors structuring fish assemblages along a bog lake successional gradient. *Ecology*, **65**, 1276–89.

Rahel, F. J. & Hubert, W. A. (1991). Fish assemblages and habitat gradients in a Rocky Mountain–Great Plains stream: biotic zonation and additive patterns of community change. *Transactions of the American Fisheries Society*, **120**, 319–22.

Rahel, F. J., Keleher, C. J. & Anderson, J. L. (1996). Potential habitat loss and population fragmentation for coldwater fish in the North Platte River drainage of the Rocky Mountains in response to climate warming. *Limnology and Oceanography*, **41**, 1116–23.

Rahel, F. J. & Nutzman, J. W. (1994). Foraging in a lethal environment: fish predation in hypoxic waters of a stratified lake. *Ecology*, **75**, 1246–53.

Regier, H. A., Magnuson, J. J., Shuter, B. J., Hill, D. K., Holmes, J. A. & Meisner, J. D. (1989). Likely effects of climate change on fish associations of the Great Lakes, In *Coping with Climate Change*, ed. J. C. Topping, Jr, pp. 234–55. Washington, DC: The Climate Institute.

Reynolds, W. W. & Casterlin, M. E. (1979). Behavioral thermoregulatory abilities of tropical coral reef fishes: a comparison with temperate freshwater and marine fishes. *Environmental Biology of Fishes*, 6, 347–9.

Rudstam, L. G. & Magnuson, J. J. (1985). Predicting the vertical distribution of fish populations: analysis of cisco, *Coregonus artedii*, and yellow perch, *Perca flavescens*. *Canadian Journal of Fisheries and Aquatic Sciences*, 42, 1178–88.

Schindler, D. W., Bayley, S. E., Parker, B. R., Beaty, K. G., Cruikshank, D. R., Fee, E. J., Schindler, E. U. & Stainton, M. P. (1996). The effects of climatic warming on the properties of boreal lakes and streams at the Experimental Lakes Area, Northwestern Ontario. *Limnology and Oceanography*, 41, 1008–17.

Schindler, D. W., Beaty, K. G., Fee, E. J., Cruikshank, D. R., DeBruyn, E. R., Findlay, D. L., Linsey, G. A., Shearer, J. A., Stainton, M. P. & Turner, M. A. (1990). Effects of climatic warming on lakes of the central boreal forest. *Science*, 250, 967–70.

Shuter, B. J. & Post, J. R. (1990). Climate, population viability, and the zoogeography of temperate fishes. *Transactions of the American Fisheries Society*, 119, 314–36.

Smith, J. B. (1991). The potential impacts of climate change on the Great Lakes. *Bulletin of the American Meteorological Society*, 72, 21–8.

Spigarelli, S. A., Romberg, G. P., Prepejchal, W. & Thommes, M. M. (1974). Body-temperature characteristics of fish at a thermal discharge on Lake Michigan. In *Thermal Ecology*, ed. J. W. Gibbons & R. R. Sharitz, pp. 119–32. AEC Symposium Series [CONF-730505].

Stefan, H. G., Hondzo, M. & Fang, X. (1993). Lake water quality modeling for projected future climate scenarios. *Journal of Environmental Quality*, 22, 417–31.

Stefan, H. G., Hondzo, M., Eaton, J. G. & McCormick, J. H. (1995). Predicted effects of global climate on fishes in Minnesota lakes. In *Climate Change and Northern Fish Populations*, ed. R. J. Beamish, pp. 57–72, *Canadian Journal of Fisheries and Aquatic Sciences, special publication* 121.

Stuntz, W. E. & Magnuson, J. J. (1976). Daily ration, temperature selection, and activity of bluegill. In *Thermal Ecology II*, ed. G. W. Esch & R. W. McFarlaine, pp. 180–4.

Tonn, W. M. & Magnuson, J. J. (1982). Patterns in the species composition and richness of fish assemblages in northern Wisconsin lakes. *Ecology*, 63, 1149–66.

Webster, K. E., Kratz, T. K., Bowser, C. J., Magnuson, J. J. & Rose, W. J. (1996). The influence of landscape position on lake chemical responses to drought in northern Wisconsin, USA. *Limnology and Oceanography*, **41**, 977–84.

Wikramanayake, E. D. & Dryden, G. L. (1993). Thermal ecology of habitat and microhabitat use by sympatric *Varanus bengalensis* and *V. salvator* in Sri Lanka. *Copeia*, *1993*, 709–14.

Wildhaber, M. L. & Crowder, L. B. (1990). Testing a bioenergetics-based habitat choice model: Bluegill (*Lepomis macrochirus*) responses to food availability and temperature. *Canadian Journal of Fisheries and Aquatic Sciences*, **47**, 1664–71.

Wildhaber, M. L. & Crowder, L. B. (1991). Mechanisms of patch choice by bluegills (*Lepomis macrochirus*) foraging in a variable environment. *Copeia*, *1991*, 445–60.

Wurtsbaugh, W. A. & Neverman, D. (1988). Post-feeding thermotaxis and daily vertical migration in a larval fish. *Nature*, **333**, 846–8.

Zahn, M. (1963). Jahreszeitliche Veranderungen der Vorzungstemperaturen von Scholle (*Pleuronectes platessa* Linne.) und Bitterling (*Rhodeus sericeus* Pallas). *Verhandlungen Deutsche Zoologische Gesellschaft*, Supplement 27, 562–80.

INDEX

Note: page numbers in *italics* refer to figures and tables